T0259563

Optical Interferometry
Second Edition

Optical Interferometry
Second Edition

By

P. Hariharan
School of Physics
University of Sydney
Sydney, Australia

ELSEVIER

AMSTERDAM • BOSTON • HEIDELBERG
LONDON • NEW YORK • OXFORD • PARIS
SAN DIEGO • SAN FRANCISCO • SINGAPORE
SYDNEY • TOKYO

Academic Press is an imprint of Elsevier

ACADEMIC PRESS

An Imprint of Elsevier

Academic Press is an imprint of Elsevier.

ACADEMIC PRESS
An Imprint of Elsevier
525 B Street, Suite 1900, San Diego, CA 92101-4495, USA
http://www.academicpress.com

Academic Press
An Imprint of Elsevier
84 Theobald's Road, London, WC1X 8RR, UK

Library of Congress Cataloging-in-Publication Data
Hariharan, P.
    Optical interferometry/by P. Hariharan.—2nd ed.
        p. cm.
    Includes bibliographical references and index.
    ISBN-13: 978-0-12-311630-7 ISBN-10: 0-12-311630-9
    1. Interferometry. I. Title.

QC411.H37 2003
535′.47′0287—dc21
ISBN-13: 978-0-12-311630-7   2003050272
ISBN-10: 0-12-311630-9

Transferred to Digital Printing 2008

Contents

Preface to the First Edition

There has been a tremendous increase in interest in optical interferometry during the last twenty years, a period in which the subject has undergone a complete transformation. The main reason for this is, of course, the development of the laser which made available, for the first time, an intense source of coherent light. Another reason has been the increasing use of digital computers to process the data obtained from an interferometer. A third reason is the availability of single-mode optical fibers which can be used to provide an optical path several metres long, with very low noise, in an extremely small space. As a result of these advances, optical interferometry is finding a remarkably wide range of applications.

The aim of this book is to present a self-contained treatment of this subject with particular emphasis on recent developments and their implications for the future. A brief historical survey leads up to three chapters covering the classical concepts of two-beam interference, coherence and multiple-beam interference. Chapter 5 then discusses interference in thin films, antireflection coatings and interference filters.

As mentioned at the outset, lasers are now being used to an increasing extent in optical interferometry; in fact, this has led to the virtual demise of the classical mercury arc. Accordingly, Chapter 6 looks at the laser as a light source and discusses techniques for obtaining a single-frequency output and for frequency stabilization.

Five chapters then deal with applications of interferometry such as length measurements, testing optical surfaces, interference spectroscopy, and Fourier-transform spectroscopy. Emphasis is placed in these chapters on techniques which have become feasible with the development of the laser, including unequal-path interferometry, fringe-counting, heterodyne and digital interferometry, fiber-optic interferometry, and nonlinear interferometry. These are followed by three chapters on holography, holographic interferometry, and speckle interferometry. A final chapter on stellar interferometry describes the intensity interferometer and techniques such as stellar speckle interferometry and speckle holography. Some useful mathematical results as well as some selected topics in optics are summarized for ready reference in five appendices.

I have tried to plan this book so that it can be used by people who would like to apply interferometric techniques in their work, as well as those who would like to learn more about interferometry. In the first instance, most topics are discussed at a level accessible to anyone with a basic knowledge of physical optics; a more detailed treatment for the serious worker then follows. Finally, the text is supplemented by a reference list of nearly 600 selected papers. Accordingly, students should find this book useful as a text, while researchers can use it as a reference work.

I am grateful to many of my colleagues for their assistance. In particular, I would like to mention W. H. (Beattie) Steel and Philip Ciddor with whom I have had many helpful discussions. I would also like to thank Colin Chidley who was responsible for the drawings.

P. Hariharan
Sydney, January 1985

Preface to the Second Edition

The 18 years since the publication of the first edition of this book have seen an explosive growth in activity in the field of optical interferometry. The use of lasers and electronic data-processing techniques is now almost universal, and the volume of literature has expanded to such an extent that it is increasingly difficult to keep up with it, and impossible to cover the entire subject in a book of reasonable size.

The aim of this second edition is, therefore, to review the growth of some selected areas of this rapidly expanding field and introduce the reader to some of these new developments. The contents reflect this shift in emphasis: by omitting the chapters in the first edition dealing with topics such as thin films and interference filters, holography and holographic interferometry, and speckle interferometry, which are not so closely related to the central theme of this book and are now the subject of books in their own right, it has been possible to describe some significant advances in the fields of interference microscopy, interferometric sensors, nonlinear interferometers, gravitational-wave detectors, and quantum effects in optical interference.

As before, a brief historical survey leads up to three chapters covering the concepts of two-beam interference, coherence, and multiple-beam interference. Chapter 5 looks at the laser as a light source and discusses techniques for obtaining a single-frequency output and for frequency stabilization, while Chapter 6 discusses electronic techniques of phase measurement. Five chapters then deal with applications of interferometry such as measurements of length, optical testing, interference microscopy, interference spectroscopy, and Fourier-transform spectroscopy. They are followed by four chapters on interferometric sensors, nonlinear interferometers, stellar interferometry, and studies of space-time and gravitation. The last three chapters discuss single-photon interference, fourth-order interference, and two-photon interferometry. Some useful mathematical results, as well as some selected topics in wave optics, are summarized in seven appendices.

This book is intended for scientists and engineers who would like to use interferometric techniques in their work, as well as students who would like to learn more about interferometry. Accordingly, most topics are introduced at a level accessible to anyone with a basic knowledge of physical optics; this introduction is followed by a more detailed treatment, which is supplemented by a bibliography and a reference list of more than 1000 selected original publications.

I have used American spelling in this edition except for the word "metre." Chapter 7 includes a description of the work by Michelson and others to establish a standard of length based on the wavelength of a monochromatic light source and cites the text of the current, internationally accepted definition of this unit. Accordingly, to avoid inconsistency, I have used the international spelling.

I am grateful to many friends and colleagues. In particular, I must mention Philip Ciddor, John Davis, Barry Sanders, Bill Tango, and Rainer Weiss; without their help, this book could not have been completed.

P. Hariharan
Sydney, April 2003

Chapter 1

Interferometry: Its Development

Almost everyone has come across interference phenomena such as the vivid colors in a soap bubble, or in an oil slick on a wet road, or the colored fringes seen in a thin air film enclosed between two glass plates when they are brought into contact. The latter are commonly known as "Newton's rings," but were, in fact, first described by Boyle and, independently, by Hooke, in the latter half of the 17th century. Their observations can be called the starting point of optical interferometry.

The development of optical interferometry extends over more than 300 years and is closely linked with the history of wave optics. The aim of this chapter is to review briefly some of the significant stages of this development, so as to put in perspective the topics which will be discussed later, in more detail, in this book.

1.1 The Wave Theory of Light

We now know that the colored fringes seen by Boyle and Hooke were produced by the interference of the light waves reflected from the two surfaces of the film. However, though Hooke did put forward a wave theory of light to explain the fringes, and this theory was expanded and put into its present form by Huyghens in 1690, it made little progress because it was opposed by Newton, who believed that a wave theory could not explain either the rectilinear propagation of light or the phenomenon of polarization. As a result, the correct explanation had to wait for almost 150 years.

The barriers to the acceptance of the wave theory were first broken by Young, when, in his Bakerian lectures in 1801 and 1803, he stated the principle of interference and demonstrated that the summation of two rays of light could give rise to darkness. However, even this demonstration did not lead to immediate acceptance of the theory. Almost no one supported Young, and the discovery by Malus, in 1809, that light could be polarized by reflection shook even Young's confidence, as he, like Huyghens, thought that light was propagated as longitudinal waves. The turning point came with

1

Fresnel's brilliant memoir on diffraction in 1818, which perfected the treatment of interference, and the discovery by Arago and Fresnel that two orthogonally polarized beams could not interfere. This led Young and Fresnel to the inevitable conclusion that light waves were transverse waves. Since only longitudinal waves can propagate in a fluid, Fresnel postulated that light waves were propagated through an elastic solid pervading all matter — the "luminiferous aether."

1.2   The Michelson–Morley Experiment

Most of the leading physicists of the 19th century supported the aether theory, even though it raised a number of questions which had no obvious answers. One such was the aberration of light discovered by Bradley in 1728, which indicated that the aether was stationary. However, theoretical calculations by Fresnel in 1818 showed that in a medium with a refractive index n moving with a velocity v, the aether should be carried along with a velocity $v(1 - 1/n^2)$. Fizeau therefore carried out an experiment in 1851 with an interferometer in which the two beams traversed two columns of running water, one beam always moving with the current, while the other moved against it. This experiment, which was repeated later by Jamin and by Michelson, showed a shift of the fringes of the expected magnitude. Based on these results, Maxwell predicted in 1880 that the movement of the earth through the aether should result in a change in the speed of light proportional to the square of the ratio of the speed of the earth to that of light. While Maxwell felt that this effect was too small to be detected experimentally, Michelson was confident that it could be observed by making use of the tremendous increase in accuracy obtained by interferometry. This led, in 1881, to Michelson's famous experiment which was designed to demonstrate the "aether drift." As things turned out, the null result obtained by Michelson led to the rejection of the concept of an aether and laid the foundations for the special theory of relativity [Shankland, 1973].

1.3   Measurement of the Metre

Other applications of interferometry followed in rapid succession. Thus, in 1896, Michelson carried out the first measurement of the length of the Pt–Ir bar which was the international prototype of the metre in terms of the wavelength of the red cadmium radiation. Although the idea of the wavelength of a monochromatic source as a natural standard of length had been suggested much earlier by Babinet and by Fizeau, it was Michelson's work which demonstrated its feasibility and led, in 1960, to the redefinition of the metre in terms of the wavelength of the orange radiation of ^{86}Kr.

1.4 Optical Testing

Another major field of application of interferometry was opened up by Twyman in 1916, when he used a modified Michelson interferometer to test optical components. This interferometer was, in turn, adapted by Linnik [1933] to permit microscopic examination of reflecting surfaces. At the same time, interferometry became a valuable tool in studies of fluid flow and combustion.

1.5 Coherence

Studies by Michelson also revealed the connection between the visibility of the fringes in his interferometer and the dimensions and spectral purity of the source. Since any thermal source can be considered as made up of many elementary radiators (atoms) which are not synchronized, the interference pattern with such a source is obtained by adding the intensities in the interference patterns formed by these incoherent elementary radiators. The gradual transition from incoherent to coherent illumination with a thermal source was demonstrated as far back as 1869 by Verdet, who showed that light from two pinholes illuminated by the sun produced an interference pattern on a screen if their separation was less than 0.05 mm. However, the first quantitative concepts of coherence were formulated by von Laue only in 1907, and it was again only after a long delay that the foundations of modern coherence theory were laid in three papers by van Cittert [1934, 1939] and Zernike [1938]. These concepts were developed in more detail, 20 years later, by Hopkins [1951, 1953] and Wolf [1954, 1955]. The discovery by Brown and Twiss [1954] of intensity-correlation effects (fourth-order correlation) led to the formulation of a general description of higher-order coherence effects by Mandel and Wolf [1965].

1.6 Interference Spectroscopy

Towards the end of the 19th century, interference techniques found their way into high-resolution spectroscopy with the development of instruments such as the Fabry–Perot etalon, the Michelson echelon, and the Lummer–Gehrcke plate. However, as improved multilayer dielectric coatings, with high values of reflectance and negligible losses, became available, the greater light-gathering power (or etendue) of the Fabry–Perot interferometer led to its rapidly replacing the other two instruments.

At this stage, a completely new approach was opened up by the development of Fourier transform spectroscopy. The origins of this technique can be traced back to 1862, when Fizeau studied the effect of the separation of the plates on Newton's rings formed with sodium light. He found that the rings almost disappeared at a separation corresponding to the passage of 490 fringes, but reached maximum contrast once again when 980 fringes had passed, showing that the sodium line was a doublet.

This method was taken a step further by Michelson in 1891, when he plotted the visibility of the fringes as a function of the optical path difference for a number of spectral lines. All of them, with the exception of the red cadmium line, exhibited a series of maxima and minima, indicating that they consisted of more than one component. A similar technique was also applied to far infrared spectroscopy ($\lambda = 100$ to 300 µm) by Rubens and Wood in 1911. However, the systematic development of Fourier-transform spectroscopy started with Fellgett [1951], who was the first to obtain a spectrum from a numerically Fourier-transformed interferogram and demonstrate the advantages of this technique. Subsequent improvements have brought this technique to the point where it reigns supreme at long wavelengths and is preferable to conventional dispersive instruments in the visible region for mapping complex spectra with the highest possible resolution.

1.7   The Laser

Throughout the first half of the 20th century, the most commonly used light source for interferometry was a pinhole illuminated by a mercury arc through a filter which isolated the green line ($\lambda = 546$ nm). Such a source gave only a very small amount of light with limited spatial and temporal coherence. The development of the laser made available, for the first time, an intense source of light with a remarkably high degree of spatial and temporal coherence and initiated a revolution in interferometry.

The origins of the laser can be traced back to 1917, when Einstein pointed out that atoms in a higher energy state, which normally radiate spontaneously, could also be stimulated to emit and revert to a lower energy state when irradiated by a wave of the correct frequency. The most remarkable feature of this process was that the emitted photon had the same frequency, polarization, and phase as the stimulating wave and propagated in the same direction.

Schawlow and Townes [1958] were the first to show that amplification by stimulated emission was possible in the visible region, and that a simple resonator consisting of two mirrors could be used for mode selection. The first practical laser was the pulsed ruby laser [Maiman, 1960]. Continuous laser action was achieved soon afterwards with the helium-neon laser, initially in the infrared [Javan, Bennett, and Herriott, 1961] and then in the visible region [White and Rigden, 1962].

The high degree of directionality and coherence of laser light were verified by Collins *et al.* [1960]. This was followed by the observation of interference fringes when the beams from two pulsed ruby lasers were superposed [Magyar and Mandel, 1963].

Lasers have removed most of the limitations of interferometry imposed by thermal sources and have made possible many new techniques.

1.8   Electronic Techniques

Another development which has revolutionized interferometry has been the increasing use of electronic techniques. This trend started with the use of photoelectric detectors with Fabry–Perot interferometers, but soon extended to such applications as fringe-counting interferometers for length measurements. Digital computers were first used in Fourier transform spectroscopy and have made it an extremely powerful tool. Digital systems, in conjunction with phase-shifting techniques, have also made possible direct measurements of the optical path difference at an array of points covering an interference pattern.

1.9 Heterodyne Techniques

The observation of beats when the beams from two lasers operating at slightly different frequencies were mixed at a photo detector [Javan, Ballik, and Bond, 1962] led to the development of a range of heterodyne techniques to replace traditional methods of interpolation. Since measurements of either the frequency or the phase of a beat can be made with very high precision, such techniques have revolutionized length interferometry. The extension of frequency measurements to the optical region also led to the redefinition of the metre in terms of the speed of light.

Light scattered from a moving particle has its frequency shifted by an amount proportional to the component of its velocity in a direction determined by the directions of illumination and viewing. Lasers made it possible to measure this frequency shift and, hence, the velocity of the particle, by detecting the beats produced by the scattered light and the original laser beam [Yeh and Cummins, 1964].

1.10 Fiber Interferometers

Another major advance followed the use of single-mode optical fibers to build analogs of conventional two-beam interferometers. Such interferometers have the advantage that very long optical paths can be accommodated in a small space. Fiber interferometers are now used widely as rotation sensors [Lin and Giallorenzi, 1979]. In addition, since the optical path through a length of such a fiber changes with pressure or temperature, fiber interferometers have found many applications as sensors for a number of physical quantities [Giallorenzi *et al.*, 1982]. Extremely high sensitivity is possible with fiber interferometers, since they have very low noise, and sophisticated detection techniques can be used with them. In addition, they can be multiplexed to make simultaneous measurements of a number of variables at a number of points.

1.11 Nonlinear Interferometers

The availability of lasers capable of producing extremely intense beams of coherent light has led to the development of interferometers using nonlinear optical elements.

Besides the measurement of third-order susceptibilities, their applications include second-harmonic interferometry [Hopf, Tomita, and Al-Jumaily, 1980], phase-conjugate interferometry [Bar-Joseph *et al.*, 1981; Feinberg, 1983] and high-speed optical switching [Lattes *et al.*, 1983].

In addition, nonlinear techniques have made possible the generation of nonclassical states of light, such as squeezed states [Walls, 1983] and entangled states, opening up completely new areas of research.

1.12 Stellar Interferometry

The idea of using interference to measure the angular dimensions of a star goes back to 1868, when Fizeau proposed an arrangement consisting of two slits placed in front of a telescope; their separation when the interference fringes crossing the image of the star disappeared could then be used to calculate the angle subtended by the star. Unfortunately, this experiment failed because the aperture of the telescope was not large enough.

However, Fizeau's proposal was taken up again by Michelson, who applied it successfully to measure the diameters of Jupiter's satellites in 1890. This success was followed in 1921 by measurements of the diameter of Betelgeuse and six other stars, using a 6-m stellar interferometer mounted on the 2.5-m (100-inch) telescope at Mt. Wilson.

An attempt by Michelson to extend this technique to longer baselines failed because of problems of mechanical stability and atmospheric turbulence, and the next advance came from a completely different approach: namely the idea of measuring the degree of correlation of the intensity fluctuations at two points in the field [Brown and Twiss, 1954]. This new method was used in the intensity interferometer [Brown *et al.*, 1964] to make measurements on 32 stars with angular diameters down to 0.0004 second of arc. Yet another approach, following the development of lasers, was based on heterodyne techniques [Johnson, Betz, and Townes, 1974; Townes, 1984]. In this method, the light from the star is mixed with light from a laser at two photodetectors, and the resulting heterodyne signals are multiplied in a correlator. The output from this correlator is a measure of the degree of coherence of the wave fields at the two photodetectors.

Other developments in this field have been the techniques of stellar speckle interferometry [Labeyrie, 1970] and speckle holography [Liu and Lohmann, 1973; Bates, Gough, and Napier, 1973], which have made it possible to determine the structure of many close groupings of stars.

More recently, modern versions of Michelson's stellar interferometer have been constructed with baselines up to 600 m [see Lawson, 1997]; techniques have also been developed to combine images from widely spaced arrays of large telescopes, to obtain the same resolution as from a single telescope with an aperture equal to the distance between the linked instruments [Baldwin *et al.*, 1996].

1.13 Space-Time and Gravitation

Laser interferometry has been used to verify the null result of the Michelson–Morley experiment to a very high degree of accuracy [Brillet and Hall, 1979] and is now about to be applied to the detection of gravitational waves from black holes and supernovae. The systems under construction for this purpose consist essentially of a Michelson interferometer in which the beam splitter and the end reflectors are attached to separate freely-suspended masses [Moss, Miller, and Forward, 1971; Weiss, 1972]. Since theoretical estimates of the intensity of bursts of gravitational radiation suggest that a sensitivity to strains of the order of a few parts in 10^{21} is needed, arm lengths of 4 km are used, and a number of ways to obtain higher sensitivity are being explored [Abramovici *et al.*, 1992; Weiss, 1999].

1.14 Quantum Effects

Interference has long been regarded as a conclusive demonstration of the wavelike nature of light. However, at low light levels, photodetectors record the annihilation of individual photons. The quantum theory of light evolved in response to the need to reconcile these two contradictory aspects.

Optical interferometry has played an increasingly important part in tests of quantum theory, ranging from experiments involving independent sources at low intensity levels which verified Dirac's famous dictum that "a photon interferes only with itself ..." [Dirac, 1958], to measurements using single-photon sources and photon-pairs. The latter include sophisticated experiments involving Bell's inequality [Bell, 1965] and experiments demonstrating the idea of a "quantum eraser." In all these cases, optical interferometry has provided evidence supporting quantum theory [see Agarwal, 1995; Hariharan and Sanders, 1996].

1.15 Future Directions

It is always an interesting, though risky, exercise to try to guess the directions of future development in any field. In the case of optical interferometry, some trends are obvious, such as the growing use of lasers and digital signal processing to increase the scope, speed, and precision of measurements. With the replacement of the human eye by electronic detectors, the advantages of using infrared wavelengths for interferometry are being exploited to an increasing extent. Other interesting possibilities have been opened up by the application of single-mode fibers and nonlinear crystals to build new types of interferometric sensors and optical switches. Optical interferometry is playing an increasingly important part in astronomy as well as in tests of relativity and quantum theory, and the possibilities of exploiting quantum interference effects are being explored. However, the history of optical interferometry, like that of any other field of science, is full of unexpected advances which have opened up new directions of research. There is every likelihood that more such surprises lie ahead.

Chapter 2

Two-Beam Interference

A beam of light is actually a propagating electromagnetic wave. If, for simplicity, we assume that we are dealing with a linearly polarized plane wave propagating in a vacuum in the z direction, the electric field E at any point can be represented by a sinusoidal function of distance and time,

$$E = a \cos[2\pi \nu(t - z/c)], \tag{2.1}$$

where a is the amplitude, ν the frequency, and c the speed of propagation of the wave. If T is the period of the vibration and ω its circular frequency,

$$T = 1/\nu = 2\pi/\omega. \tag{2.2}$$

The wavelength λ is given by the relation

$$\lambda = cT = c/\nu, \tag{2.3}$$

while the propagation constant of the wave is

$$k = 2\pi/\lambda. \tag{2.4}$$

In a medium with a refractive index n, in which the light wave propagates with a speed

$$v = c/n, \tag{2.5}$$

the wavelength of the same radiation would be

$$\lambda_n = vT$$
$$= v\lambda/c = \lambda/n. \tag{2.6}$$

9

2.1   Complex Representation of Light Waves

Equation (2.1) can also be written as

$$E = \text{Re}\{a \exp[i2\pi\nu(t - z/v)]\}, \tag{2.7}$$

where Re { } represents the real part of the expression within the braces, and i $= (-1)^{1/2}$. This representation has the advantage that the right-hand side can now be expressed as the product of spatially varying and temporally varying factors, so that

$$\begin{aligned} E &= \text{Re}\{a \exp(-i2\pi\nu z/v) \exp(i2\pi\nu t)\} \\ &= \text{Re}\{a \exp(-i\phi) \exp(i2\pi\nu t)\}, \end{aligned} \tag{2.8}$$

where

$$\begin{aligned} \phi &= 2\pi\nu z/v \\ &= 2\pi n z/\lambda. \end{aligned} \tag{2.9}$$

The product $p = nz$ is termed the optical path between the origin and the point z, and ϕ is the corresponding phase difference.

If we assume that all operations on E are linear, it is simpler to use the complex function

$$E = a \exp(-i\phi) \exp(i2\pi\nu t) \tag{2.10}$$

and take the real part at the end of the calculation. We can then rewrite Eq. (2.10) as

$$E = A \exp(i2\pi\nu t), \tag{2.11}$$

where

$$A = a \exp(-i\phi) \tag{2.12}$$

is known as the complex amplitude of the vibration.

Because of the extremely high frequencies of visible light waves ($\nu \approx 6\times 10^{14}$ Hz for $\lambda = 0.5$ μm) direct observations of the electric field are not possible. The only measurable quantity is the intensity, which is the time average of the amount of energy which, in unit time, crosses a unit area normal to the direction of the energy flow. This, in turn, is proportional to the time average of the square of the electric field.

$$\langle E^2 \rangle = \lim_{T \to \infty} \frac{1}{2T} \int_{-T}^{T} E^2 dt. \tag{2.13}$$

From Eqs. (2.1), (2.2), and (2.9) we have

$$\langle E^2 \rangle = \lim_{T \to \infty} \frac{1}{2T} \int_{-T}^{T} a^2 \cos^2(\omega t - \phi) dt$$
$$= a^2/2. \tag{2.14}$$

Since we are not interested in the absolute value of the intensity, but only in relative values over a specified region, we can ignore this factor of 1/2, as well as any other factors of proportionality, and define the optical intensity as

$$I = a^2 = |A|^2. \tag{2.15}$$

2.2 Interference of Two Monochromatic Waves

If two monochromatic waves propagating in the same direction and polarized in the same plane are superposed at a point P, the total electric field at this point is

$$E = E_1 + E_2, \tag{2.16}$$

where E_1 and E_2 are the electric fields due to the two waves. If the two waves have the same frequency, the intensity at this point is

$$I = |A_1 + A_2|^2, \tag{2.17}$$

where $A_1 = a_1 \exp(-i\phi_1)$ and $A_2 = a_2 \exp(-i\phi_2)$ are the complex amplitudes of the two waves. Accordingly,

$$I = A_1^2 + A_2^2 + A_1 A_2^* + A_1^* A_2,$$
$$= I_1 + I_2 + 2(I_1 I_2)^{1/2} \cos \Delta\phi, \tag{2.18}$$

where I_1 and I_2 are the intensities at P due to the two waves acting separately, and $\Delta\phi = \phi_1 - \phi_2$ is the phase difference between them.

If Δp is the corresponding difference in the optical paths given by Eq. (2.9), the order of interference is $N = \Delta p / \lambda$. The intensity has its maximum value $I_{\max}$ when

$$N = m, \ \Delta p = m\lambda, \ \Delta\phi = 2m\pi, \tag{2.19}$$

where m is an integer, and its minimum value $I_{\min}$ when

$$N = (2m+1)/2, \ \Delta p = (2m+1)\lambda/2, \ \Delta\phi = (2m+1)\pi. \tag{2.20}$$

A convenient measure of the contrast of the interference phenomenon is the visibility of the interference fringes, which is defined by the relation

$$\mathcal{V} = \frac{(I_{\max} - I_{\min})}{(I_{\max} + I_{\min})}. \tag{2.21}$$

In the present case, it follows from Eqs. (2.18) to (2.21) that

$$\mathcal{V} = \frac{2(I_1 I_2)^{1/2}}{(I_1 + I_2)}. \tag{2.22}$$

However, light from a thermal source, even when it comprises only a single spectral line, is not strictly monochromatic. Both the amplitude and the phase exhibit very rapid, random fluctuations, and, in the case of beams from different sources, these fluctuations are not correlated. Interference effects cannot therefore be observed between beams derived from two thermal sources. To simplify matters, we shall assume, at this stage, that the two beams are derived from the same source and the correlation between the fluctuations in the beams is complete, in which case they are said to be completely coherent. The elementary theory developed earlier for monochromatic light is then adequate to describe the effects observed.

We shall consider, in the first instance, interference between two beams; the phenomena observed can then be classified according to the method used to obtain these beams.

2.3   Wavefront Division

One way to obtain two beams from a single source is to use two portions of the original wavefront, which are then superimposed to produce interference. This method is known as wavefront division.

A simple optical system for this purpose is the arrangement known as Fresnel's mirrors. In this arrangement, as shown in Fig. 2.1, light from a point source S is reflected at two mirrors M_1 and M_2 which make a small angle ϵ with each other. Interference fringes are observed in the region where the two reflected beams overlap. Interference can be considered to take place between the light from the two virtual

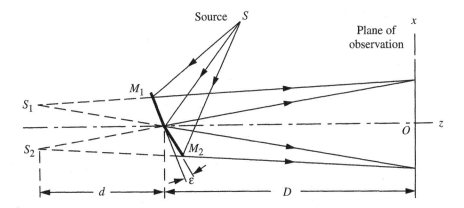

Figure 2.1. Optical system used to produce interference by wavefront division (Fresnel's mirrors).

sources S_1 and S_2, the images of S in the mirrors M_1 and M_2. If d is the distance of S from the mirrors, we have

$$S_1 S_2 = 2d\epsilon, \tag{2.23}$$

and the difference in the lengths of the optical paths traversed by the waves from these two virtual sources to a point $P(x, y)$ on the screen in the neighborhood of O is, to a first approximation,

$$p = \frac{2d\epsilon x}{(D + d)}. \tag{2.24}$$

Since successive maxima or minima in the interference pattern correspond to a change in the optical path difference

$$\Delta p = \lambda, \tag{2.25}$$

the fringes approximate to equidistant straight lines running parallel to the Y axis; their separation Δx is given by the relation

$$\Delta x = \frac{\lambda(D + d)}{2d\epsilon}. \tag{2.26}$$

Other arrangements are Young's double slit, which uses the diffracted beams from a pair of slits, and Lloyd's mirror, which uses one beam from the source and another from a mirror illuminated at near-grazing incidence.

2.4   Amplitude Division

The other method by which two beams can be obtained from a single source is by division of the amplitude over the same section of the wavefront. One way is to use a surface that reflects part of the incident light and transmits part of it.

2.4.1   Interference in a Plane-Parallel Plate

Consider a transparent plane-parallel plate illuminated, as shown in Fig. 2.2, by a point source of monochromatic light. Any point P on the same side of the plate as the source receives two beams of nearly equal amplitude from it, one reflected from the upper surface of the plate and the other from its lower surface. We can see from considerations of symmetry that the interference fringes observed in a plane parallel to the plate are circles with their centers at O', the point where this plane intersects SO, the normal to the plate.

A case of particular interest is when the plane of observation is at infinity. This is the situation when the fringes are observed in the back focal plane of a lens, as shown in Fig. 2.3. In this case the two interfering rays AL and CL' are parallel and are derived from the same incident ray. Let the thickness of the plate be d and its refractive index n_2, while that of the medium on both sides of it is n_1.

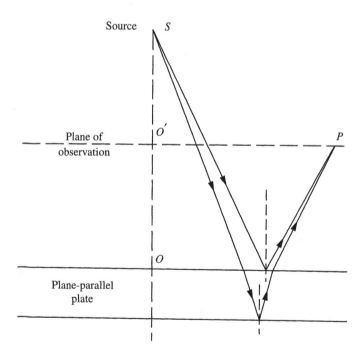

Figure 2.2. Formation of interference fringes by reflection in a plane-parallel plate.

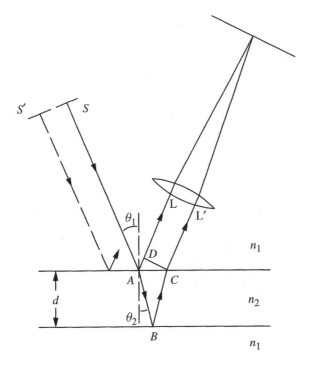

Figure 2.3. Formation of fringes of equal inclination by reflection in a plane-parallel plate.

If, then, θ_1 and θ_2 are, respectively, the angles of incidence and refraction at the upper surface, we have

$$AB = BC = d/\cos\theta_2, \tag{2.27}$$

$$AC = 2d\tan\theta_2, \tag{2.28}$$

and

$$AD = AC\sin\theta_1 = 2d\tan\theta_2\sin\theta_1. \tag{2.29}$$

Accordingly, the optical path difference between the two rays should be

$$\Delta p = n_2(AB + BC) - n_1 AD$$
$$= 2n_2 d\cos\theta_2. \tag{2.30}$$

However, we also have to take into account a phase shift of π introduced by reflection at one of the surfaces (see Appendix C.2); the effective optical path difference between the interfering wavefronts is, therefore,

$$\Delta p = 2n_2 d \cos \theta_2 \pm \lambda/2. \tag{2.31}$$

A bright fringe corresponds to the condition

$$2n_2 d \cos \theta_2 \pm \lambda/2 = m\lambda, \tag{2.32}$$

where m is an integer, while a dark fringe corresponds to the condition

$$2n_2 \cos \theta_2 \pm \lambda/2 = (2m + 1)\lambda/2. \tag{2.33}$$

Equation (2.31) shows that for a given value of d the phase difference between the wavefronts depends only on the angle θ_2. This makes it possible to use an extended monochromatic source instead of a point source. The interference fringes produced in the back focal plane of the lens by any other point S' on an extended source are identical with those produced by S, so that their visibility is unaffected. The fringes are circles centered on the normal to the plate (Haidinger's rings) and are called fringes of equal inclination.

For near-normal incidence, θ_2 is small and Eq. (2.33) can be written as

$$2n_2 d(1 - \theta_2^2/2) = m\lambda. \tag{2.34}$$

Accordingly, if there is an intensity minimum (or maximum) at the center, the radii of the dark (or bright) rings are proportional to the square roots of consecutive integers.

Similar phenomena can also be observed in transmission. In this case the directly transmitted beam interferes with the beam formed by two internal reflections. Since the net phase shift introduced by the reflections at the two surfaces of the plate is either zero or 2π, the optical path difference between the beams is

$$\Delta p = 2n_2 d \cos \theta_2. \tag{2.35}$$

The fringes are, therefore, complementary to those seen by reflection. However, since the relative amplitudes of the two beams are usually very different (1.00 and 0.05 for a glass plate in air), the visibility of the fringes is low.

2.4.2 Fizeau Fringes

We assume that the plate has only small variations of thickness and that it is illuminated at near-normal incidence with a collimated beam of monochromatic light as shown in Fig. 2.4. A lens forms an image of the plate at P. Consider two parallel incident rays which follow the paths $S_1 ABCL_1 P$ and $S_2 CL_2 P$. Since the optical paths $CL_1 P$ and $CL_2 P$ are equal, it is sufficient to consider the difference in the optical paths between C and the source. In addition, since A and C are very close together, we can neglect the variation in thickness between these points and assume that the plate has

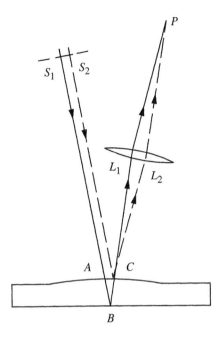

Figure 2.4. Formation of interference fringes of equal thickness (Fizeau fringes) by reflection.

a well defined thickness d over this region. The optical path difference is then, from Eq. (2.31),

$$\Delta p = 2n_2 d \cos \theta_2 \pm \lambda/2, \tag{2.36}$$

which, when θ_2 is small, reduces to

$$\Delta p = 2n_2 d \pm \lambda/2. \tag{2.37}$$

The fringes then correspond to contours of equal thickness and are known as Fizeau fringes.

In this case also, a complementary fringe pattern of low contrast is seen by transmission.

2.4.3 Interference in a Thin Film

Fringes of equal thickness can be produced in a thin film even without the use of collimated light. With an extended source, it follows from Eq. (2.36) that the optical path difference varies with the angle θ_2 as well as with the thickness d. However, for a thin film (for which d is very small), if the interference fringes are viewed with an optical system (or with the eye) focused on the film, the pupil limits the range of values of θ_2 for any point on the film, so that, for near-normal incidence ($\cos\theta_2 \approx 1$), the variation of the optical path difference over this range of angles is negligible. Under these conditions, the fringes are, effectively, fringes of equal thickness.

2.5 Localization of Fringes

So far, we have studied the phenomenon of interference under conditions which permit a relatively simple analysis. The principal restriction imposed was the use of a point source or a collimated beam of monochromatic light. We shall now examine the consequences of relaxing this restriction on the size of the source.

2.5.1 Nonlocalized Fringes

Consider the optical system shown in Fig. 2.5, which produces two images S_1 and S_2 of the point source S. Interference fringes are produced where the beams from these two secondary sources overlap. At a point P, the optical path difference between the beams is

$$\Delta p = n_1(S_2 P - S_1 P), \tag{2.38}$$

where n_1 is the refractive index of the surrounding medium. The value of Δp depends on the position of P; interference fringes corresponding to the loci of points for which Δp is constant are therefore formed over the plane containing P. Since this plane can take any position within the region of superposition of the beams, the fringes are said to be nonlocalized. Such nonlocalized fringes are always obtained with a point source. If, in addition, it is a monochromatic source, the visibility of the fringes depends only on the relative intensities of the two images S_1 and S_2.

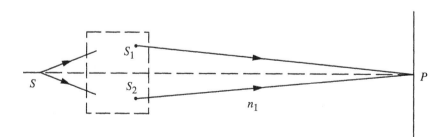

Figure 2.5. Formation of nonlocalized fringes with a point source.

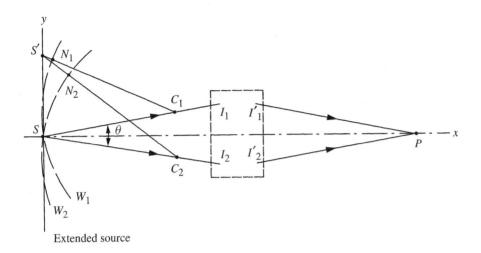

Figure 2.6. Localization of fringes with an extended source.

2.5.2 Localized Fringes

Let us now consider what happens when, as shown in Fig. 2.6, the point source S is replaced by an extended source. Such a source can be considered as an array of independent point sources, each of which produces a separate interference pattern. If the path differences at P are not the same for all these point sources, these elementary fringe patterns do not coincide, and there is a reduction in the visibility of the fringes. Since the reduction in visibility depends on the position of P, there is, in general, a position of the plane of observation for which the visibility of the fringes is a maximum; the fringes are then said to be localized in this plane.

Localization of the fringes is a normal consequence of the use of an extended source. We can study the phenomenon in more detail by examining how the optical path difference at P varies with the position of a generalized point S' in the source. For this, it is convenient to interchange the source plane and the plane of observation, and consider the situation if we were to have a point source at P. This point source would then give rise to two spherical wavefronts W_1 and W_2, passing through S, which are normal to SI_1 and SI_2, respectively. The optical path difference for these wavefronts at S is obviously given by Eq. (2.38). If now, the normals from S' to W_1 and W_2 are $S'N_1$ and $S'N_2$, respectively, these normals represent the additional paths traversed by these wavefronts before they reach S'. Accordingly, the optical path difference for the two wavefronts from S' at P is

$$\Delta p' = \Delta p + \Delta' p, \tag{2.39}$$

where $\Delta' p = n_1 (S'N_2 - S'N_1)$.

To calculate $\Delta'p$, it is convenient to choose the bisector of the angle I_1SI_2 and the normal to it as the coordinate axes SX and SY. The coordinates of C_1 and C_2, the centers of the wavefronts W_1 and W_2, are then $[R_1 \cos(\theta/2), R_1 \sin(\theta/2)]$ and $[R_2 \cos(\theta/2), -R_2 \sin(\theta/2)]$, respectively, where $R_1 = C_1S$ and $R_2 = C_2S$ are the radii of curvature of W_1 and W_2. Accordingly, since S' has the coordinates $(0, y)$, if we set $l_1 = S'N_1, l_2 = S'N_2$,

$$C_1S' = R_1 + l_1$$
$$= \{R_1^2 \cos^2(\theta/2) + [y - R_1 \sin(\theta/2)]^2\}^{1/2}, \tag{2.40}$$

and

$$C_2S' = R_2 + l_2$$
$$= \{R_2^2 \cos^2(\theta/2) + [y + R_2 \sin(\theta/2)]^2\}^{1/2}. \tag{2.41}$$

If the dimensions of the source are small compared to R_1 and R_2, terms involving powers of (y/R_1) and (y/R_2) higher than the second can be neglected, so that we have

$$l_1 = -y \sin(\theta/2) + (y^2/2R_1) \cos^2(\theta/2) \tag{2.42}$$

and

$$l_2 = y \sin(\theta/2) + (y^2/2R_2) \cos^2(\theta/2), \tag{2.43}$$

so that

$$\Delta'p = n_1(l_2 - l_1)$$
$$= n_1\{2y \sin(\theta/2) + (y^2/2)[(1/R_2) - (1/R_1)] \cos^2(\theta/2)\}, \tag{2.44}$$

or, since θ is usually small,

$$\Delta'p \approx n_1\{y\theta + (y^2/2)[(1/R_2) - (1/R_1)]\}. \tag{2.45}$$

In particular, if the source is small enough for the term involving y^2 in Eq. (2.45) to be neglected,

$$\Delta'p = n_1 y\theta. \tag{2.46}$$

The fringes then have maximum visibility at P when $\Delta'p = 0$, that is to say, when $\theta = 0$. This occurs when SI_1 and SI_2 coincide or, in other words, the two rays $I_1'P$ and $I_2'P$, which intersect at P, originate in the same ray coming from S. It follows that the interference fringes are localized on a surface which is the locus of the points of intersection of pairs of rays, each of which is derived from a single ray leaving the source.

In the vicinity of this surface of localization, $y\theta$ is negligible, and Eq. (2.45) reduces to

$$\Delta'p = n_1[(1/R_2) - (1/R_1)]y^2/2. \tag{2.47}$$

It is apparent from Eq. (2.47) that the visibility of the fringes in the plane of localization decreases as the size of the source increases. For good visibility $\Delta'p$ must be less than a small fraction of the wavelength (say, $\lambda/4$). The maximum dimensions of the source are then given by the condition that

$$n_1[(1/R_2) - (1/R_1)]y^2/2 \le \lambda/4, \tag{2.48}$$

or

$$y \le (\lambda R_1 R_2/2n_1|R_1 - R_2|)^{1/2}. \tag{2.49}$$

We shall apply these results to some specific cases in the next two sections.

2.5.3 Fringes in a Plane-Parallel Plate

In the case of a plane-parallel plate (see Section 2.4.1), when the fringes are observed in the back focal plane of a lens, $\theta = 0$, and R_1 and R_2 are infinite for all positions of S'. Accordingly, the fringes are localized at infinity, and their visibility does not decrease with an extended source.

2.5.4 Fringes in a Wedged Thin Film

Consider a thin layer of a material with a refractive index n_2 contained between two surfaces of a material with a refractive index n_1 which, as shown in Fig. 2.7, make a small angle ϵ with each other. A ray from a point source S gives rise to two reflected rays, one reflected from the upper surface and the other from the lower surface, which meet at P. It follows that, with an extended source centered on S, the fringes of equal thickness formed with such a wedge, which are straight lines running at right angles to its principal section, are localized at P. When, as in Fig. 2.7a, the points B and O are on the same side of the normal at A, the point P lies on the same side of the wedge as S. However, when, as in Fig. 2.7b, the points B and O are on opposite sides of the normal at A, the point P lies on the other side of the wedge.

A detailed analysis [Born and Wolf, 1999] shows that in both cases

$$AP \approx (an_1^2 \sin\theta \cos^2\theta)/(n_2^2 - n_1^2 \sin^2\theta), \tag{2.50}$$

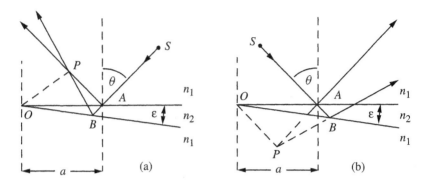

Figure 2.7. Localization of fringes in a wedged thin film.

where θ is the angle of incidence, and a is the distance of the point of incidence on the upper surface from the apex of the wedge. For an air film between thin glass plates, the effects of refraction at the upper surface may be neglected, and we can set $n_1 = n_2 = 1$. We then have

$$AP \approx a \sin \theta, \qquad (2.51)$$

and the angle OPA is very nearly a right angle.

The coordinates of P can be obtained from Fig. 2.8; they are

$$x = u \cos^2 \theta - v \sin \theta \cos \theta \qquad (2.52)$$

and

$$y = u \cos \theta \sin \theta - v \sin^2 \theta, \qquad (2.53)$$

where (u, v) are the coordinates of S. It follows, therefore, that the locus of P as A, the point of incidence, moves across the surface, is a circle with its center at $(u/2, -v/2)$ or, in other words, a circle with OS' as a diameter, where S' is the reflected image of S in the upper surface of the wedge. In the limiting case when S is at infinity, corresponding to illumination with a collimated beam, the locus of P is a plane passing through the vertex of the wedge. For normal incidence, this plane of localization coincides with the front surface of the wedge.

To determine the maximum size of the source that does not lead to an appreciable drop in the visibility of the fringes, we imagine a point source placed at P and evaluate R_1 and R_2, the radii of curvature of the wavefronts W_1 and W_2 reaching S after reflection at the upper and lower surfaces of the wedge. This calculation shows [see Born and Wolf, 1999] that, in general, W_2 has different radii in the principal section and at right angles to it. However, for near-normal incidence, this difference can be

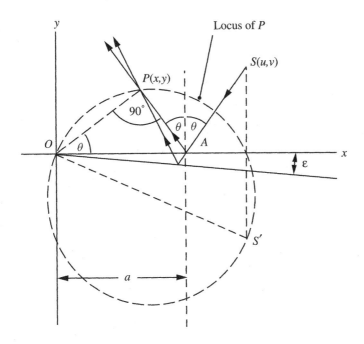

Figure 2.8. Locus of the plane of localization with changing angle of incidence.

neglected and we have $|R_2 - R_1| = 2dn_1/n_2$, where d is the thickness of the film at the point of incidence. In addition, if the source is at a relatively great distance from the film, compared to its thickness, Eq. (2.49) can be rewritten in terms of α, the angular radius of the source, as seen from P, as

$$\alpha = (1/R_1)(\lambda R_1 R_2/2n_1|R_1 - R_2|)^{1/2}$$

$$\approx (\lambda n_2/4n_1 d)^{1/2}, \tag{2.54}$$

since $R_2/R_1 \approx 1$.

2.6 Two-Beam Interferometers

For many applications, it is desirable to have an optical arrangement in which the two interfering beams travel along separate paths before they are recombined. This requirement has led to the development of a number of interferometers for specific purposes. However, apart from the Rayleigh interferometer and its variations, which make use of wavefront division, most of them are based on division of amplitude.

Three methods have been described for division of amplitude. The first uses a partially reflecting film of a metal or dielectric, commonly called a beamsplitter. Another uses a birefringent element to produce two orthogonally polarized beams. However, for these two beams to interfere, they must be derived from a single polarized beam and then brought into the same state of polarization. A third method uses a grating, or a scatter plate (a surface whose amplitude transmittance varies in a random manner), to produce one or more diffracted beams, as well as a reflected or transmitted beam.

Two-beam interferometers can also have different configurations, typified by the Michelson, Mach–Zehnder, and Sagnac interferometers.

2.7 The Michelson Interferometer

In this instrument, as shown in Fig. 2.9, light from a source S is divided at a semireflecting coating on one surface of a plane-parallel glass plate B into two beams with nearly equal amplitudes. These beams are reflected back at two plane mirrors M_1, M_2 and return to B, where they are recombined before they emerge at O.

The optical path difference p between the two arms of the interferometer is given by the difference of two summations, taken respectively over the two arms, of the

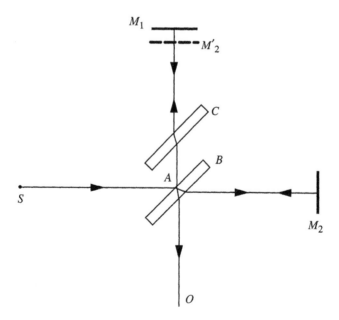

Figure 2.9. The Michelson interferometer.

products of the thickness of each medium traversed and its refractive index, so that

$$p = \sum_2 nd - \sum_1 nd. \tag{2.55}$$

The optical path difference defined by Eq. (2.55) is normally dependent on the wavelength, since all glasses have a finite dispersion $dn/d\lambda$ which is also a function of wavelength. If the optical path difference p is to be strictly independent of wavelength, it is necessary for both arms to contain the same thickness of glass having the same dispersion. Since, in the Michelson interferometer, one beam traverses the beamsplitter B only once, while the other traverses it three times, a compensating plate C of the same material and having the same thickness as B is introduced in the first beam.

Reflection at the beamsplitter B produces an image of the mirror M_2 at M_2'. The interference pattern observed is therefore the same as that produced in a layer of air bounded by the mirror M_1 and M_2', the virtual image of M_2.

2.7.1 Nonlocalized Fringes

A monochromatic point source S located at a finite distance gives rise to two virtual point sources, S_1 and S_2, which are the images of S reflected in M_1 and M_2. Nonlocalized fringes are then observed on a screen placed at O. If, as shown in Fig. 2.10a, the line $S_1 S_2$ is parallel to AO, the fringes are circular; if, as in Fig. 2.10b, $S_1 S_2$ is at right angles to AO, they are parallel straight lines.

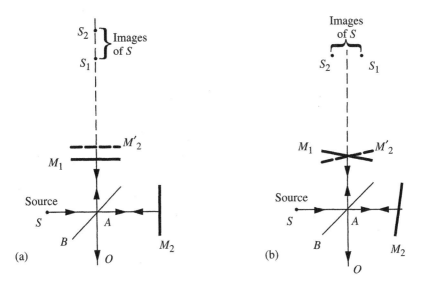

Figure 2.10. Formation of nonlocalized fringes in a Michelson interferometer.

2.7.2   Fringes of Equal Inclination

With an extended monochromatic source, circular fringes of equal inclination are formed when M_1 and M_2' are parallel. These fringes are localized at infinity and can be observed at O, either with a telescope or directly with the eye relaxed. If M_1 is moved closer to M_2', the fringes contract toward the center of the pattern where, one after the other, they vanish; at the same time, the scale of the pattern increases until, when M_1 and M_2' coincide, a uniform field is obtained.

2.7.3   Fringes of Equal Thickness

If M_1 and M_2' make a small angle with each other, and their separation is very small, fringes of equal thickness can be seen with an extended source. These fringes are equidistant straight lines parallel to the apex of the wedge and are localized at $M_1 M_2'$.

2.8   The Mach–Zehnder Interferometer

As shown in Fig. 2.11, the Mach–Zehnder interferometer uses two beamsplitters B_1, B_2, and two mirrors M_1, M_2. Light from a source S is divided at the semireflecting surface of B_1 into two beams, which, after reflection at the mirrors M_1, M_2, are recombined at the semireflecting surface of B_2. Usually, B_1, B_2 and M_1, M_2 are adjusted so that they are approximately parallel, and the paths traversed by the beams form a rectangle or a parallelogram.

We assume that the interferometer is illuminated with a collimated beam giving rise to two plane wavefronts, say W_1 and W_2, in the two arms. Let W_2' be the image of the plane W_2 in the beamsplitter B_2. The phase difference between W_1 and W_2' at a point P on W_1 is then

$$\Delta\phi = (2\pi/\lambda)nd, \qquad (2.56)$$

where d is the separation of W_1 and W_2' at P, and n is the refractive index of the medium between them. If W_1 and W_2' make a small angle with each other, a nonlocalized interference pattern is seen, consisting of equally spaced straight fringes parallel to the line of intersection of W_1 and W_2'.

A similar fringe pattern is also obtained with an extended monochromatic source, provided the separation of W_1 and W_2' is small. However, these fringes are localized, as shown in Fig. 2.12, at the meeting point O of two rays emerging from the interferometer which are derived from a single ray incident on B_1. The position of this region of localization depends on the separation of the two rays when they leave B_2, as well as on the angle between them, and can be varied by changing these parameters.

The Mach–Zehnder interferometer is a more versatile instrument than the Michelson interferometer because each of the widely separated beam paths is

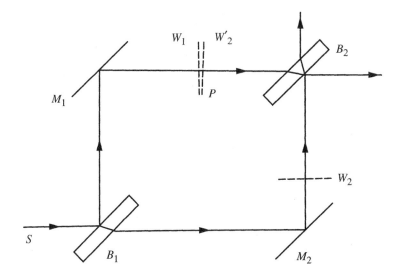

Figure 2.11. The Mach–Zehnder interferometer.

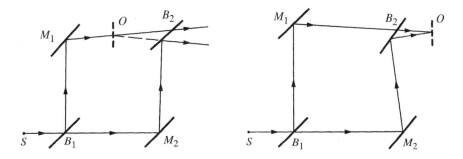

Figure 2.12. Localization of fringes in the Mach–Zehnder interferometer.

traversed only once, and the fringes can be localized in any desired plane. The Mach–Zehnder interferometer has therefore been used extensively in studies of gas flow, combustion, plasma density, and diffusion, where changes in refractive index occur that can be related to changes in pressure, temperature, or the relative concentrations of different components of a mixture [see, for example, Weinberg, 1963]. However, a problem with the Mach–Zehnder interferometer is that adjusting it, to obtain fringes of good visibility with an extended broad-band source, normally involves a number of steps [Clark, Hause, and Bennett, 1953; Panarella, 1973], since a displacement of a mirror results in a shift of the plane of localization, as well as a change in the optical path difference; the procedure can be simplified by a modified optical arrangement [Hariharan, 1969a] which decouples the two adjustments.

2.9 The Sagnac Interferometer

This type of interferometer was actually first described by Michelson [see Hariharan, 1975c] but is more commonly associated with Sagnac, who carried out an extensive series of experiments with it. In this interferometer, the two beams travel around the same closed circuit in opposite directions. As a result, the interferometer is extremely stable. In addition, since the optical paths of the two beams are always very nearly equal, it is very easy to align, even with an extended broad-band source. Two forms of this interferometer are possible, as shown in Figs. 2.13a and b, one with an even number of reflections in each path and the other with an odd number of reflections.

A major difference between the two forms is that in the one shown in Fig. 2.13b, the wavefronts are laterally inverted with respect to each other within the interferometer. As a result, the paths traversed by the two beams within the interferometer can be physically separated by a lateral displacement of the incident beam. On the other hand, in the form shown in Fig. 2.13a the two beams are always superimposed in the same sense.

The formation of interference fringes in an interferometer of the type shown in Fig. 2.13a has been analyzed in detail by Yoshihara [1968]. When B, M_1, M_2 are perpendicular to the plane of the figure, any two rays formed by the division of a single ray at the beamsplitter will always emerge parallel to one another, though they may exhibit a lateral separation. This separation can be varied by translating any one of the elements B, M_1, M_2, or by rotating any one of them about an axis perpendicular to the plane of the figure. The optical path difference between two such rays at a plane normal to them is zero for a particular angle of incidence, but increases with the deviation from this angle and the lateral separation of the rays. Consequently, when the interferometer is illuminated with an extended source, an observer sees a system of straight, vertical fringes, localized at infinity, with a spacing that is inversely proportional to the lateral separation between the emerging rays.

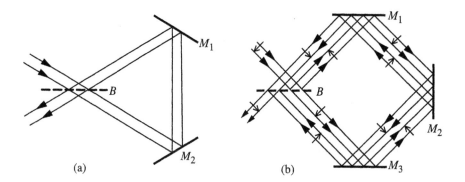

Figure 2.13. Two forms of the Sagnac interferometer.

If, however, the beamsplitter is tilted about an axis in the plane of the figure, horizontal fringes are obtained. In this case, the two emerging beams make an angle with each other, and the fringes are localized at a finite distance.

An interferometer of the form shown in Fig. 2.13b, with an odd number of reflections in each beam, is insensitive to displacements of the mirrors or the beamsplitter. However, a rotation of any of them results in the introduction of a tilt between the two rays emerging from the interferometer derived from a single ray incident on the beamsplitter. Since the lateral separation of such a pair of rays falls to zero at the mirror M_2, the fringes seen with an extended source are localized at the surface of this mirror.

2.10 Interference with White Light

With a point source of white light, a fringe system is produced for each wavelength, and the intensities of these fringe systems add at any point in the plane of observation. Since the optical path difference at the center of the plane of observation in an interferometer, such as that shown in Fig. 2.1, is zero for all wavelengths, all the fringe systems formed with different wavelengths have a maximum at this point, resulting in a white central fringe. However, because the spacing of the fringes varies with the wavelength, they rapidly get out of step as the point of observation moves away from the center of the pattern, resulting in a sequence of interference colors. These colors become less and less saturated as the path difference increases.

For interference in a thin film of air enclosed between two glass surfaces, we have to take into account the additional phase shift of π produced by reflection at one surface. As a result, the film appears black when its thickness is very small compared to the wavelength. As the thickness of the film increases, a sequence of interference colors is observed, complementary to those observed with the system shown in Fig. 2.1. Detailed calculations of the color changes in both cases have been made by Kubota [1950, 1961]. These calculations can also be applied to interferometers using amplitude division; however, in this case, we have to take into account the fact that the phase shifts for the two beams on reflection at the beamsplitter may not be the same, since, in general, these reflections do not take place under the same conditions.

2.11 Channeled Spectra

With a white light source, the visibility of the fringes decreases rapidly as the optical path difference is increased. Beyond a certain point, no interference colors are seen. However, this does not mean that interference is not taking place. To illustrate this point, consider an air film of thickness d illuminated at near-normal incidence by a point source of white light. The order of interference in this film for reflected light, for any wavelength λ, is

$$N = (2d/\lambda) + (1/2). \tag{2.57}$$

Those wavelengths which satisfy the condition

$$(2d/\lambda) = m, \tag{2.58}$$

where m is an integer, correspond to interference minima and are, therefore, missing in the light reflected from the film.

If, then, as shown in Fig. 2.14, an image of a small region of the film is projected on the slit of a spectroscope, the spectrum will be crossed by dark bands corresponding to the wavelengths defined by Eq. (2.58). For the wavelengths λ_1 and λ_2 corresponding to any two dark bands, we have

$$(2d/\lambda_1) = m_1 \tag{2.59}$$

and

$$(2d/\lambda_2) = m_2, \tag{2.60}$$

so that

$$d = \frac{\lambda_1 \lambda_2 (m_1 - m_2)}{2(\lambda_2 - \lambda_1)}. \tag{2.61}$$

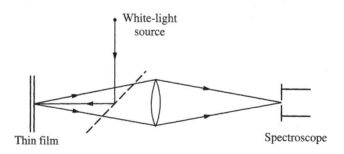

Figure 2.14. Experimental arrangement for viewing the channeled spectrum formed by interference in a thin film.

2.12 Achromatic Fringes

So far, we have assumed that the optical path difference is independent of wavelength and that, consequently, the order of interference at any point is inversely proportional to the wavelength. However, it is possible to have a situation in which the optical path difference varies with wavelength in such a manner that the order of interference is very nearly independent of the wavelength. Under these conditions, achromatic fringes are obtained.

Achromatic fringes can be produced if, as shown in Fig. 2.15, the secondary sources producing the interference pattern are the spectra formed by diffraction at a grating. In this case, the separation of the two images of the source S formed at S' and S'' by light of wavelength λ is

$$d_\lambda \approx f\lambda/\Lambda, \tag{2.62}$$

where Λ is the spacing of the lines in the grating. The spacing of the interference fringes formed in the plane of observation at a distance D from S' and S'' is then

$$\Delta x = D\Lambda/2f, \tag{2.63}$$

which is independent of the wavelength.

Very bright achromatic fringes can also be produced by introducing a thick tilted plate in a triangular-path (Sagnac) interferometer [Hariharan and Singh, 1959b].

Another method of producing achromatic fringes, or "stationary phase" fringes, is to introduce a dispersing medium in one of the paths in an interferometer. As shown in Fig. 2.16, consider a Michelson interferometer in which an additional glass plate having a thickness d and a refractive index n has been introduced in one arm. The net optical path difference between the beams at the center of the field is then

$$p = p_1 - (p_2 + nd), \tag{2.64}$$

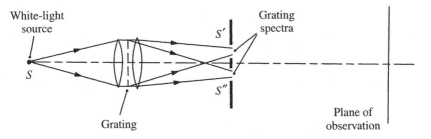

Figure 2.15. Experimental arrangement for producing achromatic interference fringes.

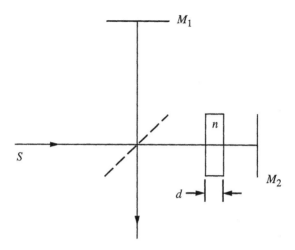

Figure 2.16. Production of achromatic interference fringes in a Michelson interferometer.

where p_1 and p_2 are the paths in air in the two arms, and the interference order is

$$N = (p/\lambda)$$
$$= (1/\lambda)[p_1 - (p_2 + nd)]. \qquad (2.65)$$

Achromatic fringes are obtained when $(dN/d\lambda) = 0$, that is to say, when

$$p_1 - \{p_2 + d[n - \lambda(dn/d\lambda)]\} = 0. \qquad (2.66)$$

A comparison of Eqs. (2.64) and (2.66) shows that achromatic fringes can be obtained only with a finite optical path difference between the beams. With an extended source we have circular fringes of equal inclination localized at infinity. The setting for achromatic fringes is best identified by viewing the channeled spectrum formed with a white light source. When the phase is stationary at some wavelength, the fringes at that point of the spectrum run parallel to the dispersion, or, if M_1 and M_2 are parallel, are spread out so that the interference order passes through a flat maximum [Hariharan and Singh, 1959a].

As will be shown later (see Section 3.3) the term $[n - \lambda(dn/d\lambda)]$ is the group refractive index for the glass plate. Accordingly, Eq. (2.66) is the condition that the difference of the group optical paths in the two arms is zero; this is also the condition that the interferometer is compensated for a range of wavelengths with a finite optical path difference between the beams [Steel, 1962].

2.13   Standing Waves

When a collimated beam of monochromatic light is incident normally on a plane mirror, interference between the incident wave and the reflected wave results in a system of standing waves. The existence of these standing waves was demonstrated in 1890 by Wiener, who also showed that the electric vector was zero, corresponding to a node, at the reflecting surface.

For oblique incidence, the amplitude of the standing wave is a maximum when the incident light is polarized with its electric vector perpendicular to the plane of incidence, and does not then depend on the angle of incidence. If, however, the incident light is polarized in the plane of incidence, the standing wave disappears when the angle of incidence is 45°, since the electric vectors of the incident wave and the reflected wave are then at right angles and cannot interfere.

2.14   Interferential Color Photography

Interferential color photography, demonstrated by Lippmann in 1891, is based on recording these standing waves [see Fournier, 1991, 1994].

If a high-resolution photographic plate is placed with the emulsion side in contact with a mirror and illuminated normally with monochromatic light of wavelength λ, a series of layers of reduced silver are formed in the photographic emulsion, running parallel to its surface. These layers correspond to the antinodal planes and are separated by a distance

$$\Lambda = \lambda/2n, \tag{2.67}$$

where n is the refractive index of the emulsion. Since the thickness of the photographic emulsion is much greater than the separation of the layers, a large number of layers are formed in the emulsion. When the processed photographic plate is illuminated with white light, the reflected components from all these layers, having a wavelength λ, emerge in phase, so that their amplitudes add. As a result, the plate selectively reflects a narrow band of wavelengths centered on the wavelength to which it was exposed.

If, however, the exposure is made with polychromatic light, the interference pattern that is recorded represents the superposition of the individual patterns that would be generated by the monochromatic components of the incident light. The envelope of the interference fringes recorded in the emulsion is then the temporal Fourier transform of the spectrum of the incident light (see Appendix A.1). The contrast of the fringe system is a maximum near the surface and falls off rapidly with increasing depth in the emulsion. When the plate is illuminated with white light, the reflected light has the same spectral distribution as the light to which the plate was originally exposed.

It follows that if an image of a multicolored object is formed on the photographic plate, each point on it reproduces the spectral distribution to which it was exposed. Lippmann photography can, therefore, accurately reproduce a continuous spectrum, unlike conventional color photography, where the colors formed are obtained by mixing three primaries.

Because it is virtually impossible to copy Lippmann photographs, they offer advantages as a unique security device [Bjelkhagen, 1999].

Chapter 3

Coherence

A detailed study of the effects due to the finite size and spectral bandwidth of a source requires the more powerful tools provided by coherence theory, which is essentially a statistical description of the properties of the radiation field in terms of the correlation between the vibrations at different points in the field. Comprehensive treatments of coherence theory are to be found in reviews by Mandel and Wolf [1965] and by Troup and Turner [1974], as well as in books by Beran and Parrent [1964], Peřina [1971], Mandel and Wolf [1995] and Born and Wolf [1999].

3.1 Quasi-Monochromatic Light

As mentioned in Chapter 2, light from a thermal source is not strictly monochromatic. Accordingly, to evaluate the electric field due to a thermal source emitting over a range of frequencies, we have to sum the fields due to the individual monochromatic components. From Eqs. (2.1) and (2.9) the net electric field is given by the integral

$$V^{(r)}(t) = \int_0^\infty a(v) \cos[2\pi v t - \phi(v)] \mathrm{d}v, \qquad (3.1)$$

where $a(v)$ and $\phi(v)$ are the amplitude and phase, respectively, of a monochromatic component of frequency v. Since $V^{(r)}(t)$ is a real function, our first step is to develop a complex representation which is a generalization of that used earlier for monochromatic light. For this we make use of the associated function (also real)

$$V^{(i)}(t) = \int_0^\infty a(v) \sin[2\pi v t - \phi(v)] \mathrm{d}v. \qquad (3.2)$$

We can then define a complex function

$$V(t) = V^{(r)}(t) + iV^{(i)}(t)$$

$$= \int_0^\infty a(v) \exp\{i[2\pi vt - \phi(v)]\}dv. \tag{3.3}$$

This complex function is called the analytic signal associated with the real function $V^{(r)}(t)$; its properties have been discussed in detail by Born and Wolf [1999]. In particular, it can be shown that if $v(v)$ is the Fourier transform of $V^{(r)}(t)$, so that

$$V^{(r)}(t) = \int_{-\infty}^\infty v(v) \exp(-i2\pi vt)dv, \tag{3.4}$$

we have

$$V(t) = 2 \int_0^\infty v(v) \exp(-i2\pi vt)dv. \tag{3.5}$$

It can also be shown that

$$\langle V(t)V^*(t) \rangle = 2\langle [V^{(r)}(t)]^2 \rangle, \tag{3.6}$$

so that, if we ignore a factor of $(1/2)$, as we have done in Eq. (2.15), the optical intensity due to a quasi-monochromatic source is

$$I = \langle V(t)V^*(t) \rangle. \tag{3.7}$$

It follows that if the operations on $V^{(r)}(t)$ are linear, it is possible to replace it by $V(t)$ and take the real part at the end of the calculation.

3.2 Waves and Wave Groups

Monochromatic light corresponds to an infinitely long train of waves, all of which have the same wavelength or frequency. The superposition of infinite trains of waves of slightly different frequencies (as from a quasi-monochromatic source) results in the formation of wave groups.

For convenience, we take $E = a\cos(\omega t - kz)$, where $\omega = 2\pi v$ and $k = 2\pi/\lambda$, to represents a monochromatic wave propagating along the z axis. A quasi-monochromatic beam consisting of a number of such waves whose angular frequencies lie within a very small interval $\pm\Delta\omega$ about a mean angular frequency $\bar{\omega}$ can then be represented by the relation

$$V(z, t) = \int_{\bar{\omega}-\Delta\omega}^{\bar{\omega}+\Delta\omega} a(\omega)\exp[-i(\omega t - kz)]d\omega. \tag{3.8}$$

If $\bar{k}$ is the propagation constant for an angular frequency $\bar{\omega}$, Eq. (3.8) can be rewritten as

$$V(z, t) = A(z, t)\exp[-i(\bar{\omega}t - \bar{k}z)], \tag{3.9}$$

where

$$\begin{aligned} A(z, t) &= \int_{\bar{\omega}-\Delta\omega}^{\bar{\omega}+\Delta\omega} a(\omega)\exp\{-i[(\omega - \bar{\omega})t - (k - \bar{k})z)]\}d\omega \\ &\approx \int_{\bar{\omega}-\Delta\omega}^{\bar{\omega}+\Delta\omega} a(\omega)\exp\{-i[(\omega - \bar{\omega})[t - (dk/d\omega)z]\}d\omega, \end{aligned} \tag{3.10}$$

provided $(dk/d\omega)$ does not vary appreciably over the range $\bar{\omega} \pm \Delta\omega$.

Equation (3.9) represents a wave with an angular frequency $\bar{\omega}$ whose amplitude varies periodically at a much lower frequency. The modulation envelope effectively divides the carrier wave into groups whose length is inversely proportional to the modulation frequency and, hence, inversely proportional to the spectral bandwidth of the light.

3.3 Phase Velocity and Group Velocity

Equation (3.9) shows that the planes of constant phase in an amplitude modulated wave are propagated with a velocity

$$v = \bar{\omega}/\bar{k}, \tag{3.11}$$

known as the phase velocity. The phase velocity is also the velocity of propagation of the carrier wave. On the other hand, Eq. (3.10) shows that a plane of constant amplitude is propagated with a velocity

$$v_g = d\omega/dk, \tag{3.12}$$

which is known as the group velocity.

To evaluate the group velocity, we make use of the fact that for a monochromatic wave $\omega = kv$. Equation (3.12) then gives

$$v_g = v + k(\mathrm{d}v/\mathrm{d}k)$$
$$= v - \lambda(\mathrm{d}v/\mathrm{d}\lambda). \tag{3.13}$$

In a nondispersive medium, $\mathrm{d}v/\mathrm{d}\lambda = 0$, and the group velocity is the same as the phase velocity. However, in a dispersive medium, in which the phase velocity depends on the frequency, the group velocity differs from the phase velocity.

We also have, corresponding to the group velocity, a group refractive index

$$n_g = c/v_g, \tag{3.14}$$

which can be evaluated by expanding Eq. (3.14) as a Taylor series; we then have

$$n_g = n - \lambda(\mathrm{d}n/\mathrm{d}\lambda)$$
$$= n + v(\mathrm{d}n/\mathrm{d}v). \tag{3.15}$$

Since, in most dispersive media, the phase velocity increases with the wavelength, $v_g < v$ and $n_g > n$. Values of n and n_g for some typical media are presented in Table 3.1.

Experimental measurements of the velocity of light from the time of flight give the group velocity and not the phase velocity, since a peak on the envelope, which identifies a particular wave train, travels with the group velocity. This distinction disappears only when the medium has zero dispersion, since then the group and phase velocities are the same. On the other hand, in the case of interference phenomena, the order of interference is derived from the phase index, even with quasi-monochromatic radiation; however, as will be shown later (see Section 3.11), the setting for maximum visibility of the fringes, with dispersive media in the paths, involves their group indices.

Table 3.1. Phase and group indices of refraction for some materials ($\lambda = 589$ nm)

Material	n	n_g
Air	1.000 276	1.000 284
Water	1.333	1.351
Crown glass	1.517	1.542
Carbon disulfide	1.628	1.727

3.4 The Mutual Coherence Function

The quantum theory shows that even for waves originating from a single point on a thermal source, the amplitude and phase exhibit rapid, irregular fluctuations. For waves originating from different points on a source of finite size, these fluctuations are completely uncorrelated. As a result, the wave fields at any two points illuminated by such a source, or at the same point at different instants of time, will, in general, exhibit only partial correlation.

To evaluate this correlation, consider the optical system shown in Fig. 3.1, which is similar to that used in Young's interference experiment. In this system, a quasi-monochromatic source of finite size illuminates a screen containing two pinholes P_1 and P_2, and the light leaving these pinholes forms an interference pattern at a second screen.

Let the analytic signals corresponding to the wave fields produced by the source at P_1 and P_2 be $V_1(t)$ and $V_2(t)$, respectively. P_1 and P_2 can then be considered as two secondary sources, so that the analytic signal at a point P in the interference

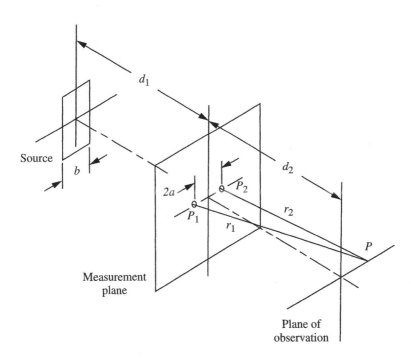

Figure 3.1. Measurement of the coherence of the radiation field produced by a quasi-monochromatic source of finite size.

pattern is

$$V_P(t) = K_1 V_1(t - t_1) + K_2 V_2(t - t_2), \tag{3.16}$$

where $t_1 = r_1/c$ and $t_2 = r_2/c$ are the times taken for the waves from P_1 and P_2 to travel to P, and K_1 and K_2 are parameters whose magnitude is determined by geometrical factors such as the size of the pinholes and the distance from the screen to P.

Now, since the interference field is stationary (or, in other words, independent of the time origin selected), Eq. (3.16) can be rewritten as

$$V_P(t) = K_1 V_1(t + \tau) + K_2 V_2(t), \tag{3.17}$$

where $\tau = t_2 - t_1$. The average intensity at P is then

$$
\begin{aligned}
I_P &= \langle V_P(t) V_P^*(t) \rangle \\
&= |K_1|^2 \langle V_1(t + \tau) V_1^*(t + \tau) \rangle + |K_2|^2 \langle V_2(t) V_2^*(t) \rangle \\
&\quad + K_1 K_2^* \langle V_1(t + \tau) V_2^*(t) \rangle + K_1^* K_2 \langle V_1^*(t + \tau) V_2(t) \rangle \\
&= |K_1|^2 I_1 + |K_2|^2 I_2 + 2|K_1 K_2| \, \mathrm{Re} \, \{\Gamma_{12}(\tau)\},
\end{aligned}
\tag{3.18}
$$

where I_1 and I_2 are the intensities at P_1 and P_2, respectively, and

$$\Gamma_{12}(\tau) = \langle V_1(t + \tau) V_2^*(t) \rangle \tag{3.19}$$

is known as the mutual coherence function of the wave fields at P_1 and P_2.

The mutual coherence function has the same dimensions as intensity. If we normalize $\Gamma_{12}(\tau)$, we obtain the dimensionless quantity

$$\gamma_{12}(\tau) = \Gamma_{12}(\tau)/(I_1 I_2)^{1/2}, \tag{3.20}$$

which is called the complex degree of coherence of the wave fields at P_1 and P_2.

Now, $|K_1|^2 I_1$ and $|K_2|^2 I_2$ are the intensities at P due to the sources P_1 and P_2 acting separately. If we denote these intensities by $I_{P(1)}$ and $I_{P(2)}$, respectively, and make use of Eq. (3.20), Eq. (3.18) can be rewritten as

$$I_P = I_{P(1)} + I_{P(2)} + 2[I_{P(1)} I_{P(2)}]^{1/2} \, \mathrm{Re} \, \{\gamma_{12}(\tau)\}. \tag{3.21}$$

This is the general law of interference for partially coherent light.

Equation (3.21) shows that to determine the intensity at P, we must know the intensities due to P_1 and P_2 acting separately, and, in addition, the real part of the complex degree of coherence. For this, we express the complex degree of coherence as the product of a modulus and a phase factor, so that

$$\gamma_{12}(\tau) = m_{12}(\tau) \exp\{i[\alpha_{12}(\tau) - 2\pi\bar{\nu}\tau]\}, \tag{3.22}$$

where $m_{12}(\tau) = |\gamma_{12}(\tau)|$, $\alpha_{12}(\tau)$ is the phase difference between the waves incident at P_1 and P_2, and $\bar{\nu}$ is the mean frequency.

The quantity $m_{12}(\tau)$ is known as the degree of coherence. It can be shown from Eq. (3.18) and the Schwarz inequality that it satisfies the condition

$$0 \le m_{12}(\tau) \le 1. \tag{3.23}$$

The wave fields at P_1 and P_2 are said to be coherent if $m_{12} = 1$, and incoherent if $m_{12} = 0$. In all other cases, they are partially coherent. It should be noted that the first limiting case, $m_{12} = 1$, is attainable only for strictly monochromatic radiation and, hence, does not exist in reality; it can also be shown that a nonzero radiation field for which $m_{12} = 0$, for all values of τ and any pair of points in space, cannot exist.

If we make use of Eq. (3.22), Eq. (3.21) becomes

$$I_P = I_{P(1)} + I_{P(2)} + 2[(I_{P(1)}I_{P(2)})]^{1/2}m_{12}(\tau)\cos[\alpha_{12}(\tau) - 2\pi\bar{\nu}\tau]. \tag{3.24}$$

With quasi-monochromatic light, both $m_{12}(\tau)$ and $\alpha_{12}(\tau)$ are slowly varying functions of τ when compared with $\cos 2\pi\bar{\nu}\tau$ and $\sin 2\pi\bar{\nu}\tau$. As a result, when P moves across the plane of observation, the spatial variations in I_P are essentially due to the changes in $2\pi\bar{\nu}\tau$ due to the changes in the value of $(r_2 - r_1)$. In addition, if the openings at P_1 and P_2 are sufficiently small, we can assume that $I_{P(1)}$ and $I_{P(2)}$ are constant over an appreciable region surrounding P. The intensity distribution over this region then consists of a uniform background on which fringes are superposed, corresponding to the variations of the term $\cos[\alpha_{12}(\tau) - 2\pi\bar{\nu}\tau]$.

From Eqs. (2.21) and (3.24), the visibility of these fringes is

$$\mathcal{V} = \frac{2(I_{P(1)}I_{P(2)})^{1/2}}{I_{P(1)} + I_{P(2)}}m_{12}(\tau), \tag{3.25}$$

which, when the two beams have the same intensity, reduces to

$$\mathcal{V} = m_{12}(\tau). \tag{3.26}$$

In this case, the visibility of the fringes gives the degree of coherence. Equation (3.24) also shows that the fringes are displaced, relative to those that would be formed if the waves from P_1 and P_2 had the same phase, by an amount proportional to the term $\alpha_{12}(\tau)$.

Another useful function which we may define at this stage is the cross-power spectrum $g_{12}(\nu)$, which is the Fourier transform of $\Gamma_{12}(\tau)$ and is therefore given by the relation

$$g_{12}(\nu) = \int_{-\infty}^{\infty} \Gamma_{12}(\tau) \exp(i2\pi\nu\tau) d\tau. \qquad (3.27)$$

Since $\Gamma_{12}(\tau)$ is also an analytic signal, it has no negative-frequency components, and we can write the inverse transform as

$$\Gamma_{12}(\tau) = \int_{0}^{\infty} g_{12}(\nu) \exp(-i2\pi\nu\tau) d\nu. \qquad (3.28)$$

For quasi-monochromatic light, the cross-power spectrum $g_{12}(\nu)$ has a value differing significantly from zero only over a very small frequency range on either side of the mean frequency $\bar{\nu}$. Accordingly, Eq. (3.28) can be rewritten as

$$\Gamma_{12}(\tau) = \exp(-i2\pi\nu\tau) \int_{-\bar{\nu}}^{\infty} j_{12}(\nu') \exp(-i2\pi\nu'\tau) d\nu'$$

$$= \exp(-i2\pi\bar{\nu}\tau) J_{12}(\tau), \qquad (3.29)$$

where $\nu' = \nu - \bar{\nu}$, $j_{12}(\nu') = g_{12}(\nu)$, and $J_{12}(\tau)$, the Fourier transform of $j_{12}(\nu')$, is known as the mutual intensity.

Since, for quasi-monochromatic light, $j_{12}(\nu')$ differs significantly from zero only over a narrow range of values of ν' around $\nu' = 0$, $J_{12}(\tau)$ contains only low-frequency components and is, therefore, a slowly varying function of τ. Accordingly, if the time delay τ between the interfering beams is small enough that $|(\nu - \bar{\nu})\tau| \ll 1$ over the range of frequencies for which $g_{12}(\nu)$ is not equal to zero, $|\Gamma_{12}(\tau)|$, $|\gamma_{12}(\tau)|$, and $\alpha_{12}(\tau)$ are very nearly equal to $|\Gamma_{12}(0)|$, $|\gamma_{12}(0)|$, and $\alpha_{12}(0)$, respectively. We can then set

$$J_{12} = \Gamma_{12}(0), \qquad (3.30)$$

$$\mu_{12} = \gamma_{12}(0), \qquad (3.31)$$

and

$$\beta_{12} = \alpha_{12}(0) \qquad (3.32)$$

and rewrite Eq. (3.24) in the form

$$I_P = I_{P(1)} + I_{P(2)} + 2[I_{P(1)}I_{P(2)}]^{1/2}|\mu_{12}|\cos(\beta_{12} - 2\pi\bar{\nu}\tau).\tag{3.33}$$

3.5 Spatial Coherence

In the limiting case just discussed, when the difference in the optical paths is small enough that effects due to the finite bandwidth radiated by the source can be neglected, we are concerned essentially with the spatial coherence of the field. To evaluate the spatial coherence of the field, we consider two points P_1 and P_2 in the plane of observation illuminated, as shown in Fig. 3.2, by a source S, whose dimensions are small compared to its distance from P_1 and P_2. If, then, $V_{i1}(t)$ and $V_{i2}(t)$ are the analytic signals at P_1 and P_2 due to a small element dS_i located at $S(x_i, y_i, 0)$ on the source, the analytic signals at these points due to the whole source are

$$V_1(t) = \sum_{i=1}^{N} V_{i1}(t)\tag{3.34}$$

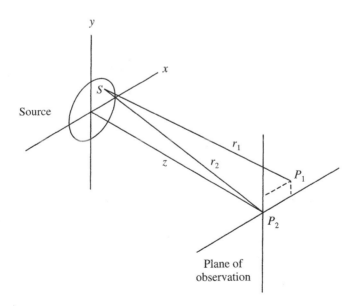

Figure 3.2. Calculation of the coherence of the fields at two points P_1, P_2 illuminated by an extended source.

and

$$V_2(t) = \sum_{i=1}^{N} V_{i2}(t), \tag{3.35}$$

respectively. Accordingly, the mutual intensity is

$$J_{12} = \langle V_1(t) V_2^*(t) \rangle$$

$$= \sum_{i=1}^{N} \langle V_{i1}(t) V_{i2}^*(t) \rangle + \sum_{i \neq j}^{N} \sum_{j}^{N} \langle V_{i1}(t) V_{j2}^*(t) \rangle. \tag{3.36}$$

However, since the fluctuations of the wave fields from different elements are statistically independent,

$$\sum_{i \neq j}^{N} \sum_{j}^{N} \langle V_{i1}(t) V_{j2}^*(t) \rangle = 0, \tag{3.37}$$

so that

$$J_{12} = \sum_{i=1}^{N} \langle V_{i1}(t) V_{i2}^*(t) \rangle. \tag{3.38}$$

Now, the analytic signals at P_1 and P_2 due to the element dS_i can be written as

$$V_{i1}(t) = (1/r_1)\, a_i(t - r_1/c) \exp[-i2\pi\, \bar{\nu}(t - r_1/c)] \tag{3.39}$$

and

$$V_{i2}(t) = (1/r_2)\, a_i(t - r_2/c) \exp[-i2\pi\, \bar{\nu}(t - r_2/c)], \tag{3.40}$$

where $a_i(t)$ is the complex amplitude of the wave field at the element dS_i at time t, and r_1 and r_2 are the distances from dS_i to P_1 and P_2, respectively. Accordingly,

$$V_{i1}(t) V_{i2}^*(t) = (1/r_1 r_2)\, a_i(t - r_1/c)\, a_i(t - r_2/c)$$

$$\times \exp[-i2\pi\, \bar{\nu}(r_2 - r_1)/c]. \tag{3.41}$$

As shown in Section 3.1, with a quasi-monochromatic source emitting only over a range of frequencies $\bar{\nu} \pm \Delta\nu$, $a_i(t)$ varies slowly enough that if the time interval

$(r_2-r_1)/c$ is small compared to $1/\Delta \nu$, we can take $a_i(t-r_1/c) \approx a_i(t-r_2/c) \approx a_i(t)$. We can, therefore, write

$$V_{i1}(t)V_{i2}^*(t) = (1/r_1r_2)|a_i(t)|^2 \exp[-i2\pi\bar{\nu}(r_2-r_1)/c]. \qquad (3.42)$$

To obtain the mutual intensity due to the whole source, Eq. (3.42) is integrated over the area of the source. If $I(S_i)$ is the local value of the intensity at the element dS_i,

$$I(S_i)dS_i = |a_i(t)|^2 \qquad (3.43)$$

and

$$J_{12} = \int_S (1/r_1r_2)I(S) \exp[-i\bar{k}(r_2-r_1)]dS, \qquad (3.44)$$

where $\bar{k} = 2\pi\bar{\nu}c = 2\pi/\bar{\lambda}$.

Now, if I_1 and I_2 are the intensities at P_1 and P_2 due to the source,

$$I_1 = \int_S [I(S)/r_1^2]dS \qquad (3.45)$$

and

$$I_2 = \int_S [I(S)/r_2^2]dS. \qquad (3.46)$$

Hence, from Eqs. (3.44), (3.45), and (3.46), the complex degree of coherence of the fields at P_1 and P_2 is

$$\mu_{12} = (I_1I_2)^{-1/2} \int_S (1/r_1r_2)I(S) \exp[-i\bar{k}(r_2-r_1)]dS. \qquad (3.47)$$

It is apparent from the formal similarity of Eq. (3.47) to the Fresnel–Kirchhoff integral (see Appendix B) that the evaluation of the complex degree of coherence is equivalent to the calculation of the complex amplitude in a diffraction pattern. This result was first established by van Cittert [1934] and later obtained in a simpler way by Zernike [1938], and is now known as the van Cittert–Zernike theorem; it may be stated as follows:

> Imagine that the source is replaced by a screen with a transmittance for amplitude at any point proportional to the intensity at this point in the source, and that this screen is illuminated by a spherical wave converging to a fixed point P_2. The complex degree of coherence μ_{12} which exists between the vibrations at P_1 and P_2 is then proportional to the complex amplitude at P_1 in the diffraction pattern.

Equation (3.47) can be simplified if the dimensions of the source, and the separation of P_1 and P_2, are extremely small compared to the distances of P_1 and P_2 from the source. For convenience, we will assume that the point P_2 is located on the z axis at $(0, 0, z)$, while $P_1(x, y, z)$ is free to move over the plane of observation. If, then, $(x_s, y_s, 0)$ are the coordinates of the point S on the source,

$$r_1^2 = (x - x_s)^2 + (y - y_s)^2 + z^2, \tag{3.48}$$

and, since x_s, y_s, x, y are all very small compared to z,

$$r_1 \approx z + [(x - x_s)^2 + (y - y_s)^2]/2z. \tag{3.49}$$

Similarly, for the point P_2,

$$r_2 \approx z + (x_s^2 + y_s^2)/2z. \tag{3.50}$$

Accordingly, from Eqs. (3.49) and (3.50),

$$r_2 - r_1 \approx -[(x^2 + y^2)/2z] + [(xx_s + yy_s)/z]. \tag{3.51}$$

If we define new coordinates in the source plane such that

$$\xi = x_s/z,$$
$$\eta = y_s/z, \tag{3.52}$$

and set

$$\phi_{12} = -k(x^2 + y^2)/2z, \tag{3.53}$$

the complex degree of coherence of the field defined by Eq. (3.47) can be written as

$$\mu_{12} = (I_1 I_2)^{1/2} \exp(i\phi_{12}) \int \int_S (1/z^2) I(\xi, \eta) \exp[i\bar{k}(x\xi + y\eta)] d\xi \, d\eta. \tag{3.54}$$

However,

$$I_1 \approx I_2 \approx \int \int_S (1/z^2) I(\xi, \eta) d\xi \, d\eta, \tag{3.55}$$

so that Eq. (3.54) becomes

$$\mu_{12} = \frac{\exp(i\phi_{12}) \int \int_S I(\xi, \eta) \exp[i\bar{k}(x\xi + y\eta)] d\xi \, d\eta}{\int \int_S I(\xi, \eta) d\xi \, d\eta}. \tag{3.56}$$

The complex degree of coherence is therefore given, in this case, by the normalized Fourier transform of the intensity distribution over the source.

A case of interest is a rectangular source. Since the fringes obtained in Young's interference experiment are parallel straight lines, it is possible to use a line source instead of a point source, provided that its long dimension is parallel to the fringes. The field due to each point on the line source is incoherent with respect to the fields due to all the other points, but the fringes produced by the individual point sources are displaced with respect to each other only along their length and, therefore, effectively coincide. However, with a rectangular source having a width b, as shown in Fig. 3.1, the intensity distribution across the source is given by the expression

$$I(x_s) = \text{rect}\,(x_s/b),\qquad(3.57)$$

where

$$\text{rect}\,(x) = \begin{cases} 1, & \text{when } |x| \le 1/2, \\ 0, & \text{when } |x| > 1/2. \end{cases}\qquad(3.58)$$

Accordingly, from Eqs. (3.56) and (3.26), the visibility of the fringes is

$$\mathcal{V} = |\text{sinc}\,(2ab/\lambda d_1)|,\qquad(3.59)$$

where sinc $x = (1/\pi x)\sin(\pi x)$.

It is apparent that when b is small enough that $2ab/\lambda d_1 \ll 1$, the visibility of the fringes is close to unity. As b is increased, the visibility of the fringes decreases and becomes zero when $b = \lambda d_1/2a$. Beyond this point the sinc function is negative, so that the fringes reappear, but with reversed contrast.

Another interesting case is that of a circular source of radius r. In this case, the visibility of the fringes is

$$\mathcal{V} = 2J_1(u)/u,\qquad(3.60)$$

where $J_1(u)$ is the Bessel function of the first kind of order one, and $u = 4\pi ra\lambda d_1$. Experiments showing the variation of the complex degree of coherence with the effective radius of such a source have been described by Thompson and Wolf [1957] and by Hariharan and Singh [1961].

3.6 Temporal Coherence

In the other limiting case, when the dimensions of the source are very small (so that it is, effectively, a point source) but it radiates over a range of wavelengths, we are concerned with the temporal coherence of the field. In this case, the complex degree of coherence depends only on τ, the difference in the transit times from the source to P_1 and P_2, and the mutual coherence function is merely the autocorrelation function

$$\Gamma_{11}(\tau) = \langle V(t + \tau)V^*(t)\rangle. \tag{3.61}$$

The degree of temporal coherence of the field is then

$$\gamma_{11}(\tau) = \frac{\langle V(t + \tau)V^*(t)\rangle}{\langle V(t)V^*(t)\rangle}. \tag{3.62}$$

From Eqs. (3.25) and (3.62), it is apparent that the degree of temporal coherence can be obtained from the visibility of the interference fringes, as the optical path difference between the interfering wavefronts is varied. This result leads to the concepts of the coherence time and the coherence length of the radiation.

3.7 Coherence Time and Coherence Length

If we make use of the Wiener–Khinchin theorem (see Appendix A.2), it follows from Eq. (3.61) that $W(v)$, the frequency spectrum of the radiation, is given, in this case, by the Fourier transform of the mutual coherence function, so that

$$W(v) \leftrightarrow \Gamma_{11}(\tau), \tag{3.63}$$

while the complex degree of coherence is

$$\gamma_{11}(\tau) = \mathcal{F}\{W(v)\}/\int_{-\infty}^{\infty} W(v)dv. \tag{3.64}$$

Consider, now, a source which radiates over a range of frequencies Δv centered on a mean frequency $\bar{v}$. The frequency spectrum of the radiation is then given by the function

$$W(v) = \text{rect}[(v - \bar{v})/\Delta v]. \tag{3.65}$$

Accordingly, from Eq. (3.64), the complex degree of coherence of the radiation is

$$\gamma_{11}(\tau) = \text{sinc}(\tau \Delta v). \tag{3.66}$$

This is a damped oscillating function whose first zero occurs at a time difference $\Delta \tau$ given by the relation

$$\Delta \tau \Delta v = 1. \tag{3.67}$$

This time interval $\Delta\tau$ is known as the coherence time of the radiation; its coherence length is defined as

$$\Delta l = c\Delta\tau = c/\Delta\nu, \tag{3.68}$$

where c is the speed of light. From Eqs. (2.3) and (3.68), we also have

$$\Delta l \approx (\bar{\lambda}^2/\Delta\lambda), \tag{3.69}$$

where $\bar{\lambda}$ is the mean wavelength of the radiation, and $\Delta\lambda$ is the range of wavelengths emitted by the source.

3.8 Coherence in the Space-Frequency Domain

As we have seen, the mutual coherence function $\Gamma_{12}(\tau)$ takes into account both the spatial and temporal correlation properties of the light field. However, with an extended source emitting over a finite spectral range, it is useful to study correlations in the space-frequency domain [Wolf, 1981, 1982, 1986].

To do this we apply the van Cittert–Zernike theorem to the field for each spectral component and integrate the results over the bandwidth of the radiation. We thereby obtain the integrated mutual coherence function

$$\bar{\Gamma}(P_1, P_2, \tau) = \int_0^\infty J_{12}(\nu) \exp(-\mathrm{i}2\pi\nu\tau)\mathrm{d}\nu, \tag{3.70}$$

where

$$J_{12}(\nu) = \int_S (1/r_1 r_2) I(S, \nu) \exp[-\mathrm{i}\bar{k}(r_2 - r_1)]\mathrm{d}S, \tag{3.71}$$

and $I(S, \nu)$ is the intensity distribution over the source at a frequency ν.

We can then use Eq. (3.70) to define the cross-spectral density function [Mandel and Wolf, 1976]

$$W(P_1, P_2, \nu) = \int_{-\infty}^\infty \Gamma(P_1, P_2, \tau)e^{2\pi\mathrm{i}\nu\tau}\mathrm{d}\tau. \tag{3.72}$$

The cross-spectral density function and the (integrated) mutual coherence function form a Fourier-transform pair, since

$$\overline{\Gamma}(P_1, P_2, \tau) = \int_0^\infty W(P_1, P_2, \nu)e^{2\pi\mathrm{i}\nu\tau}\mathrm{d}\nu. \tag{3.73}$$

When P_1 and P_2 coincide, the cross-spectral density function reduces to the power spectrum of the light field.

If we normalize the cross-spectral density, we obtain the function

$$\mu(P_1, P_2, \nu) = \frac{W(P_1, P_2, \nu)}{[W(P_1, \nu)]^{1/2}[W(P_2, \nu)]^{1/2}}, \tag{3.74}$$

which is called the degree of spectral coherence at frequency ν of the light field at the points P_1 and P_2. Its modulus satisfies the condition

$$0 \le \mu(P_1, P_2, \nu) \le 1. \tag{3.75}$$

If, now, we consider the Young's interferometer shown in Fig. 3.1, it can be shown that the spectral density at the point P in the plane of observation is

$$\begin{aligned}
W(P, \nu) = {} & W_{P(1)}(P, \nu) + W_{P(2)}(P, \nu) \\
& + 2[W_{P(1)}(P, \nu)]^{1/2}[W_{P(2)}(P, \nu)]^{1/2} \\
& \times \mathrm{Re}\,[\mu(P_1, P_2, \nu)e^{-2\pi i\nu(r_1-r_2)/c}].
\end{aligned} \tag{3.76}$$

This formula, which may be called the spectral interference law, shows that, in general, the spectral density (power spectrum) $W(P, \nu)$ of the field at P is not just the sum of the spectral densities of the fields due to the two beams reaching P from the two pinholes P_1 and P_2, but also depends on the spectral degree of coherence $\mu(P_1, P_2, \nu)$ of the light emerging from these two pinholes [Wolf, 1987].

In particular, it follows that even when the two beams have the same spectral distribution, that is to say when

$$W_{P(2)}(P, \nu) = W_{P(1)}(P, \nu), \tag{3.77}$$

in which case Eq. (3.76) reduces to

$$\begin{aligned}
W(P, \nu) = {} & 2W_{P(1)}(P, \nu) \\
& + 2[W_{P(1)}(P, \nu)]\,\mathrm{Re}\,\{\mu(P_1, P_2, \nu)e^{-2\pi i\nu(r_1-r_2)/c}\},
\end{aligned} \tag{3.78}$$

the spectral distribution of the light field obtained by superposing the two beams may differ from the spectral distributions of the light fields due to the individual beams. The only situation where no modification of the spectral distribution is to be expected is where no correlation exists between the fields due to the two beams.

The formulation of higher-order coherence functions in the space-frequency domain has been discussed by Agarwal and Wolf [1993].

3.9 Nonclassical Light

The normalized correlation function of two measurements of the intensity may be termed the classical degree of higher-order coherence. However, it is also possible to define an analogous function, the quantum degree of higher-order coherence, in terms of the photon counts. It has been shown [Loudon, 1980] that the classical degree of higher-order coherence has to satisfy three inequalities which do not apply to its quantum analog. Light whose degree of higher-order coherence falls outside the range allowed by classical theory can be said to be nonclassical light.

Nonclassical light can be generated by various techniques including a two-photon cascade and parametric down-conversion (see Section 16.3).

3.10 Effects in Two-Beam Interferometers

While the most important features of the interference patterns observed with two-beam interferometers can often be obtained quite simply from geometric considerations, as outlined in Chapter 2, the application of coherence theory makes possible a more exact analysis and leads to a better understanding of some of the effects observed. Such a treatment has been developed by Steel [1965, 1983] and is outlined in this section.

We start by defining two reference planes in the interferometer. One is, as shown in Fig. 3.3, the source plane $\mathcal{S}$, while the other is the plane of observation $\mathcal{O}'$.

On looking through the interferometer from the source side, two images of the plane of observation are seen. As shown in Fig. 3.4, these images make an angle ϵ with each other called the tilt. In addition, O_1 and O_2, the two images of the origin O, appear at different distances, z and $z + \Delta z$. The distance Δz between O_1 and O_2 along the line of sight is called the shift. Finally, the two optical paths from C, the center of the source, to O may differ by an amount p_0 corresponding to a time difference, or delay, $\tau = p_0/c$. In a simple interferometer made up of plane mirrors, the shift Δz is equal to the optical path difference p_0. However, this may not be so if the paths contain focusing systems or refractive media, since the distances z and $z + \Delta z$ to O_1 and O_2 are not actual distances, but correspond to the radii of curvature of the two wavefronts from O when they reach the source plane.

It is also possible, as shown in Fig. 3.5, for the images of the plane of observation to be shifted and rotated with respect to each other and even to be of different sizes. As a result, P_1, P_2, the two images of an arbitrary point in the plane of observation will exhibit a lateral separation. The vector distance $\mathbf{s} = P_1 P_2$ is then called the shear for P. If P_1 and P_2 have position vectors $\mathbf{u}_1$ and $\mathbf{u}_2$ with respect to O_1 and O_2, the images of O, the origin of the plane of observation, we then have

$$\mathbf{s} = \mathbf{s}_0 + \mathbf{u}_2 - \mathbf{u}_1, \tag{3.79}$$

where $\mathbf{s}_0$ is the shear for the origin.

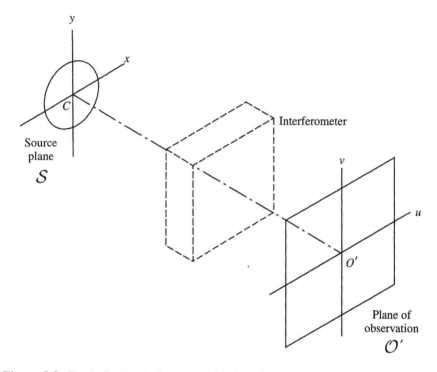

Figure 3.3. Equivalent optical system of an interferometer.

The complementary representation of the interferometer is shown in Fig. 3.6. On looking through the interferometer from the plane of observation, two images of the source are seen, whose centers C'_1 and C'_2 appear to be at distances z' and $z' + \Delta z'$. By analogy with the shift Δz defined earlier, $\Delta z'$ can be called the source shift; similarly, $\mathbf{s}'$, the lateral separation of S'_1 and S'_2, the images of a point S on the source, can be called the source shear for this point. We then have

$$\mathbf{s}' = \mathbf{s}'_0 + \mathbf{x}'_2 - \mathbf{x}'_1, \tag{3.80}$$

where $\mathbf{s}'_0$ is the lateral separation of C'_1 and C'_2, and $\mathbf{x}'_1$ and $\mathbf{x}'_2$ are the position vectors of S'_1 and S'_2 with respect to C'_1 and C'_2.

To evaluate the optical path difference between the two beams from any point S on the source to any point P in the plane of observation, we first calculate the optical path difference p_0 for the central ray from C to O'; this is the difference of two summations, taken over each of the two optical paths, of the products of the thickness of individual elements and their refractive index, and it can be written as

$$p_0 = \sum_2 nd - \sum_1 nd. \tag{3.81}$$

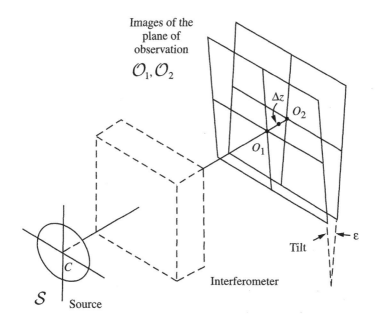

Figure 3.4. Images of the plane of observation seen from the source side of an interferometer [Steel, 1983].

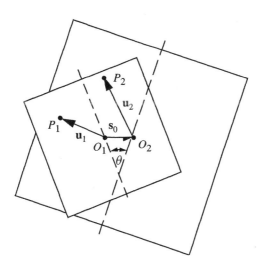

Figure 3.5. Sheared images of the plane of observation seen from the source [Steel, 1983].

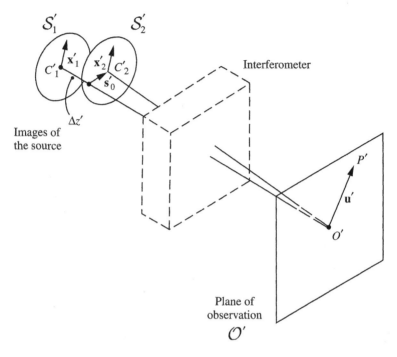

Figure 3.6. Complementary representation of an interferometer, showing the two images of the source seen from the plane of observation [Steel, 1983].

The additional optical path difference to a point P' in the plane of observation from a point S on the source, as a function of $\mathbf{x}$, the position vector of S, is then

$$\Delta p_x \approx -\mathbf{s} \cdot \mathbf{x}/z - \Delta z |\mathbf{x}|^2/2z^2. \tag{3.82}$$

Similarly, the additional optical path difference to a point S in the source plane from a point P' in the plane of observation, as a function of $\mathbf{u}'$ the position vector of P', is

$$\Delta p_u \approx -\mathbf{s}' \cdot \mathbf{u}'/z' - \Delta z' |\mathbf{u}'|^2/2z'^2. \tag{3.83}$$

Accordingly, the total path difference between the pair of points S and P' is

$$\Delta p(\mathbf{x}', \mathbf{u}') = p_0 + \Delta p_x + \Delta p_u. \tag{3.84}$$

It follows that a two-beam interferometer can be described by six parameters. Two of these are the tilt and the delay. In addition, we have, looking through the interferometer from the source side, the shift and the shear. Finally, we have, looking through the interferometer from the plane of observation, the source shift and the source shear. Diagrams corresponding to Figs. 3.4 and 3.6 can be used to evaluate these parameters for any particular configuration. They are constants for many interferometers but, in some cases, they may vary across the source or the plane of observation and are, then, functions of the position vectors $\mathbf{x}$ or $\mathbf{u}'$. These parameters completely determine the value of $\gamma(\mathbf{u}_1, \mathbf{u}_2, \tau)$, the complex degree of coherence of the wave fields at any point in the plane of observation.

The intensity distribution in the plane of observation is then given by Eq. (3.21), the general equation for interference, which can now be written in the form

$$I(\mathbf{u}') = I_1 + I_2 + 2(I_1 I_2)^{1/2} \, \mathrm{Re}\, \{\gamma(\mathbf{u}_1, \mathbf{u}_2, \tau)\}. \tag{3.85}$$

3.11 Source-Size Effects

With monochromatic light, or for small optical path differences with quasi-monochromatic light, we can replace $\gamma(\mathbf{u}_1, \mathbf{u}_2, \tau)$ in Eq. (3.85) by μ_{12}, which can be evaluated from the van Cittert–Zernike theorem (see Section 3.5). We have, in this case,

$$\mu_{12} = \frac{\int_{-\infty}^{\infty} I(\mathbf{x}) \exp[i\phi(\mathbf{x})]\mathrm{d}\mathbf{x}}{\int_{-\infty}^{\infty} I(\mathbf{x})\mathrm{d}\mathbf{x}}, \tag{3.86}$$

where $I(\mathbf{x})$ is the intensity distribution across the source and, from Eq. (3.75),

$$\phi(\mathbf{x}) = \bar{k}[-\mathbf{s} \cdot \mathbf{x}/z - \Delta z|\mathbf{x}|^2/2z^2]. \tag{3.87}$$

Equation (3.87) shows that with a point source (for which $\mathbf{x} = 0$), the shear and shift have no effect on the visibility of the fringes. However, with a source of finite size, the fringes will have maximum visibility only where both the shear $\mathbf{s}$ and the shift Δz are close to zero; this is the region of localization which we have found earlier by a purely geometrical analysis (see Section 2.5.2).

As the plane of observation moves away from this region, the visibility of the fringes decreases until they vanish completely when $\mu_{12} = 0$. An interesting result, not apparent from a simple geometrical analysis, is the existence of secondary regions of localization [Hariharan, 1969b]. These regions arise because, beyond the point at which $\mu_{12} = 0$, it is possible for its value to become finite again, though negative, and pass through a minimum. The fringes then reappear, but with a phase reversal, a dark fringe taking the place of a bright fringe.

Equations (3.86) and (3.87) also show that, to obtain fringes of good visibility with a finite shear, the source must be smaller than a certain size. From the van Cittert–Zernike theorem, this limiting size corresponds to the aperture of a lens which can barely resolve two points separated by a distance equal to the shear.

3.12   Spectral Bandwidth Effects

With a source that also emits over a finite spectral bandwidth, Eqs. (3.70) and (3.85) can be used to evaluate the intensity distribution in the fringe pattern, as long as the paths traverse nondispersive media. Conversely, the spectral energy distribution of the source can be evaluated from the Fourier transform of the self-coherence function $\Gamma_{11}(\tau)$. This was the method first used by Michelson to determine the structure of isolated spectral lines; it is also the basis of the modern technique of Fourier-transform spectroscopy (see Chapter 11).

If, however, the beam paths include dispersive media, the delay becomes a function of the frequency, and Eq. (3.85) is no longer applicable. In this case, it has been shown that, provided the group velocity can be taken to be constant over the spectral range involved, we can replace τ in Eq. (3.85) by the group delay τ_g [Pancharatnam, 1963]. The group delay is defined, by analogy with Eq. (3.81), as

$$\tau_g = \sum_2 n_g d/c - \sum_1 n_g d/c, \qquad (3.88)$$

where the n_g are the group refractive indices given by Eq. (3.15) for the media in the interferometer. For the fringe visibility to be a maximum, the group delay must be zero; as we have seen earlier in Section 2.12, this is the condition for achromatic fringes.

3.13   Spectral Changes Due to Coherence

As discussed in Section 3.8, changes in the spatial coherence of the fields at an extended source can lead to changes in the spectrum of the radiated field [Wolf, 1987]. An experiment demonstrating shifts of spectral lines produced by partially correlated sources was performed by Gori *et al.* [1988], and it was shown by Kandpal, Vaishya, and Joshi [1989] that the spectrum of light diffracted by an aperture could be affected by the size of the aperture.

James and Wolf [1991a, 1991b, 1998] have shown that the conditions in which appreciable spectral changes are produced are, to some extent, complementary to those required for the formation of an interference pattern. The production of an extended interference pattern requires light with a narrow spectral bandwidth, while light with a broad bandwidth is necessary to observe significant spectral changes.

An analysis of the spectral changes produced by interference in a Mach–Zehnder interferometer has been made by Agarwal and James [1993]. These changes have been confirmed experimentally by Rao and Kumar [1994].

3.14   Polarization Effects

To make the preceding analysis complete, we also have to evaluate the state of polarization of the beams leaving the interferometer. We note, at the outset, that if the light entering the interferometer is unpolarized or partially polarized, it can be regarded as made up of two orthogonally polarized components. Each of these components can be treated separately, and their intensities can be added finally.

If, then, the state of polarization of one of these orthogonal components is specified by the Jones vector A_0 (see Appendix D), the Jones vectors A_1, A_2 of the two beams derived from it, when they leave the interferometer, are given by the relations

$$A_1 = M_1 A_0 \tag{3.89}$$

and

$$A_2 = M_2 A_0, \tag{3.90}$$

where M_1 and M_2 are matrices representing the combined effects of the elements in each of the arms. The difference in the states of polarization of the two beams can then be specified by an angle ψ defined by the relation

$$\cos \psi = \langle |A_1^\dagger A_2| \rangle, \tag{3.91}$$

where $A_1^\dagger$ is the Hermitian conjugate of A_1. The intensity in the interference pattern is, accordingly,

$$I(\mathbf{u}') = I_1 + I_2 + 2(I_1 I_2)^{1/2} \, \text{Re} \, \{\gamma(\mathbf{u}_1, \mathbf{u}_2, \tau)\} \cos \psi, \tag{3.92}$$

where

$$I_1 = \langle A_1^\dagger A_1 \rangle, \tag{3.93}$$

and

$$I_2 = \langle A_2^\dagger A_2 \rangle. \tag{3.94}$$

A comparison with Eq. (3.85) shows that the visibility of the fringes is reduced by the factor $\cos \psi$. The generalization of this formula to quasi-monochromatic be&ms has been discussed by Pancharatnam [1963].

If the polarization states of the two beams leaving the interferometer are to be the same for unpolarized light (or, in other words, for any incident polarization), we require M_1 and M_2 to be identical. Each of these matrices is the product of diagonal matrices representing reflection at the mirrors and matrices which represent the change of orientation (if any) of the plane of incidence from one mirror to the next. Even if the two arms contain similar optical elements, the order in which they are traversed is not the same. As a result $\mathbf{M}_1$ and $\mathbf{M}_2$ will not be identical unless the matrix products commute; this requires all the component matrices to be diagonal matrices. The interferometer is then said to be compensated for polarization.

A simple case in which the interferometer is compensated for polarization is when the normals to all the mirrors and beamsplitters are in the same plane. Other possible cases have been discussed by Steel [1964b]. However, compensation may not be complete if the interferometer contains elements such as a beamsplitter coated with a thin metal film, for which the phase shifts on reflection differ for incidence from the air side and the glass side [Chakraborty, 1970, 1973].

Compensation for polarization is not possible in interferometers in which elements such as cube corners are used. The resulting effects, which can be quite complex, have been studied by Leonhardt [1972, 1974, 1981]. In such cases, it is possible to obtain fringes with good visibility only by using two suitably oriented polarizers in the interferometer, one just after the source and the other just before the plane of observation.

Chapter 4

Multiple-Beam Interference

In Chapter 2, when we studied interference phenomena in plates and thin films, we neglected the contributions of multiple reflected beams. This simplification is justifiable only when the reflectance of the surfaces is low. We shall now study the fringe patterns produced when the reflectance of the surfaces is high, and the effects of multiple reflected beams must be taken into account.

4.1 Fringes in a Plane-Parallel Plate

Consider a plane-parallel plate of refractive index n on which, as shown in Fig. 4.1, a plane wave of monochromatic light of unit amplitude is incident. As a result of multiple reflections within the plate, the ray SA_1, which corresponds to this incident wave, gives rise to a series of transmitted rays whose intensities fall off progressively. Since these rays are parallel to one another, the interference pattern produced with an extended source will be localized at infinity.

The optical path difference between successive transmitted waves can be evaluated from Eq. (2.31): the corresponding phase difference is

$$\phi = (4\pi/\lambda)nd \cos\theta_2. \tag{4.1}$$

If, then, the complex reflection and transmission coefficients for amplitude at the two surfaces are r_1, t_1 and r_2, t_2, respectively, for a wave incident on the surfaces from the surrounding medium, and r_1', t_1', r_2', t_2', for a wave incident on them from within the plate, the complex amplitudes of the transmitted waves $C_1 D_1, C_2 D_2, C_3 D_3 \ldots$ can be written, as shown in Fig. 4.1, as $t_1 t_2' \exp(-i\phi/2)$, $t_1 t_2' r_1' r_2' \exp(-i3\phi/2)$, $t_1 t_2' r_1'^2 r_2'^2 \exp(-i5\phi/2) \ldots$. Accordingly, the resultant complex amplitude due to all

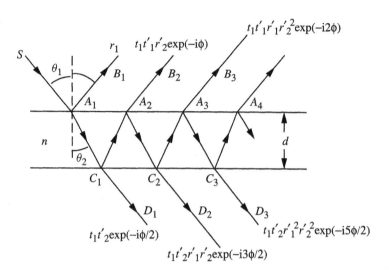

Figure 4.1. Multiple-beam interference in a plane-parallel plate.

the transmitted waves is

$$A(\phi) = t_1 t_2' \exp(-i\phi/2)[1 + r_1' r_2' \exp(-i\phi) + r_1'^2 r_2'^2 \exp(-i2\phi) + \cdots]$$

$$= \frac{t_1 t_2' \exp(-i\phi/2)}{1 - r_1' r_2' \exp(-i\phi)}. \tag{4.2}$$

The complex reflection coefficients r_1' and r_2' can be written in the form $r_1' = |r_1'| \exp(-i\beta_1')$ and $r_2' = |r_2'| \exp(-i\beta_2')$, where β_1' and β_2' are the phase shifts on reflection at the surfaces. The intensity in the interference pattern is, therefore,

$$I(\psi) = \frac{T_1 T_2}{1 + R_1 R_2 - 2(R_1 R_2)^{1/2} \cos \psi}, \tag{4.3}$$

where $T_1 = |t_1|^2$, $T_2 = |t_2'|^2$, $R_1 = |r_1'|^2$, $R_2 = |r_2'|^2$ are, respectively, the transmittances and reflectances for intensity of the two surfaces and $\psi = \phi + \beta_1' + \beta_2'$. In the special case when $R_1 = R_2 = R$, and $T_1 = T_2 = T$, Eq. (4.3) reduces to

$$I(\psi) = \frac{T^2}{1 + R^2 - 2R \cos \psi}$$

$$= \frac{T^2}{(1 - R)^2 + 4R \sin^2(\psi/2)}. \tag{4.4}$$

This is known as the Airy formula.

A further simplification is possible if we are dealing with reflection at interfaces between transparent media. In this case, if the media above and below the plate are the same, the sum of the phase shifts due to successive reflections at the two surfaces, $\beta_1' + \beta_2' = 0$, or 2π, and can be neglected.

The intensity in the fringe pattern is a maximum when $\sin(\psi/2) = 0$, and a minimum when $\sin(\psi/2) = 1$. The maximum intensity is

$$I_{max} = T^2/(1 - R)^2, \tag{4.5}$$

while the minimum intensity is

$$I_{min} = T^2/(1 + R)^2. \tag{4.6}$$

If we define a function G such that

$$G = 4R/(1 - R)^2, \tag{4.7}$$

we can write Eq. (4.4) as

$$I(\psi) = \frac{I_{max}}{1 + G\sin^2(\psi/2)} \tag{4.8}$$

and Eq. (4.6) as

$$I_{min} = \frac{I_{max}}{1 + G}. \tag{4.9}$$

It follows from Eq. (4.5) that if $T + R = 1$, $I_{max} = 1$. This is not the case when thin metal coatings are used to obtain a higher reflectance, since such coatings always exhibit absorption, and $T + R < 1$. However, as we will see later, a reflectance approaching unity can now be obtained with multilayer dielectric films, which have negligible losses and for which $T + R \approx 1$. Under these conditions, the peak intensity in the fringes is always close to unity. The intensity distribution in the fringe pattern is then given by the relation

$$I(\psi) = \frac{1}{1 + G\sin^2(\psi/2)} \tag{4.10}$$

and is shown in Fig. 4.2 as a function of the phase difference ψ for different values of the reflectance R. As R increases, the intensity of the minimum decreases, and the maxima become sharper. Near normal incidence, the pattern observed in transmitted light consists of very narrow bright circular fringes on an almost completely dark background. Such a device, consisting either of a plane-parallel plate with reflecting coatings on its two faces, or of two plates with reflecting coatings separated by an air space, is known as a Fabry–Perot interferometer, or Fabry–Perot etalon.

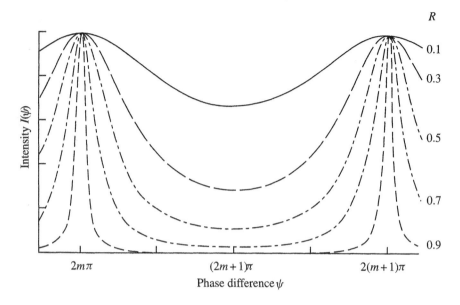

Figure 4.2. Intensity distribution in multiple-beam fringes of equal inclination formed in transmitted light for different values of the reflectance.

We can define W, the half-width of the fringes, as the interval between two points on either side of a maximum at which the intensity is equal to half its maximum value. At these points, we have, from Eq. (4.8),

$$\frac{I_{\max}}{2} = \frac{I_{\max}}{1 + G \sin^2(\psi/2)}, \tag{4.11}$$

or

$$\sin(\psi/2) = \pm G^{-1/2}. \tag{4.12}$$

Since G is large, $\sin(\psi/2)$ is small, so that we can set $\sin(\psi/2) \approx (\psi/2)$. The half-width of the fringe then corresponds to a change in ψ given by the relation

$$W = 4G^{-1/2}. \tag{4.13}$$

The ratio of the separation of adjacent fringes to their half-width is called their finesse. Since the separation of the fringes corresponds to a change in ψ of 2π, their finesse is given by the relation

$$F = \frac{2\pi}{W} = \frac{\pi G^{1/2}}{2} = \frac{\pi R^{1/2}}{1 - R}. \tag{4.14}$$

4.2 Fringes by Reflection

In addition to the fringes produced by the transmitted beams, a fringe system is produced by the reflected beams. The resultant complex amplitude due to all the reflected waves is, from Fig. 4.1,

$$A(\phi) = r_1 + t_1 t_1' r_2' \exp(-i\phi)[1 + r_1' r_2' \exp(-i\phi) + r_1'^2 r_2'^2 \exp(-i2\phi) + \cdots]$$

$$= r_1 + \frac{[t_1 t_1' r_2' \exp(-i\phi)]}{[1 - r_1' r_2' \exp(-i\phi)]}. \tag{4.15}$$

Now, if we consider the simplest case of reflection at interfaces between dielectric layers, $r_1 = -r_1'$. In addition, from the Stokes relations (see Appendix C.2),

$$r_1^2 + t_1 t_1' = 1. \tag{4.16}$$

Accordingly, Eq. (4.15) can be simplified and written as

$$A(\phi) = \frac{-r_1' + r_2' \exp(-i\phi)}{1 - r_1' r_2' \exp(-i\phi)}. \tag{4.17}$$

In this case also, if the media above and below the plate are the same, the phase shift due to two reflections within the plate must be either 0 or 2π and can be neglected. The intensity in the fringe pattern is, therefore,

$$I(\phi) = \frac{R_1 + R_2 - 2(R_1 R_2)^{1/2} \cos \phi}{1 + R_1 R_2 - 2(R_1 R_2)^{1/2} \cos \phi}. \tag{4.18}$$

In the special case when $R_1 = R_2$, Eq. (4.18) reduces to

$$I(\phi) = \frac{2R(1 - \cos \phi)}{1 + R^2 - 2R \cos \phi}$$

$$= \frac{4R \sin^2(\phi/2)}{(1 - R)^2 + 4R \sin^2(\phi/2)}$$

$$= \frac{G \sin^2(\phi/2)}{1 + G \sin^2(\phi/2)}, \tag{4.19}$$

where, as defined by Eq. (4.7), $G = 4R/(1 - R)^2$.

In this case, the pattern consists of narrow dark fringes on a nearly uniform bright background and is complementary to the pattern formed by the transmitted beams.

4.3 Fringes of Equal Thickness

Very narrow fringes of equal thickness can be observed between two highly reflecting surfaces enclosing a thin film. The formation of such fringes has been analyzed by Brossel [1947].

We consider, as shown in Fig. 4.3, a wedge-shaped film whose faces make an angle ϵ with each other, illuminated by a monochromatic plane wave W_0 incident at an angle θ. The transmitted light then contains, in addition to the directly transmitted wave W_0, a family of plane waves W_1, W_2, ... formed by multiple reflections at the film. The angle between successive members of the family is 2ϵ, and their amplitude decreases in geometric progression.

For simplicity, we will assume that the surfaces are interfaces between dielectrics. The phase shifts on reflection can then be neglected, and all the members of the family will be in phase at O.

If, now, we consider a point $P(x, y)$, the optical path to the wave which has undergone $2m$ reflections is

$$
\begin{aligned}
p_m &= nPM_m \\
&= n[x\cos(\theta + 2m\epsilon) + y\sin(\theta + 2m\epsilon)],
\end{aligned}
\tag{4.20}
$$

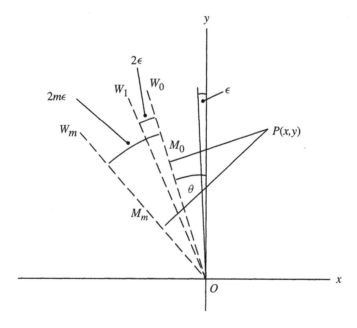

Figure 4.3. Formation of multiple-beam fringes of equal thickness in a wedged film.

where n is the refractive index of the film. Accordingly, the optical path difference between the directly transmitted wave and the wave which has undergone $2m$ reflections is

$$\Delta p_m = p_m - p_0$$
$$= n\{x[\cos(\theta + 2m\epsilon) - \cos\theta] + y[\sin(\theta + 2m\epsilon) - \sin\theta]\}. \tag{4.21}$$

Equation (4.21) can be simplified for normal incidence, in which case $\theta = 0$, and the fringes are localized at the film. The point of observation P is then located in the film ($x = 0$), and we have

$$\Delta p_m = ny \sin 2m\epsilon. \tag{4.22}$$

If, then, the thickness of the film at the point of observation $P(0, y)$ is d, we have

$$y = d/\epsilon, \tag{4.23}$$

and Eq. (4.22) can be expanded to give

$$\Delta p_m = (nd/\epsilon) \sin 2m\epsilon$$
$$= (nd/\epsilon)[2m\epsilon - (2m\epsilon)^3/3 + \cdots]. \tag{4.24}$$

Equation (4.24) shows that, with a wedged film, the optical path difference between successive waves is not exactly constant, because of the presence of higher order terms. If these higher order terms are to be neglected, this additional optical path difference must not exceed $\lambda/4$; the condition for this is

$$8m^3\epsilon^2 nd/3 < \lambda/4. \tag{4.25}$$

This condition usually imposes severe restrictions on the angle of the wedge and the thickness of the film.

Typically, for $\lambda = 633$ nm, if we have (say) 1 fringe per mm in an air wedge, so that $\epsilon = 3.16 \times 10^{-4}$, and surfaces with coatings having a reflectance of 0.9, corresponding to about 50 effective beams, Eq. (4.25) gives a maximum value of d of about 5 μm.

If the thickness of the film is greater than the limit set by Eq. (4.25), the intensity distribution is no longer given by the Airy formula. Numerical calculations [Kinosita, 1953] show that the maximum intensity of the fringes drops, and their width increases; in addition, they may, as shown in Fig. 4.4, suffer a displacement away from the apex of the wedge and become asymmetrical with secondary maxima on the thicker side.

Fringes of equal thickness are also seen in reflected light. If there are no absorption losses, the intensity distribution in these fringes is complementary to that seen by transmission.

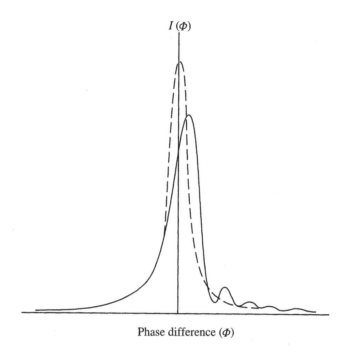

Figure 4.4. Intensity distribution in multiple-beam fringes formed in a wedged film (solid line) compared with that given by the Airy formula (broken line) [Kinosita, 1953].

An important application of multiple-beam fringes of equal thickness is in the study of surface structure [Tolansky, 1955, 1961]. The surface to be studied is coated with a highly reflecting layer of silver, and a small reference flat which has a semitransparent silver film on its front face is placed in contact with it. The reflection fringes can then be viewed by means of a microscope with a vertical illuminator.

4.4   Fringes of Equal Chromatic Order

We have seen in Section 2.11 that when a thin film illuminated with white light is viewed through a spectroscope, a channeled spectrum is obtained. Such fringes can also be produced by multiple-beam interference.

Consider the setup shown in Fig. 4.5 in which a film is illuminated at normal incidence with a collimated beam of white light, and an image of the film is formed on the slit of a spectrograph. The transmitted amplitude is a maximum at wavelengths

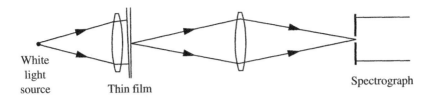

Figure 4.5. Optical system used to observe fringes of equal chromatic order (FECO fringes) in a thin film.

which satisfy the condition

$$(4\pi/\lambda_m)nd + \beta_1 + \beta_2 = 2m\pi, \tag{4.26}$$

where β_1 and β_2 are the phase shifts on reflection at the two surfaces, and m is an integer. If the reflectance of the surfaces is high, the spectrum is crossed by narrow bright fringes separated by relatively broad, dark regions.

If the thickness of the film is constant, the fringes are straight lines parallel to the slit. On the assumption that the phase shifts on reflection at the surfaces do not vary with wavelength, we then have from Eq. (4.26),

$$(1/\lambda_m) - (1/\lambda_{m+1}) = 1/2nd. \tag{4.27}$$

On the other hand, if the thickness of the film varies along the section imaged on the slit, the values of λ_m along each fringe will reflect these variations. Such fringes are known as fringes of equal chromatic order, or FECO fringes [Tolansky, 1945], and have been used widely for the study of surface structure.

To implement this technique, fringes are formed between the surface under study and a flat surface. If, then, d_1 and d_2 are the thicknesses of the film at two points and the wavelengths corresponding to the fringe maximum of order m at these points are $\lambda_1(m)$ and $\lambda_2(m)$, respectively, we have, from Eq. (4.26),

$$\Delta d = d_2 - d_1$$
$$= (m/2)[\lambda_2(m) - \lambda_1(m)], \tag{4.28}$$

provided that the phase shifts on reflection do not change appreciably over this range of wavelengths. With a spectrograph having a linear dispersion, the outline of the fringe corresponds directly to the profile of the surface. To obtain absolute values of the surface deviation Δd, the interference order can be evaluated from Eq. (4.27) using the values of the wavelengths corresponding to two adjacent orders, say m and $(m + 1)$, for the same point on the surface.

FECO fringes have several advantages over fringes of equal thickness for the study of surface structure. The most important is that the wedge angle between the

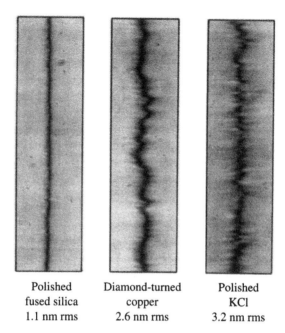

Polished	Diamond-turned	Polished
fused silica	copper	KCl
1.1 nm rms	2.6 nm rms	3.2 nm rms

Figure 4.6. Measurement of the residual irregularities of polished surfaces using fringes of equal chromatic order (courtesy J. M. Bennett, Michelson Laboratory).

surfaces can be virtually eliminated, so that highly reflecting coatings can be used effectively to obtain fringes with extremely high finesse. As shown in Fig. 4.6, surface irregularities < 1 nm in height can be measured with this technique [Bennett and Bennett, 1967].

4.5   Fringes of Superposition

Although it is not possible to observe interference fringes with white light in a thick, plane-parallel plate, it is possible to obtain fringes by superposing two plates of very nearly the same thickness, so that the optical path differences introduced in one set of beams by reflection in the first plate are matched by the optical path differences introduced in the other set of beams by reflection in the second plate. Such fringes were first observed by Brewster in 1817 and are called fringes of superposition.

To understand how such fringes are formed, consider two plane-parallel plates M_1 and M_2 of equal thickness, which, as shown in Fig. 4.7, make a small angle ϵ with each other. Any ray incident on M_1 will give rise to a series of rays by reflection within M_1; each of these will then give rise to a series of rays by reflection within M_2. For simplicity, we select two rays B_1 and B_2 which originate from a single ray

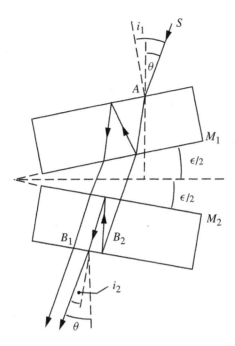

Figure 4.7. Formation of fringes of superposition.

incident on M_1, B_1 being reflected at the two surfaces of M_1 and directly transmitted through M_2, while B_2 is transmitted through M_1 and reflected at the two surfaces of M_2.

Let r_1 and r_2 be the angles of refraction at the two plates. The optical path difference between these two rays is then

$$\Delta p = 2nd(\cos r_1 - \cos r_2), \tag{4.29}$$

where n is the refractive index of the plates and d is their thickness. Since r_1 and r_2 are small, we can rewrite Eq. (4.29) as

$$\Delta p \approx 2nd[(1 - r_1^2/2) - (1 - r_2^2/2)]$$
$$\approx nd(r_2^2 - r_1^2). \tag{4.30}$$

If the incident ray SA makes an angle θ with the normal to the bisector of the angle between M_1 and M_2, the angles of incidence on M_1 and M_2 are, respectively,

$$i_1 = \theta + \epsilon/2 \tag{4.31}$$

and

$$i_2 = \theta - \epsilon/2. \tag{4.32}$$

However, $i_1 \approx nr_1$ and $i_2 \approx nr_2$. Accordingly, Eq. (4.30) reduces to

$$\Delta p = 2\theta\epsilon d/n. \tag{4.33}$$

Equation (4.33) shows that the optical path difference between these two beams is proportional to θ and drops to zero when $\theta = 0$. Straight-line fringes parallel to the apex of the wedge and localized at infinity can be seen with an extended white light source, though their visibility is low because they are superposed on a uniform bright background produced by the other beams, for which the optical path differences are too large for them to interfere.

From Eq. (4.33), the condition for an intensity maximum is

$$2\theta\epsilon d/n = m\lambda, \tag{4.34}$$

where m is an integer. The angular separation of the fringes is, therefore,

$$\theta_m - \theta_{m+1} = n\lambda/2\epsilon d, \tag{4.35}$$

and increases as ϵ, the angle between the plates, decreases. When the plates are parallel ($\epsilon = 0$), a uniform interference field is obtained.

A more complete analysis of this phenomenon requires us to take into account all the beams formed by multiple reflections at M_1 and M_2. In addition, we have to consider the general case in which the thicknesses of M_1 and M_2 are no longer equal.

We first consider M_1 alone and assume that a monochromatic wave of wavelength λ and unit amplitude is incident on it. If the reflectance and transmittance for amplitude for both its faces are r_1 and t_1, respectively, for a wave incident on them from the surrounding medium, and r_1' and t_1' for a wave incident from within, the complex amplitude of the transmitted wave is (see Section 4.1)

$$A_1(\lambda) = t_1 t_1' \exp(-i\phi_1/2)[1 + r_1'^2 \exp(-i\psi_1) + r_1'^4 \exp(-i2\psi_1) + \cdots]$$

$$= T_1 \exp(-i\phi_1/2) \sum_{l=0}^{\infty} R_1^l \exp(-il\phi_1), \tag{4.36}$$

where T_1 and R_1 are the transmittance and reflectance, for intensity, of the faces, ϕ_1 is the phase delay on transmission through the plate, ψ_1 is the phase difference between successive beams, and l is an integer. The intensity of the transmitted wave is therefore

$$I_1(\lambda) = T_1^2 \sum_{l=0}^{\infty} \sum_{m=0}^{\infty} R_1^l R_1^m \exp[-i(l - m)\psi_1]. \tag{4.37}$$

Equation (4.37) can be simplified if we set $l - m = a$; we then have

$$I_1(\lambda) = T_1^2 \sum_{l=0}^{\infty} R_1^{2l} \left\{ 1 + \sum_{a=1}^{\infty} R_1^a [\exp(ia\psi_1) + \exp(-ia\psi_1)] \right\}$$

$$= [T_1^2/(1 - R_1^2)] \left(1 + 2 \sum_{a=1}^{\infty} R_1^a \cos a\psi_1 \right). \tag{4.38}$$

In the same manner, if we consider M_2 alone, the intensity of the transmitted wave would be

$$I_2(\lambda) = [T_2^2/(1 - R_2^2)] \left(1 + 2 \sum_{b=1}^{\infty} R_2^b \cos b\psi_2 \right), \tag{4.39}$$

where T_2 and R_2 are, respectively, the transmittance and reflectance, for intensity, of its faces, and ψ_2 is the phase difference between successive beams. Equations (4.38) and (4.39) represent the effective transmittances of M_1 and M_2. Accordingly, if we neglect the effects of beams reflected backward and forward between M_1 and M_2, the intensity of the wave transmitted through them is given by the product of $I_1(\lambda)$ and $I_2(\lambda)$,

$$I_{1,2}(\lambda) = [T_1^2 T_2^2/(1 - R_1^2)(1 - R_2^2)]$$

$$\times \left(1 + 2 \sum_{a=1}^{\infty} R_1^a \cos a\psi_1 \right) \left(1 + 2 \sum_{b=1}^{\infty} R_2^b \cos b\psi_2 \right). \tag{4.40}$$

With a source emitting over a wide spectral bandwidth, the intensity of the transmitted wave is the sum of the intensities of the monochromatic components. If we assume that the incident light has a spectral energy distribution $I(\lambda)$, and that the reflectance and transmittance of the surfaces do not vary with wavelength, the total transmitted intensity is

$$I_{1,2} = [T_1^2 T_2^2/(1 - R_1^2)(1 - R_2^2)]$$

$$\times \int I(\lambda) \left[1 + 2 \sum_{a=1}^{\infty} R_1^a \cos a\psi_1 + 2 \sum_{b=1}^{\infty} R_2^b \cos b\psi_2 \right.$$

$$+2\sum_{a=1}^{\infty}\sum_{b=1}^{\infty}R_1^a R_2^b \cos(a\psi_1 + b\psi_2)$$

$$+2\sum_{a=1}^{\infty}\sum_{b=1}^{\infty}R_1^a R_2^b \cos(a\psi_1 - b\psi_2)\Bigg] d\lambda. \tag{4.41}$$

If the spectral bandwidth of the source is broad enough that fringes cannot be seen with either of the plates taken singly, it means that over this range of wavelengths ψ_1 and ψ_2 vary by amounts that are large compared to 2π. As a result, the terms involving $\cos a\psi_1$, $\cos b\psi_2$, and $\cos(a\psi_1 + b\psi_2)$ in Eq. (4.41) change sign many times, and their contribution to the integral can be neglected. Equation (4.41), therefore, reduces to

$$I_{1,2} = [T_1^2 T_2^2/(1 - R_1^2)(1 - R_2^2)]$$

$$\times \int I(\lambda)\left[1 + 2\sum_{a=1}^{\infty}\sum_{b=1}^{\infty}R_1^a R_2^b \cos(a\psi_1 - b\psi_2)\right] d\lambda. \tag{4.42}$$

Further consideration of Eq. (4.42) shows that the contribution of the terms involving $\cos(a\psi_1 - b\psi_2)$ is also negligible, except in the particular case when ψ_1/ψ_2 is close to g/f, where g and f are two small integers. In this case, those terms involving $m(f\psi_1 - g\psi_2)$, where m is an integer, remain, and Eq. (4.42) can be written as

$$I_{1,2} = [T_1^2 T_2^2/(1 - R_1^2)(1 - R_2^2)]$$

$$\times \int I(\lambda)\left\{1 + 2\sum_{m=1}^{\infty}(R_1^f R_2^g)^m \cos[m(f\psi_1 - g\psi_2)]\right\} d\lambda. \tag{4.43}$$

The series within the braces in Eq. (4.43) is similar to that on the right-hand side of Eq. (4.38); the latter is the expression for the intensity of a monochromatic wave transmitted by a single plate and its sum is given by Eq. (4.10). Accordingly, we have

$$1 + 2\sum_{m=1}^{\infty}(R_1^f R_2^g)^m \cos[m(f\psi_1 - g\psi_2)]$$

$$= \frac{[1 - (R_1^f R_2^g)^2]}{(1 - R_1^f R_2^g) + 4R_1^f R_2^g \sin^2[(f\psi_1 - g\psi_2)/2]} \tag{4.44}$$

and

$$I_{1,2} = \{T_1^2 T_2^2[1 - (R_1^f R_2^g)^2]/(1 - R_1^2)(1 - R_2^2)(1 - R_1^f R_2^g)^2\}$$

$$\times \int I(\lambda)\{1 + G \sin^2[(f\psi_1 - g\psi_2)/2]\}^{-1}d\lambda, \tag{4.45}$$

where $G = 4R_1^f R_2^g/(1 - R_1^f R_2^g)^2$.

The intensity distribution defined by Eq. (4.45) is equivalent to the superposition of a number of monochromatic fringe patterns similar to that defined by Eq. (4.8). The zero-order maxima of all these patterns, which must satisfy the condition $(f\psi_1 - g\psi_2) = 0$, coincide. In the vicinity of this zero-order maximum, the intensity distribution is similar to that in a thin film whose surfaces have a reflectance $R_1^f R_2^g$. Accordingly, the visibility of the fringes decreases for higher values of f and g.

From Eq. (4.35), the angular separation of the fringes is

$$\theta_m - \theta_{m+1} = n\lambda/2\epsilon f d_1$$
$$= n\lambda/2\epsilon g d_2. \tag{4.46}$$

Because their separation is proportional to the wavelength, successive maxima on either side of the zero-order maximum get increasingly out of step, and the visibility of the pattern falls off. With white light, the fringes on either side of the white zero-order fringe are colored; these colors progressively become less and less distinct and, finally, give way to uniform illumination. Some applications of such fringes of superposition have been discussed by Cagnet [1954].

4.6 Three-Beam Fringes

With two-beam interference, it is difficult to measure the position of a fringe by visual estimation to better than 1/20 of the interfringe distance. However, visual intensity matching of two uniform interference fields, one of which contains a small phase step, makes it possible to detect a change in the optical path of $\lambda/1000$ [Kennedy, 1926; Hariharan and Sen, 1960a]. Zernike [1950] therefore proposed the following simple technique, using three beams, which permits a photometric setting on a system of interference fringes.

As shown schematically in Fig. 4.8, two plane waves of equal amplitude a_0 are used as reference wavefronts. These two wavefronts make angles $\pm\epsilon$ with the plane of observation and intersect at a point O in this plane. The complex amplitudes due to these wavefronts at a point P in the plane of observation at a distance x from O are then

$$a_1 = a_0 \exp(-ik\epsilon x) \tag{4.47}$$

and

$$a_2 = a_0 \exp(ik\epsilon x), \tag{4.48}$$

and they produce an interference pattern consisting of equally spaced, parallel, straight fringes.

A third wavefront which is parallel to the plane of observation is now superposed on the first two. We assume that this wave also has an amplitude a_0, and that its optical

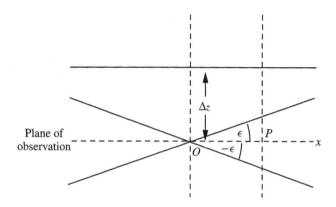

Figure 4.8. Formation of three-beam fringes [Zernike, 1950].

path differs by an amount Δz from the optical paths of the first two waves to the point O. The resultant complex amplitude at P can then be written as

$$A = a_0 \exp(-i\psi) + a_0 \exp(i\psi) + a_0 \exp(-i\phi), \qquad (4.49)$$

where $\psi = k\epsilon x$ and $\phi = k\Delta z$, and the resultant intensity is

$$I = I_0[3 + 2\cos 2\psi + 4\cos \phi \cos \psi], \qquad (4.50)$$

where $I_0 = a_0^2$.

Curves of the intensity distribution in the fringes when $\phi = 2m\pi$, $2m\pi + \pi/2$, and $(2m+1)\pi$ are presented in Fig. 4.9. These curves show that the introduction of the third wavefront results in a modulation of the intensity of the fringes. The intensities of adjacent maxima are, in general, unequal except when ϕ is an odd multiple of $\pi/2$.

We can therefore use a photometric setting criterion which involves adjusting the phase of the third wavefront by means of a compensator until all the fringes have the same intensity. To evaluate the precision with which this setting can be made, consider the situation when $\phi = (2m + 1)\pi/2 + \Delta\phi$ where $\Delta\phi \ll \pi/2$. The intensities at adjacent maxima are then given by the relations

$$I_m = I_0(5 - 4\Delta\phi) \qquad (4.51)$$

and

$$I_{m+1} = I_0(5 + 4\Delta\phi). \qquad (4.52)$$

The relative difference in their intensities is, therefore,

$$\Delta I/I = (8/5)\Delta\phi. \qquad (4.53)$$

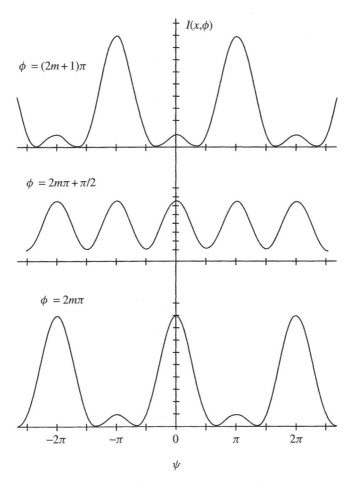

Figure 4.9. Intensity distribution in three-beam fringes for different values of ϕ, the phase difference between the third beam and the other two beams [Hariharan and Sen, 1959a].

If we assume that a difference of 5% in the intensities of adjacent fringes can just be detected visually, the setting error is

$$\Delta\phi \approx 2\pi/200, \tag{4.54}$$

and measurements of the optical path can be made with a precision

$$\Delta p \approx \lambda/200. \tag{4.55}$$

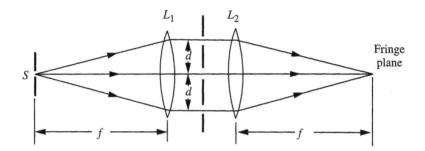

Figure 4.10. Zernike's three-beam interferometer.

In the optical arrangement used by Zernike [1950], which is shown in Fig. 4.10, the three waves are produced by division of a plane wavefront at a screen containing three equidistant parallel slits. The two outer slits, whose center lines are separated by a distance $2d$, provide the reference beams, while the beam from the middle slit is used for measurements.

In this arrangement, the optical paths of all the three beams are equal at a point on the axis in the back focal plane of the lens L_2. However, at a plane located at a distance z from L_2, the optical path of the middle beam will be longer by an amount

$$\Delta p = (d^2/2)[(1/f) - (1/z)]. \tag{4.56}$$

There will, therefore, be two positions of the plane of observation that correspond to values of Δp of $\pm\lambda/4$. Any small optical path difference introduced in the middle beam can be measured from the shift in these positions.

This interferometer is easy to set up and very stable. It has been used to calibrate phase-retardation plates as well as to measure very small changes in the optical thickness of a specimen [Vittoz, 1956]. However, because the beams are obtained by wavefront division, the amplitudes of the individual beams are not uniform over the field. As a result, the photometric setting has to be made using only the fringes on either side of the central fringe. In addition, since a very narrow slit source must be used, the intensity of the fringes is low.

A brighter fringe system can be obtained if the source slit is replaced by a grating whose period is equal to the spacing of the fringes [Maréchal, Lostis, and Simon, 1967]. Both problems are eliminated if the three beams are produced by amplitude division, using an optical system similar to that employed in the Jamin interferometer [Hariharan and Sen, 1959a].

4.7 Double-Passed Fringes

In three-beam interference, the intensity distribution in a two-beam fringe pattern is modulated by the wavefront whose phase is to be measured. A similar modulation of

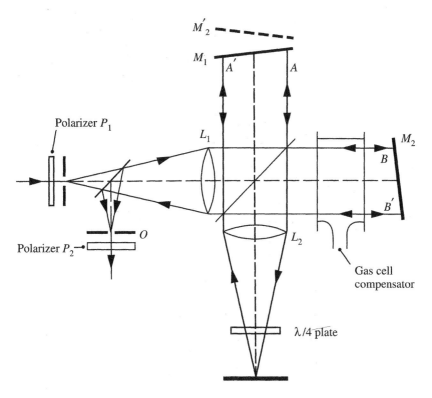

Figure 4.11. Double-passed Michelson interferometer [Hariharan and Sen, 1960b].

the fringes can also be obtained by reflecting the beams emerging from a two-beam interferometer back through it [Hariharan and Sen, 1960b].

To study the formation of these fringes, consider a Michelson interferometer illuminated with collimated light, in which, as shown in Fig. 4.11, the beams emerging from the interferometer are reflected back through it by a mirror placed at the focus of the lens L_2. The double-passed fringes are viewed, by means of an auxiliary beamsplitter, at O. The beams reflected back from the interferometer after only a single pass are eliminated by the polarizer P_2, whose axis is at right angles to P_1, while the double-passed beams, which have traversed the $\lambda/4$ plate twice, are transmitted freely. A gas-cell compensator in one beam is used to vary the optical path difference.

Four beams are formed in this case as a result of the double passage through the interferometer; their total optical paths may be designated as p_{11}, p_{12}, p_{21}, and p_{22} corresponding to the paths traversed on the outward and return journeys. If therefore, M_1 and M_2', the image of M_2 in the beamsplitter, make an angle 2ϵ with each other and are separated by a distance d at the center of the field, the optical paths of the double-passed beams at a point P at a distance x from the center of the field can be written as

$$p_{11} = p_0 - 2d, \tag{4.57}$$

$$p_{12} = p_0 + 4x\epsilon, \tag{4.58}$$

$$p_{21} = p_0 - 4x\epsilon, \tag{4.59}$$

$$p_{22} = p_0 + 2d. \tag{4.60}$$

The resultant complex amplitude at P can then be written as

$$A = A_0[\exp(i2\phi) + \exp(-i2\psi) + \exp(i2\psi) + \exp(-i2\phi)], \tag{4.61}$$

where $\phi = kd$ and $\psi = kx\epsilon$, and the resultant intensity at P is

$$I = I_0[\cos 2\phi + \cos 2\psi]^2. \tag{4.62}$$

In this case also, the intensity of adjacent fringes is equal when

$$\phi = (2m + 1)\pi/2, \tag{4.63}$$

but the precision of measurements is

$$\Delta p = \lambda/1000. \tag{4.64}$$

The formation of double-passed fringes in other types of interferometers has been discussed by Hariharan and Sen [1960c, 1961a]; they have also made a detailed analysis of possible sources of errors [Hariharan and Sen, 1961b].

Chapter 5

Lasers

As we have seen, many types of interferometers require a source of coherent light. The closest approximation to such a source was, for many years, a pinhole illuminated, through a narrow-band filter, by a mercury vapor lamp. However, such a thermal source has two major drawbacks. One is its very low intensity; the other is the limited spatial and temporal coherence of the light. The high intensity and high degree of spatial and temporal coherence of laser light eliminate most of the problems associated with the use of thermal sources and make the laser an almost ideal light source for interferometry [Hariharan, 1987a].

The principles of operation of lasers have been described in detail in several books [see Siegman, 1971; Svelto, 1989; Silfvast, 1996].

5.1   Lasers for Interferometry

Some of the lasers that are commonly used for interferometry are listed in Table 5.1.

Helium–neon (He-Ne) lasers are widely used for interferometry because they are inexpensive and easy to operate. The most common output wavelengths are 0.63 and 0.54 μm.

Argon-ion (Ar^+) lasers normally require water cooling and a wavelength-selecting prism, but are useful in applications which require higher powers or multiple wavelengths.

Carbon-dioxide (CO_2) lasers can be operated on any one of a number of lines in the 9-μm and 10.6-μm bands. They are very useful for measurements over long distances.

Because of the very large number of dyes available, dye lasers can be operated at a number of wavelengths. In addition, they can be tuned continuously over a range of wavelengths (typically, about 70 nm) [see Schäfer, 1973].

Table 5.1. Lasers for interferometry

Laser Type	Wavelength (μm)	Output
He-Ne	3.39, 1.15, 0.63, 0.61, 0.54	0.5–25 mW
Ar^+	0.51, 0.49	0.5–a few W
CO_2	~10.6, ~9.0	few W–few kW
Dye	~1.08–~0.41	~10–~100 mW
Diode	~0.87, ~0.75, ~0.67	few mW–few W
Ruby	0.69	0.6–10 J
Nd:YAG	1.06	0.1–0.15 J
(cw)		few mW–few W

Diode lasers [Thompson, 1980] are cheap, are very compact, and use little power. They can be tuned over a limited range by varying the injection current or the temperature. The output beam is divergent and astigmatic, but packages are available with additional optics to produce a collimated beam.

A laser can also be built using as the active medium a solid rod of a suitably doped crystal or glass which is irradiated with light from a flash lamp. The laser output takes the form of a pulse or a series of pulses lasting a few microseconds. Commonly used materials are ruby ($\lambda = 694$ nm) and neodymium-doped yttrium aluminum garnet (Nd:YAG) ($\lambda = 1.06$ μm), the output from which can be converted to visible light by means of a frequency-doubling crystal [see Koechner, 1976]. Diode-pumped Nd:YAG lasers can give a cw output.

Very short pulses of light (≈ 15 ns duration) can be produced by using a Q switch in the laser cavity. Such a Q-switched ruby laser can be used for interferometric studies of high-speed gas flows [Tanner, 1966, 1967]. Shorter pulses, with a duration less than 500 ps, produced by a pulsed nitrogen laser ($\lambda = 337$ nm), have been used for interferometric studies of plasmas [Schmidt, Salzmann, and Strowald, 1975]. Even shorter pulses can be produced by a mode-locked Ti–sapphire laser.

5.2   Laser Modes

The two mirrors in a laser constitute a resonant cavity whose length L is much greater than its lateral dimensions. The modes of such a resonant cavity can be defined as stationary configurations of the electromagnetic field within the cavity which satisfy the boundary conditions.

The simplest form of resonant cavity is the plane-parallel, or Fabry–Perot, resonator. The modes of such a resonator correspond to two systems of plane waves propagating in opposite senses along the axis of the cavity, and meeting the condition that the optical path for each round trip is an integral number of wavelengths.

The resonant frequencies are then defined by the relation

$$v = q(c/2L), \tag{5.1}$$

where q is an integer. Plane-parallel resonators were used in the earliest lasers, but they are difficult to align and only marginally stable. A better configuration consists of two spherical mirrors with the same radius of curvature b, separated by a distance $L = b$. Since the focal points of the two mirrors coincide, this arrangement is called a confocal resonator.

5.2.1   Modes of a Confocal Resonator

The modes of a confocal resonator have been analyzed by Boyd and Gordon [1961], who have shown that any eigenfunction (a spatial mode) must satisfy the condition that it is its own Fourier transform. The eigenfunctions that satisfy this condition are the products of Hermite polynomials H_l, H_m and a Gaussian function.

Since all these modes have negligible electric and magnetic fields along the axis, they are identified by a designation TEM_{lmq}, where the indices l and m correspond to the orders of the Hermite polynomials H_l and H_m, and q is an integer.

The lowest-order mode, which corresponds to $l = 0$, $m = 0$ (the TEM_{00} mode), is of particular interest since its eigenfunction is

$$E_{00}(x, y) = \exp[-\pi(x^2 + y^2)/L\lambda]. \tag{5.2}$$

The field at the mirrors has a Gaussian profile, and its amplitude drops to $(1/e)$ of its maximum value at a distance w from the axis, known as the spot size, where

$$w = (\lambda L/\pi)^{1/2}. \tag{5.3}$$

This mode has the lowest losses and does not exhibit any phase reversals in the electric field across the beam, so that the output beam exhibits complete spatial coherence. For this reason, lasers used as light sources for interferometry should operate in the TEM_{00} mode.

The confocal resonator has the disadvantage that the spot size is too small to make effective use of the available cross-section of the laser medium. The most commonly used resonator configuration consists, therefore, either of two concave mirrors with a radius of curvature greater than the resonator length, or of a plane mirror in conjunction with a long-radius concave mirror. Such a resonator gives a larger spot size than a confocal resonator and can, at the same time, be aligned easily.

The theory of such generalized spherical resonators has been analyzed by Boyd and Kogelnik [1962] and by Kogelnik and Li [1966] by reducing the problem to that of an equivalent confocal resonator.

5.2.2 Longitudinal Modes

The active medium in a gas laser has a significant gain over a finite frequency range due to various line-broadening mechanisms such as collisions between atoms and Doppler broadening. The latter is the principal effect at low gas pressures, as in the He-Ne laser, while the former becomes dominant at higher gas pressures. Sustained oscillation is possible at any frequency at which the gain in the active medium exceeds the losses in the resonator. Even for the TEM_{00} mode, the cavity has a number of resonant frequencies which are given, for a confocal resonator, by the expression

$$\nu_q = (c/2L)(q + 1/2), \tag{5.4}$$

where q is an integer. Accordingly, if more than one of these resonant frequencies falls within the gain profile of the medium, as shown in Fig. 5.1a, the laser will oscillate in a number of longitudinal modes.

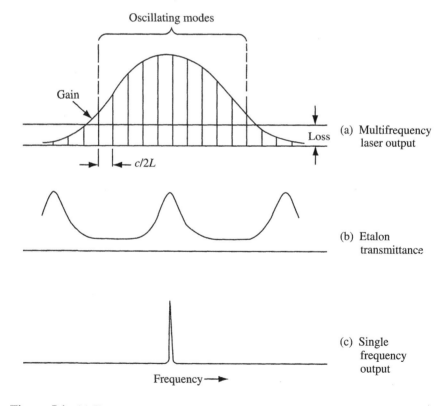

Figure 5.1. (a) Frequency spectrum of a gas laser oscillating in a single transverse mode, but in a number of longitudinal modes. (b) Transmittance of an intracavity etalon. (c) Single-frequency output obtained with an intracavity etalon.

With low-power gas lasers, in which the mirrors are sealed directly onto the ends of the discharge tube, mode competition results in adjacent longitudinal modes having orthogonal polarizations. In higher-power lasers, the tube containing the active medium is sealed with windows set at the Brewster angle, and separate mirrors are used. Losses for the TM wave (the wave polarized in the plane of incidence) are then close to zero, while losses for the TE wave are high enough to prevent oscillation. As a result, all the longitudinal modes are polarized in the same plane.

The width of the individual modes depends on the mechanical stability of the cavity structure as well as on the losses and is typically about 3 MHz in the visible region. With a laser oscillating in a single longitudinal mode, the coherence length of the output could therefore be as much as 100 m. However, the existence of more than one longitudinal mode reduces the coherence length severely. If we assume, that, as a first approximation, the output spectrum of the beam can be represented by Q equally spaced delta functions, so that it can be written as

$$S(\nu) = \sum_{i=q}^{i=q+Q-1} \delta(\nu - \nu_i),\tag{5.5}$$

the degree of temporal coherence for an optical path difference p is

$$\gamma_{11}(p) = \left| \frac{\sin(Q\pi p/2L)}{Q\sin(\pi p/2L)} \right|.\tag{5.6}$$

If $Q > 1$, this is an oscillatory function and the coherence length is defined conventionally by its first zero, which occurs when

$$p = 2L/Q.\tag{5.7}$$

Interference fringes will be obtained once again as the optical path difference is increased beyond this point, but with reversed contrast.

5.2.3 Single-Frequency Operation

The simplest way to force a gas laser to operate in a single longitudinal mode is to use a very short cavity. The frequency difference between the longitudinal modes can then be made greater than the width of the gain profile (about 1.7 GHz for a He-Ne laser), so that only one longitudinal mode is sustained. However, the gain available with such a small length of the active medium is limited, so that the power output is very low.

If single-mode operation at higher power levels is required, some method of longitudinal-mode selection has to be applied to the laser. One technique described by Smith [1965] involves the use of an auxiliary resonant cavity. A simpler method of obtaining single-frequency operation is to use an etalon in the laser cavity. This is usually a plane-parallel plate of fused silica [Hercher, 1969], though alignment is easier with an etalon having concentric spherical surfaces [Hariharan, 1982]. Oscillation is then possible only on a mode common to both cavities, as shown in Figs. 5.1b and c, for which the combined losses are low.

Single-frequency operation of diode lasers is fairly easy since, with proper design, the number of longitudinal modes decreases rapidly as the injection current is increased [Streifer, Burnham, and Scifres, 1977; Aiki et al., 1978; Kajimura et al., 1979; Henry, 1991]. Above a critical value of the current, a single-frequency output is obtained.

A narrower line width as well as a more stable output frequency can be obtained by antireflection coating the end faces of the chip and operating it in an external cavity with two low-finesse Fabry–Perot etalons as wavelength-selecting elements [Voumard, 1977].

5.3   Comparison of Laser Frequencies

The coherence time for light from a laser can be long enough that it is possible to observe beats between light waves from two lasers operating on the same transition [Javan, Ballik, and Bond, 1962]. Measurements of the beat frequency and its variance over time can be used to estimate the frequency stability of a laser.

To observe such beats, the beams from the two lasers are combined at a semi-transparent mirror and allowed to fall on a photodetector. The resultant electric field at the detector can then be represented by the real vibration

$$E(t) = E_1(t) + E_2(t), \tag{5.8}$$

where $E_1(t)$ and $E_2(t)$ are the real vibrations corresponding to the two superimposed fields. These vibrations can be written as

$$E_1(t) = a_1 \cos(2\pi \nu_1 t + \phi_1), \tag{5.9}$$

$$E_2(t) = a_2 \cos(2\pi \nu_2 t + \phi_2), \tag{5.10}$$

where a_1 and a_2 are the amplitudes, ν_1 and ν_2 are the frequencies, and ϕ_1 and ϕ_2 are the phases of the two waves.

With a square-law detector, the output current $i(t)$ is proportional to $[E(t)]^2$ so that

$$\begin{aligned} i(t) &= [E_1(t) + E_2(t)]^2 \\ &= (1/2)(a_1^2 + a_2^2) + (1/2)[\cos 2(2\pi \nu_1 t + \phi_1) + \cos 2(2\pi \nu_2 t + \phi_2)] \\ &\quad + a_1 a_2 \cos[2\pi (\nu_1 + \nu_2)t + \phi_1 + \phi_2] \\ &\quad + a_1 a_2 \cos[2\pi (\nu_1 - \nu_2)t + \phi_1 - \phi_2]. \end{aligned} \tag{5.11}$$

The second and third terms on the right-hand side of Eq. (5.11) correspond to oscillatory components at frequencies of $2\nu_1$, $2\nu_2$, and $(\nu_1 + \nu_2)$ which are too high for the detector to follow. Accordingly Eq. (5.11) reduces to

$$i(t) = (1/2)(a_1^2 + a_2^2)$$
$$+ a_1 a_2 \cos[2\pi(\nu_1 - \nu_2)t + \phi_1 - \phi_2]. \tag{5.12}$$

The output from the detector consists of a steady current, on which is superposed an oscillatory component at the beat frequency $(\nu_1 - \nu_2)$; since this beat frequency is, typically, in the radio-frequency domain, it can be observed.

We can also interpret Eq. (5.12) as implying the formation of a set of moving interference fringes, the number of fringes passing any point on the detector in unit time being equal to the beat frequency. Obviously, for the beats to be detected, it is necessary for the dimensions of the detector to be small compared to the interfringe distance, or in other words,

$$d \ll \lambda / \sin\theta, \tag{5.13}$$

where d is the diameter of the sensitive area of the detector and θ is the angle between the interfering beams. Siegman [1966] has shown that there is a trade-off between the angular field of view of the detector and its effective area, their product being equal to λ^2.

Beats can even be observed with a single laser if it is oscillating in more than one axial mode [Herriot, 1962]. The frequency separation of these modes and, therefore, the beat frequency is given, as we have seen earlier, by the relation

$$\Delta\nu = c/2L, \tag{5.14}$$

where c is the speed of light and L is the length of the laser cavity. With a typical He-Ne laser having a cavity length of 250 mm, this beat frequency is about 600 MHz and can be measured either with a spectrum analyzer, using a fast photodiode as the detector, or by direct frequency counting.

The phenomenon of beats is easily explained in classical terms on the basis that the amplitude and phase of each beam do not vary appreciably over the coherence time $1/\Delta\nu$, where $\Delta\nu$ is the laser line width. A more detailed analysis made by Mandel [1964] shows that even with two light beams derived from completely independent sources, the correlation between the intensities at two space-time points is a periodic function of their separation, indicating the presence of transient interference effects. These effects cannot be observed with light from thermal sources, for which the degeneracy parameter δ (the average number of photons in the same spin state falling on a coherence area in the coherence time) is less than 10^{-3}. However, they become observable with laser beams for which $\delta > 1$.

5.4   Frequency Stabilization

Many interferometric measurements require a source whose frequency is very stable. In addition, for measurements of length (see Chapter 7) this frequency must be precisely known. The output from a free-running gas laser, even when it is oscillating in a single longitudinal mode, does not meet these requirements to better than about 1 part in 10^6. The reason for this is that the output frequency is inversely proportional to the optical distance between the end mirrors, which can vary because of thermal or mechanical effects. As a result, although such a laser can, in principle, produce an output with a bandwidth less than 500 Hz, the mean frequency of this output can vary over the much wider bandwidth of the gain profile. Because of this, some method of stabilizing the output frequency of a laser is necessary. Some of the commonly used methods will be described briefly here.

5.4.1   Stabilization with a Reference Etalon

The simplest way of stabilizing the output frequency of a laser is by locking its output wavelength to the transmission peak of a reference Fabry-Perot etalon. With a diode laser this can be a short length of a single-mode optical fiber [Wolfelschneider and Kist, 1984].

5.4.2   Polarization-Stabilized Laser

A simple method of stabilizing the output frequency of a He-Ne laser involves using a laser tube with internal mirrors, whose length is chosen so that it operates simultaneously on just two longitudinal modes. Normally, these two modes are orthogonally polarized and their directions of polarization remain fixed. It is therefore possible to divide the beam from the rear mirror of the laser at a polarizing beamsplitter, as shown in Fig. 5.2, and to compare the intensities of the two modes. The difference of the two intensities is used as the error signal in a servo amplifier that controls either the discharge current [Balhorn, Kunzmann, and Lebowsky, 1972] or the current in a heater wound on the plasma tube [Bennett, Ward, and Wilson, 1973; Gordon and Jacobs, 1974], so that the intensities of the two modes are held equal. The frequencies of the two modes are then symmetrically positioned about the center frequency of the gain profile. One of the modes can be selected by means of a polarizer in the main beam to obtain a single-frequency output.

Measurements with such lasers show that their frequency variations can be held to less than ± 50 kHz for short periods and ± 5 MHz, or 1 part in 10^8, over quite long periods [Ciddor and Duffy, 1983]. However, care must be taken to avoid systematic offsets due to optical feedback arising from external reflections [Brown, 1981].

5.4.3   Stabilized Transverse Zeeman Laser

As mentioned earlier, a low-power He-Ne laser with internal mirrors usually oscillates in two or three longitudinal modes which are linearly polarized in orthogonal

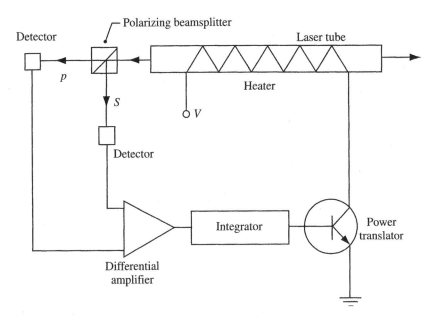

Figure 5.2. Frequency stabilization system for a He-Ne laser using two orthogonally polarized modes [Gordon and Jacobs, 1974].

directions. Morris, Ferguson, and Warniak [1975] found that the application of a transverse magnetic field to such a laser results in the modes splitting. If the value of the magnetic field is properly chosen, adjacent components of two modes can be made to overlap near the line center, while the outer components move to the wings of the gain curve and are suppressed. As a result the laser oscillates on a single axial mode composed of two orthogonally polarized waves. These two components exhibit a small frequency difference due to the magnetically induced birefringence of the gas in the laser tube, and this frequency difference depends on the position of the mode within the gain profile. This property has been exploited to stabilize the output frequency of the laser.

As shown in Fig. 5.3, the two frequencies in the back beam of the laser are mixed with a polarizer, and the beat frequency is detected with a phototransistor. The output from a frequency-to-voltage converter circuit is then used to control the length of the cavity through a servo amplifier and a heating coil on the laser tube. The properties of lasers using this method of frequency stabilization have been studied in detail by Ferguson and Morris [1978] and by Umeda, Tsukiji, and Takasaki [1980]. Brown [1981] has shown that the frequency variations are of the same order as those cited earlier for polarization stabilization but, since stabilization is effected by measurements of a beat frequency, this system is less influenced by optical feedback; however, problems can arise due to stray magnetic fields.

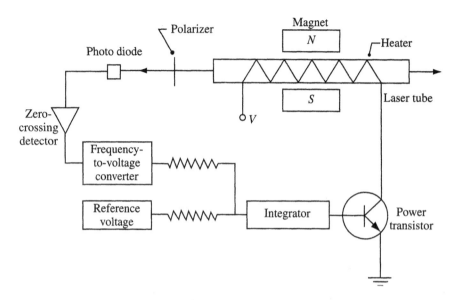

Figure 5.3. Frequency stabilization system for a He-Ne laser using a transverse magnetic field [Morris, Ferguson, and Warniak, 1975].

5.4.4 Stabilization on the Lamb Dip

When a laser oscillates, the number of atoms available in the upper (or excited) state drops, so that the available gain is reduced. However, this reduction is not uniformly spread over the whole gain profile. With a laser operating at a single frequency, only those atoms whose Doppler-shifted frequencies match the laser frequency within narrow limits will interact with the radiation field. As a result, there is a local dip in the gain profile; this effect is known as *hole burning*. If the laser frequency is not at the center of the gain profile, two holes are burned in the gain profile, as shown in Fig. 5.4, because of the interaction of waves traveling in opposite directions along the axis of the resonator with two groups of atoms having equal but opposite Doppler shifts. However, if the laser frequency coincides with the center of the gain profile, these two holes coalesce to produce a single deeper hole. Under these conditions, saturation of the stimulated emission produces a drop in the output power known as the Lamb dip. The width of the Lamb dip is about a tenth of the Doppler width for the He-Ne laser and it has been used, by modulating the laser frequency and monitoring the output with a phase-sensitive detector, to stabilize the frequency of a He-Ne laser to about 1 part in 10^8 [Rowley and Wilson, 1972].

5.4.5 Stabilization by Saturated Absorption

Stabilization on the Lamb dip has been superseded by a similar technique based on saturated absorption. When a cell containing gas molecules with an absorption

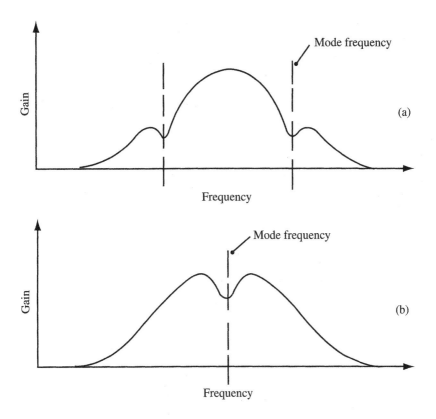

Figure 5.4. Hole burning due to saturation of stimulated emission: (a) mode frequency offset from the center of the gain profile, and (b) mode frequency at the center of the gain profile.

line at a suitable frequency is placed within a laser cavity, two groups of molecules with equal but opposite velocity components will normally interact with the two counterpropagating waves. However, when the laser frequency coincides with the unshifted absorption frequency, the same group of molecules will interact with the two counterpropagating waves, and there is a dip in the observed absorption due to saturation of the absorption. The width of this dip is limited only by the lifetimes of the energy states involved and the interaction time, and is typically less than 10^{-3} of the Doppler width.

Frequency stabilization by this technique was first proposed by Lee and Skolnick [1967], and later achieved by Hall [1968] using the 3.39-μm line of a He-Ne laser with CH_4 as the absorber. Malyshev *et al.* [1980] have reported that this line is reproducible to 5 parts in 10^{13}, and its frequency has been measured with an accuracy of ± 3 parts in 10^{11} [Knight *et al.*, 1980].

In the visible region, both $^{127}I_2$ and $^{129}I_2$ have absorption lines that coincide with the He-Ne laser line at 633 nm [Hanes and Dahlstrom, 1969; Knox and Pao, 1970]. Because the saturated absorption peaks in iodine are very small ($<1\%$) and are superposed on a sloping background, it is necessary to use a third-harmonic servo system to eliminate systematic frequency shifts due to this slope [Wallard, 1972; Hanes, Baird, and De Remigis, 1973]. A detailed description of such an iodine-stabilized He-Ne laser operating at a wavelength of 633 nm has been given by Layer [1980]. This laser has an output whose frequency is stable to 3 parts in 10^{13} over 1000 s and reproducible to better than 5 parts in 10^{10}. This technique of frequency stabilization has been the subject of intensive research [Brillet and Cerez, 1981], and lasers using it provide a number of wavelengths which are used for interferometric measurements of length (see Section 7.9).

5.4.6   Stabilization by Saturated Fluorescence

The most common infrared laser is the CO_2 laser. A widely used method of frequency stabilization for this laser is an interesting variation of the technique of saturated absorption. Low-pressure CO_2 in an auxiliary cell absorbs weakly at 10.6 μm, because the lower laser level has a small thermal population, and, as a result, fluoresces in the 4.3 μm band, which connects the upper laser level to the ground state. Saturation of the absorption results in the fluorescence efficiency exhibiting a dip at the line center. If the frequency of the laser is modulated, the resultant signal can be used very effectively to stabilize the laser center frequency [Freed and Javan, 1970]. This technique has the advantage that it permits stabilization on any one of a number of transitions in the 10.6 μm band on which the CO_2 laser can oscillate. A frequency stability of 1 part in 10^{12} can be obtained, with a reproducibility of about 2 parts in 10^{10}.

5.4.7   Diode Lasers

A considerable amount of work has been done on frequency stabilization of diode lasers. The most common technique for locking a diode laser to an absorption line uses wavelength modulation to generate a signal approximating the derivative of absorption as a function of wavelength. The zero crossing of the first derivative, which occurs at the line center, and the linear region around the zero crossing are used to generate a feedback signal to drive the laser toward line center [Ikegami, Sudo, and Sakai, 1995].

Diode lasers have been stabilized on the lines of rubidium at 780 nm and 795 nm. In addition, water vapor has a large number of absorption lines in the 810 to 840 nm region, as well as around 944 nm and 791 nm, offering the possibility of stabilizing a diode laser at several neighboring wavelengths [Koch et al., 2001; Grohe et al., 2001].

5.5 Laser Beams

The beam from a laser oscillating in the TEM_{00} mode has a Gaussian intensity profile and its diameter is often much smaller than the aperture of the interferometer. It is then necessary to expand the beam to obtain a reasonably uniform intensity distribution over the field. This can be done by a microscope objective which brings the beam to a focus, followed, where a collimated beam is required, by a lens of an appropriate focal length. The expanded beam may exhibit diffraction patterns (spatial noise) due to dust on the surfaces of the microscope objective, but these patterns can be eliminated by a pinhole placed at the focus of the objective which acts as a spatial filter and blocks the diffracted light. If the diameter of the pinhole is slightly greater than the central maximum of the image (the Airy disc), the loss of light is negligible.

With a diode laser, because of the small dimensions of the active region, a single spatial mode has a nearly Gaussian profile. However, the divergence of the output beam is quite large and is greater in one direction than the other, typically $5° \times 50°$. Interferometric measurements of the astigmatism of the output wavefront have been made by Tatsuno and Arimoto [1981] and Creath [1985].

Laser light can also give rise to problems because of its high spatial and temporal coherence. If a diffuser is introduced in the beam to obtain an extended diffuse source, speckle (see Appendix G) can be a nuisance. Speckle can be minimized when observing or photographing interference fringes by using a moving diffuser or a combination of two moving diffusers [Lowenthal and Arsenault, 1970].

A more serious problem is reflected light from the various surfaces in the optical paths in the interferometer. Because this stray light is coherent with the main beam, but has traversed an additional optical path d, its amplitude a_s adds vectorially to the amplitude a of the main beam, as shown in Fig. 5.5, resulting in a phase shift (assuming $a_s \ll a$)

$$\Delta\phi \approx (a_s/a) \sin \phi, \tag{5.15}$$

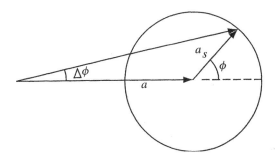

Figure 5.5. Phase shift of a beam produced by coherent light that has traversed an additional optical path.

where $\phi = (2\pi/\lambda)d$. Spatial or temporal variations in either the amplitude of the stray light or its phase relative to that of the main beam will give rise to noise in the interferogram.

Feedback due to light reflected back from the interferometer into the laser must also be avoided, since changes in the amplitude or the phase of the reflected light can cause changes in the output or even (see Section 5.4.2) the frequency of the laser [Brown, 1981].

Finally, it is very important, when working with lasers, to see that adequate precautions are taken to avoid damage to the retina, including the use of protective eye wear, where necessary (see Sliney and Wolbarsht, 1980; ANSI Standard Z 136.1-1993).

Chapter 6

Electronic Phase Measurements

Another development which has revolutionized interferometry has been the increasing use of electronic techniques for direct measurements of phase differences in interference patterns with very high accuracy. Photodetector arrays, in conjunction with digital computers, have made it possible to extend these techniques to map optical phase differences at a network of points covering an interference pattern in a very short time. As a result, traditional methods of recording and analyzing interference patterns have been largely replaced by electronic techniques.

6.1 Photoelectric Settings

Photoelectric detection has been used for many years to obtain increased precision in interferometric measurements. One of the earliest techniques was a null method in which a small modulation of the optical path difference was introduced [Baird, 1954; Bruce and Hill, 1961]. The fringe pattern was imaged on a small aperture, and the transmitted flux was analyzed into its harmonic components. The disappearance of the component at the modulation frequency defined a setting on an intensity maximum or minimum.

The precision attainable with this technique has been discussed by Hill and Bruce [1962, 1963], Hanes [1963], and Ciddor [1973], who have shown that the major limitations with thermal radiation arise from its limited coherence and low intensity. With a laser, the precision of measurements is limited, in practice, by the mechanical stability of the interferometer.

6.2 Fringe Counting

Another technique, which was first described by Peck and Obetz [1953], but has now become increasingly important, is electronic fringe counting. Basically, this involves an optical system giving two uniform interference fields, in one of which there is an additional phase difference of $\pi/2$ between the interfering beams. The two fields can be produced conveniently by using a beamsplitter with a multilayer semireflecting metal coating [Raine and Downs, 1978]; the interference pattern formed by the normal output beams and that formed by the beams going back to the source then meet this condition. Two detectors viewing these two fields provide two signals in quadrature which can drive a bidirectional counter. The same signals can also be used to determine the fractional interference order. One way of doing this is to apply these signals to the horizontal and vertical deflection amplifiers of an oscilloscope. A circular pattern is obtained, which is traversed once each time the phase difference between the beams changes by 2π. Fringe counting interferometers of this type have been described by Gilliland et al. [1966] and by Matsumoto, Seino, and Sakurai [1980]. Fringe-counting systems have also been used in free-fall instruments for absolute measurements of gravitational acceleration (g) [Zumberge, Rinker, and Faller, 1982; Arnautov et al., 1983], and more recently for measurements of the Newtonian gravity constant (G) [Schwarz et al., 1999].

An alternative method of fringe counting is based on polarization coding [Dyson et al., 1972]. In this technique, the two beams emerging from the interferometer are linearly polarized at right angles to each other and traverse a $\lambda/4$ plate oriented at 45°, which converts one beam into right-handed, and the other into left-handed, circularly polarized light. If the amplitudes of the two beams are the same, they add to produce a linearly polarized beam whose plane of polarization rotates through 360° for a change in the optical path difference between the beams of two wavelengths. The rotation of the plane of polarization can be followed by a polarizer controlled by a servo system [Roberts, 1975]. The effect of dc drifts can be minimized by modulating the output intensity by a Pockels cell or a Faraday cell and using a phase-sensitive detector. A detailed analysis of the possible errors in such a system (also known as an *optical screw*) has been made by Hopkinson [1978].

A drawback of both these fringe counting systems is that the inherent low-frequency intensity fluctuations in a laser can affect their operation. This problem can be avoided with a system that uses two different optical frequencies [Dahlquist, Peterson, and Culshaw, 1966; Dukes and Gordon, 1970]. The two optical frequencies are generated by a single He-Ne laser, which is forced to oscillate simultaneously on two frequencies, ν_1 and ν_2, separated by a constant difference of about 2 MHz, by the application of an axial magnetic field. The beam from the laser consisting of these two waves, which are circularly polarized in opposite senses, goes through a $\lambda/4$ plate, which converts them to orthogonal linear polarizations, and a beam-expanding telescope before entering the interferometer.

In the interferometer, as shown schematically in Fig. 6.1, a portion of the beam is split off initially and, after going through a polarizer which mixes the two frequencies, is incident on a detector D_R. The output from this detector, which is at the beat frequency $(\nu_2 - \nu_1)$, constitutes a reference frequency. The main beam goes to a polarizing beamsplitter at which one frequency (say ν_1) is reflected to a fixed cube corner C_1, while the other, ν_2, is transmitted to a movable cube corner C_2 which

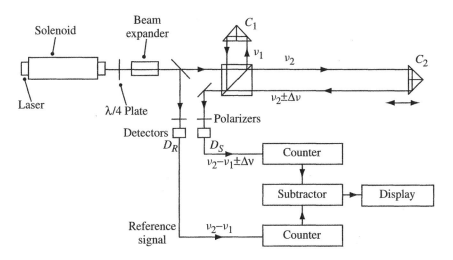

Figure 6.1. Fringe counting interferometer using a two-frequency laser [after Dukes and Gordon, 1970]. © Copyright 1970 Hewlett-Packard Company. Reproduced with permission.

constitutes the measuring element. Both frequencies then return along a common axis and, after passing through a polarizer, are incident on another detector D_S.

When both the cube corners are stationary, the output signal from the detector D_S is also at the beat frequency $(\nu_2 - \nu_1)$. However, if the cube corner C_2 is moved slowly, the frequency ν_2 is shifted up or down because of the Doppler effect. Typically, a velocity of 0.1 m/s causes a shift $\Delta\nu$ of approximately 300 kHz. As a result, the frequency at the output from the detector D_S becomes $(\nu_2 - \nu_1 \pm \Delta\nu)$.

The outputs from the two photodetectors D_R and D_S go to a differential counter. One frequency drives the counter forward, while the other drives it backward. If the cube corner C_2 is stationary, the two frequencies are equal and no net count accumulates. However, if C_2 is moved, a net positive or negative count is produced, which gives the change in optical path in wavelengths of light.

This type of interferometer can be used for accurate measurements over paths as long as 60 m [Liu and Klinger, 1979; Bobroff, 1993]. Simultaneous measurements over a number of axes can be made with a system using a two-longitudinal-mode He-Ne laser yielding a higher output power by down-converting the high intermode beat frequency [Kim and Kim, 2002].

6.3 Heterodyne Interferometry

In this method, a small frequency shift is introduced in one beam of the interferometer [Crane, 1969; Lavan, Cadwallender, and De Young, 1975]. Systems which have been used for this purpose include a rotating $\lambda/4$ plate and a rotating grating. However, the most common method is to use two acousto-optic modulators operated at slightly different frequencies.

The output current from a photodetector is then, from Eq. (5.12),

$$i(x, y, t) = (1/2)[a_1^2(x, y) + a_2^2(x, y)]$$
$$+ a_1(x, y)a_2(x, y)\cos[2\pi(\nu_1 - \nu_2)t - \Delta\phi(x, y)], \qquad (6.1)$$

where ν_1, ν_2 are the two optical frequencies, $a_1(x, y)$ and $a_2(x, y)$ are the amplitudes of the two waves, $\phi_1(x, y)$ and $\phi_2(x, y)$ are their phases, and $\Delta\phi(x, y) = \phi_1(x, y) - \phi_2(x, y)$.

It follows from Eq. (6.1) that the output from a photodetector at this point is modulated at a frequency $(\nu_1 - \nu_2)$, and the phase of this modulation corresponds to the original phase difference between the interfering wavefronts. The phase of the modulation can be measured electronically with respect to a reference signal derived either from a second detector to which the optical paths remain unchanged, or from the source driving the modulator. This technique is capable of very high precision and has been used for a variety of measurements [Ohtsuka and Sasaki, 1974; Lavan et al., 1976; Kristal and Peterson, 1976].

The light source normally used for heterodyne interferometry (a frequency-stabilized single-mode He-Ne laser) has an output less than 1 mW, which is not adequate for multiple-axis and high-target-speed interferometry. This requirement can be met by a system using a 5-mW He-Ne laser operating in three longitudinal modes, along with super-heterodyne detection [Yokoyama et al., 2001]. An added benefit is that the sensitivity is doubled.

As can be seen, the heterodyne technique replaces measurements of the positions of fringes by measurements in the frequency domain, which can be made with very high precision.

6.4   Phase-Locked Interferometry

In this technique, the phase is sinusoidally modulated with a small amplitude while an additional reference phase $\alpha(t)$ is introduced, either by moving a mirror [Johnson, Leiner, and Moore, 1979; Matthews, Hamilton, and Sheppard, 1986] or by using a diode laser as the light source and controlling the injection current [Suzuki, Sasaki, and Maruyama, 1989; Suzuki et al., 1991].

The time varying output from a point detector is then filtered to obtain a signal proportional to the sine of the sum of the original phase $\phi(t)$ and the additional reference phase $\alpha(t)$. The value of this signal is held at zero by means of a feedback loop, so that the value of $\alpha(t)$ is a direct measure of the value of $\phi(t)$. The effects of external disturbances can be eliminated by making simultaneous measurements at two points: a fixed point which gives information on any disturbances and a point which is scanned along the surface of the object. The surface profile is then obtained by subtracting the former from the latter [Suzuki *et al.*, 1992].

6.5 Computer-Aided Fringe Analysis

For many years, the chief disadvantage of interferometry as a tool for quantitative measurements was the labor involved in analyzing an interferogram. Digital techniques are now used widely to process images of the fringes for very rapid and very precise measurements [Robinson and Reid, 1993; Dorrio and Fernández, 1999].

Computer-aided analysis of fringe patterns is possible either by plotting the loci of the centers of the fringes [Womack *et al.*, 1979], using a television camera linked to a digital computer, or by directly measuring the phase difference between the two interfering wavefronts at an array of points covering the field. The latter approach is preferable, since it is capable of higher precision and is not affected by variations in the illumination over the field. The techniques described for this purpose fall into two broad groups.

One approach is based on point-by-point measurements using mechanical or optical scanning of the fringe pattern. Measurements of the phase difference between the interfering wavefronts can be made either by a null-detection technique using a small phase modulation of one of the beams [Pearson *et al.*, 1976; Johnson, Leiner, and Moore, 1977], or by heterodyne techniques [Sommargren and Thompson, 1973; Massie, Nelson, and Holly, 1979]. The time involved in making such measurements can be reduced by using an image dissector camera to scan the interference pattern [Mottier, 1979; Massie, 1980].

Real-time measurements can be made by using two orthogonally polarized beams and a system of polarizing beamsplitters and retarders at the output of the interferometer to generate four signals in quadrature [Smythe and Moore, 1984].

6.5.1 Fourier-Transform Techniques

A tilt introduced in one of the beams (say, along the x direction) generates background fringes corresponding to a spatial carrier frequency ξ_0 [Takeda, Ina, and Kobayashi, 1982]. This spatial carrier is modulated by the local variations $\Delta\phi(x, y)$ in the phase difference between the beams, so that the intensity in the interference pattern is given by the relation

$$I(x, y) = a(x, y) + b(x, y)\cos[2\pi\xi_0 x + \Delta\phi(x, y)]$$

$$= a(x, y) + c(x, y)\exp(2\pi i\xi_0 x) + c^*(x, y)\exp(-2\pi i\xi_0 x), \quad (6.2)$$

where $c(x, y) = (1/2)b(x, y)\exp[i\Delta\phi(x, y)]$.

The Fourier transform of the intensity with respect to x is then [Macy, 1983]

$$\mathcal{I}(\xi, y) = A(\xi, y) + C(\xi - \xi_0, y) + C^*(\xi + \xi_0, y), \quad (6.3)$$

where ξ is the spatial frequency in the x direction. If the spatial carrier frequency ξ_0 is sufficiently high, the term $C(\xi - \xi_0, y)$ can be isolated and processed to obtain $C(\xi, y)$. The inverse Fourier transform of $C(\xi, y)$ then yields $c(x, y)$, from which the phase difference at any point can be calculated using the relation

$$\Delta\phi(x, y) = \arctan\left[\frac{\text{Im}\{c(x, y)\}}{\text{Re}\{c(x, y)\}}\right]. \tag{6.4}$$

Fourier transform techniques are most useful where it is possible to produce straight, equally spaced carrier fringes, and the local phase variations are small as, for example, in studies of plasmas [Nugent, 1985; Bone, Bachor, and Sandemann, 1986; Toyooka, Nishida, and Takezaki, 1989]. The application of such techniques to fringe analysis has been reviewed by Takeda [1990] and Kujawinska and Wojciak [1991].

It has also been shown that this technique can be extended to closed-fringe patterns [Garcia-Marquez, Malacara-Hernandez, and Servin, 1998; Ge *et al.*, 2001]. A continuous wavelet-transform ridge-extraction technique can be used to analyze lower-quality images [Tomassini *et al.*, 2001].

6.6   Phase-Shifting Interferometry

Phase-shifting techniques permit simultaneous measurements on a number of points, using a charge-coupled-detector (CCD) array [see Creath, 1988]. This approach also has the advantage that a better signal-to-noise (S/N) ratio is possible at low light levels, since the signal at each element is integrated over the whole measurement period. Figure 6.2 is a schematic of a system used for phase-shifting interferometry.

In one method [Bruning *et al.*, 1974], the optical path difference is changed in equal steps by means of a mirror mounted on a PZT, and the corresponding values of the intensity at each point in the interference pattern are recorded. If the complex amplitudes of the reference and test wavefronts are written as

$$a(x, y) = a_0 \exp(-ikp) \tag{6.5}$$

and

$$b(x, y) = b_0 \exp[-ikW(x, y)], \tag{6.6}$$

where $W(x, y)$ represents the profile of the test wavefront, the intensity in the interference pattern is

$$I(x, y, p) = a_0^2 + b_0^2 + 2a_0 b_0 \cos k[W(x, y) - p]. \tag{6.7}$$

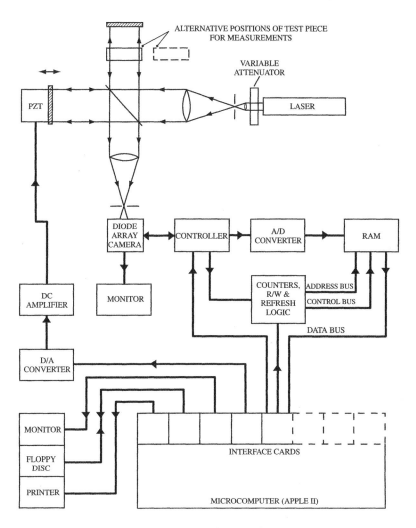

Figure 6.2. Schematic of a system for phase-shifting interferometry [Hariharan, Oreb, and Leistner, 1984].

An alternative representation for Eq. (6.7) is

$$I(x, y, p) = f + g_1 \cos kp + h_1 \sin kp, \qquad (6.8)$$

where the coefficients f, g_1, and h_1 are functions of x and y. This is a Fourier series consisting only of the dc term and the first harmonics. Accordingly, the coefficients at any point can be found by measuring the intensity $I(x, y, p)$ for values of p given by

$$p = p_j = j\lambda/2m, \qquad (6.9)$$

where $j = 1, 2, \ldots, mq$, and m and q are integers. We then have

$$g_1 = (2/mq) \sum_{j=1}^{mq} I(x, y, p_j) \cos k p_j$$

$$= 2a_0 b_0 \cos k W(x, y), \tag{6.10}$$

and

$$h_1 = (2/mq) \sum_{j=1}^{mq} I(x, y, p_j) \sin k p_j$$

$$= 2a_0 b_0 \sin k W(x, y), \tag{6.11}$$

so that

$$W(x, y) = (1/k) \arctan(h_1/g_1). \tag{6.12}$$

While as many as 100 readings ($m = 25$, $q = 4$) have been used, a typical implementation of this technique used an eight-step staircase modulation ($m = 8$, $q = 1$) of the optical path difference.

A simpler method [Carré, 1966] involves only four measurements of the intensity at a point corresponding to three equal increments of the phase. It is then possible to calculate the original phase difference between the beams as well as the magnitude of the phase shifts [Cheng and Wyant, 1985b].

A further simplification is possible if the phase shifts are known, in which case only three measurements of the intensity are necessary. Typically, one value of the phase shift can be zero, while the other two are $\pi/2$ and π [Frantz, Sawchuk, and von der Ohe, 1979; Dörband, 1982]. Alternatively, it is possible to use phase shifts of $\pm 2\pi/3$ [Hariharan, Oreb, and Brown, 1982].

However, the most commonly used algorithm for calculations of the phase difference is one using four frames of intensity data recorded with phase shifts of 0, $\pi/2$, π, and $3\pi/2$. In this case,

$$\tan \phi = \frac{I(3\pi/2) - I(\pi/2)}{I(0) - I(\pi)}. \tag{6.13}$$

In another method [Wyant, 1975; Stumpf, 1979; Schaham, 1982], the optical path difference between the interfering wavefronts is made to vary linearly with time. The intensity at any point $P(x, y)$ in the interference pattern can then be written as a function of time,

$$I(x, y) = I_0(x, y)\{1 + \mathcal{V} \sin[(2\pi t/T) + \phi(x, y)]\}, \tag{6.14}$$

where $I_0(x, y)$ is the average intensity at P, $\mathcal{V}$ is the visibility of the fringes, T is the period of the resulting modulation, and the phase difference between the two interfering wavefronts at P, when $t = 0$, is $\phi(x, y) - \pi/2$.

To determine $\phi(x, y)$, the integrated output from each of the elements is read out at the end of four equal intervals covering one such period. If these outputs are taken as A, B, C, and D, respectively, it follows that

$$\tan \phi(x, y) = (A - C)/(B - D). \tag{6.15}$$

This system requires the variation of the optical path with time to be strictly linear, though the effect of nonlinearities can be reduced by increasing the number of integration periods per cycle to eight.

Since measurements can be made rapidly, and the data can be stored, the effects of air currents and mechanical instability can be reduced by averaging a number of observations. Similarly, it is possible to eliminate systematic errors due to imperfections in the interferometer by subtracting from the readings made with the test piece a set of readings made without it or with a standard test piece [Hariharan, Oreb, and Leistner, 1984; Greivenkamp and Bruning, 1992].

6.6.1 Error-Correcting Algorithms

With the algorithms described in the previous section, systematic errors can arise in the measured value of the phase difference from several causes. The most important of these are: (1) miscalibration of the phase shifter, (2) nonlinearity and hysteresis of the phase shifter, (3) nonlinearity of the photodetector, and (4) deviations of the intensity distribution in the fringes from a sinusoid due to multiply reflected beams [Schwider *et al.*, 1983; Hariharan, 1987b; Creath, 1991]. All these effects produce errors involving terms of the form $\cos n\phi(x, y)$ and $\sin n\phi(x, y)$, where n is an integer: that is to say, sinusoidal functions of $\phi(x, y)$ with spatial frequencies that are integer multiples of the interference fringe frequency [Kinstaetter *et al.*, 1988; van Wingerden, Frankena, and Smorenburg, 1991; Joenathan, 1994].

One way to eliminate phase-shift errors is by using a general least-squares algorithm in which the phase shifts are also considered as unknowns and identified through an iterative numerical procedure [Okada, Sato, and Tsujiuchi, 1991; Kim, Kang, and Han, 1997]. Another is to use algorithms with a larger number of phase shifts, the simplest being one using five frames of intensity data recorded with nominal phase increments of $\pi/2$ [Hariharan, Oreb, and Eiju, 1987].

The characteristics of these algorithms can be evaluated by an analysis in the Fourier domain [Freischlad and Koliopoulos, 1990; Larkin and Oreb, 1992; Malacara, Servin, and Malacara, 1998]. Several algorithms using a larger number of phase shifts have been developed, using the Fourier-transform approach [Hibino, Oreb, and Farrant, 1995; Hibino *et al.*, 1997]. Other algorithms have been developed using an averaging technique, first introduced by Schwider *et al.* [1993] and extended by

Schmit and Creath [1995, 1996] and Zhang, Lalor, and Burton [1999], as well as a data-windowing technique [de Groot, 1995].

Surrel [1993] showed that while an m-frame algorithm, with phase increments of $2\pi/m$, was quite sensitive to miscalibration of the phase shifter, an $(m + 1)$-frame algorithm (with one redundant frame) offered insensitivity to phase-shifter miscalibration as well as the possibility of using many frames to improve the signal-to-noise ratio. An alternative approach to the design of efficient algorithms, based on the use of a characteristic polynomial, was also proposed by Surrel [1996, 1997b, 1998], who showed that to obtain insensitivity to harmonics up to the mth harmonic, the minimum number of frames required was $2m + 2$, with phase increments of $2\pi/(m + 2)$.

The susceptibility of error-compensating algorithms to random noise has been analyzed by Hibino [1997] and Surrel [1997a], while Zhao [1997] has evaluated the errors due to intensity quantization with an m-frame algorithm. Data-processing techniques for suppressing phase errors due to vibration have also been developed [Huntley, 1998].

General methods for generating algorithms insensitive to various sources of error have been described by Phillion [1997], Wei, Chen, and Wang [1999], and Zhu and Gemma [2001].

6.7   Techniques of Phase Shifting

The most common method of introducing phase shifts is by mounting one of the mirrors of the interferometer on a piezoelectric transducer (PZT) to which appropriate voltages are applied. However, two other techniques have advantages for specific applications.

6.7.1   Frequency Shifting

An alternative method of phase shifting is by using a diode laser whose output frequency can be varied by varying the drive current. In this case, the interferometer is set up initially with a known optical path difference p between the beams. A frequency change $\Delta \nu$ then introduces an additional phase difference

$$\Delta \phi = (2\pi p/\nu)\Delta \nu \qquad (6.16)$$

between the two wavefronts [Ishii, Chen, and Murata, 1987]. The phase is calculated using the output from a CCD array sensor that integrates the intensity at each point over four successive intervals of one-quarter period [Chen, Ishii, and Murata, 1988].

Dynamic phenomena can be monitored by synchronizing the variation of the injection current with the field pulse of a CCD camera and storing, on a videotape, a series of interferograms recorded with phase increments of $\pi/2$ at video rates.

Four successive phase-shifted interferograms can then be used to evaluate the phase distribution at any time [Onodera and Ishii, 1999b].

A problem is that a change in the drive current also affects the intensity of the laser beam, resulting in errors [Hariharan, 1989]. One way to avoid this problem is by using photothermal modulation. In this technique, the radiation from an external pump laser is focused onto the active region of the diode laser used as the source [Klimcak and Camparo, 1988]. The resulting changes in the temperature and the charge carrier density produce a frequency shift with a much smaller change in intensity [Anderson and Jones, 1992]. A combination of both techniques can be used to produce frequency modulation free from intensity modulation [Suzuki *et al.*, 1999b].

The phase shift can be precisely controlled and the effects of external disturbances minimized by a feedback-control system [Yamaguchi, Liu, and Kato, 1996; Yoshino and Yamaguchi, 1998; Yokota, Asaka, and Yoshino, 2001; Suzuki, Zhao, and Sasaki, 2001].

6.7.2 Polarization Techniques

Yet another way of shifting the phase of a beam is by means of a cyclic change in its state of polarization (the Pancharatnam phase; see Appendix E.2).

With a linearly polarized beam, it is possible to use a $\lambda/2$ plate (H) mounted between two $\lambda/4$ plates (Q) as a phase shifter [Pancharatnam, 1956]. The two $\lambda/4$ plates have their optic axes fixed at an azimuth of 45°, while the $\lambda/2$ plate can be rotated. Rotation of the $\lambda/2$ plate through an angle θ shifts the phase of the linearly polarized output beam by 2θ. With two orthogonally polarized beams, it is possible to use a simpler system consisting of a $\lambda/4$ plate, with its axis at 45°, followed by a rotatable polarizer [Hariharan and Sen, 1960e]. Rotation of the polarizer through an angle θ introduces a phase difference of 2θ between the two beams. Some other systems which can be used as phase shifters with polarized beams have been reviewed by Kothiyal and Delisle [1985].

Because the Pancharatnam phase is a geometric phase and, therefore, a topological phenomenon, it is, in principle, independent of the wavelength. As a result, the phase shift with a QHQ system, even with simple retarders, is very nearly achromatic [Hariharan and Ciddor, 1994; Hariharan, Larkin, and Roy, 1994]. Switchable achromatic phase shifts of ±90° and ±120° can be obtained by replacing the $\lambda/2$ plate by two ferroelectric liquid-crystal cells [Hariharan and Ciddor, 1999]. Achromatic phase shifters using a rotating polarizer have been discussed by Helen, Kothiyal, and Sirohi [1998].

6.7.3 Simultaneous Measurements

In order to make measurements under adverse conditions, or measurements of dynamic events, at least three phase-shifted interferograms need to be recorded simultaneously. The use of beamsplitters [Smythe and Moore, 1984] requires separate cameras, involving problems of synchronization. An alternative is to use a grating, in which case a single camera is sufficient [Kwon, 1984; Hettwer, Kranz, and Schwider, 2000].

6.8   Sinusoidal Phase Modulation

In this method, the phase of the reference beam is modulated by a sinusoidally vibrating mirror, and a CCD image sensor is used to detect the time-varying intensity distribution in the interference pattern. The profile of the test surface is then derived from the frequency components of the detected signal [Sasaki and Okazaki, 1986].

In a variation of this method, the phase is modulated sinusoidally, and a video camera integrates the time-varying intensity at each point of the interference pattern over a quarter of the period of the sinusoidal modulation. If the modulation parameters are properly chosen, the phase at each point can then be calculated from four successive patterns obtained in a single modulation period [Sasaki, Okazaki, and Sakai, 1987].

This method can be implemented conveniently by using a diode laser as the light source and modulating the injection current [Sasaki, Takahashi, and Suzuki, 1990]. It has the advantage of being relatively insensitive to temperature fluctuations and low-frequency vibration.

Chapter 7

Measurements of Length

One of the main applications of interferometry is in accurate measurements of length. This essentially involves the determination of the mean interference order over a specified area of the interference pattern. Such length measurements are usually carried out on material standards of two types: line standards, consisting of a set of fine lines engraved on a polished surface, and end standards, consisting of a bar with polished, flat, parallel ends.

7.1   Line Standards

Measurements on line standards, such as scales, are carried out with an interference comparator (see the review by Baird [1963]). The scale to be calibrated is mounted on a movable carriage which also carries one of the mirrors of a Michelson interferometer. Successive graduations on the scale are located by a fixed photoelectric microscope, while the displacement of the mirror is evaluated from the corresponding change in the interference order [Ciddor and Bruce, 1967].

7.2   End Standards

Measurements of end standards are commonly made with a Kösters interferometer. As shown in Fig. 7.1, this is basically a Michelson interferometer using collimated light and incorporating a dispersing prism to select a single spectral line from the source. The end standard is wrung on to a metal platen, which is used as one of the mirrors in the interferometer. The image of the reference mirror is positioned halfway between the platen and the top of the end standard to minimize the optical path differences, and the difference in the interference orders for the two fields is determined. The

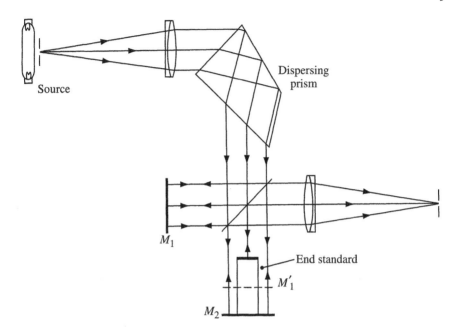

Figure 7.1. Kösters interferometer for end standards.

length of the end standard can then be evaluated using the method of exact fractions (see Section 7.4).

The value of the length obtained by this method includes the thickness of a contacting layer between the face of the platen and the surface of the gauge which depends on a number of factors. Uncertainties due to this cause can be eliminated by using an interferometric arrangement in which both ends of the gauge are viewed directly without the need for an auxiliary platen [Hariharan and Sen, 1959b; Khavinson, 1999].

7.3   The Integral Interference Order

A problem in all interferometric measurements of length is that if the interference order with monochromatic light of wavelength λ is, say, $(m+\epsilon)$, where m is an integer and ϵ is a fraction, observations on the fringes only yield the value of the fraction ϵ. Various methods have been used to find m, the integral interference order. A direct method is to count the fringes which pass a given point in the field while one of the mirrors of the interferometer is moved over the distance to be measured; this method can be used very effectively with photoelectric fringe counting techniques (see Section 6.2).

Another method involves the use of the zero-order white-light fringe to judge the equality of the two optical paths in the interferometer; this criterion can be used when comparing two end standards of very nearly the same length. Yet another method is optical multiplication. If two Fabry–Perot interferometers are placed in series, the spacings of the plates in the two interferometers can be adjusted using fringes of superposition (see Section 4.5), so that they are in an integral ratio. This technique makes it possible to compare a long optical path with a shorter, known optical path. However, the most commonly used procedure is the method of exact fractions.

7.4 Exact Fractions

In this method, the interferometer is illuminated successively with different wavelengths $\lambda_1, \lambda_2 \ldots \lambda_n$, and the fractional orders of interference $\epsilon_1, \epsilon_2 \ldots \epsilon_n$ are measured for these wavelengths. If p is the optical path difference, we then have

$$p = (m_1 + \epsilon_1)\lambda_1, \tag{7.1}$$

$$p = (m_2 + \epsilon_2)\lambda_2, \tag{7.2}$$

$$\ldots \ldots \ldots \ldots$$

$$p = (m_n + \epsilon_n)\lambda_n, \tag{7.3}$$

where $m_1, m_2 \ldots m_n$ are the corresponding integral orders. Since the value of p is already known within fairly close limits from mechanical measurements, an approximate estimate of the integral order can be made for one of the wavelengths, say λ_1. A range of integral values for m_1, centered on this value, is then taken, and the corresponding values of $\epsilon_2 \ldots \epsilon_n$ are calculated from Eqs. (7.1) to (7.3). The correct value of m_1 is then that for which the calculated values of all the excess fractions agree with the measured values to within the experimental uncertainty.

The method of exact fractions can be applied very effectively with a CO_2 laser, since it can generate a number of lines whose wavelengths are known with a high degree of accuracy [Bourdet and Orszag, 1979; Gillard, Buholz, and Ridder, 1981; Gillard and Buholz, 1983]. A set of analytical equations amenable to automatic computation has been developed by Tilford [1977] that permit the length to be calculated by a series of approximations in terms of the wavelengths and the measured fractional interference orders. These equations can also be used to choose suitable wavelengths for the range to be covered.

7.5 The Refractive Index of Air

A problem closely related to the measurement of length is the measurement of the refractive index of air. Unless a length measurement is carried out in a vacuum, the value obtained for the optical path length has to be divided by the refractive index of the air within the interferometer to obtain the true length.

The first detailed measurements of the refractive index of air were made by Barrell and Sears [1939] using two Fabry–Perot interferometers of almost the same length in series. To start with, the shorter one was filled with air while the other was evacuated. The first interferometer was then slowly evacuated and the change in the interference order was measured from observations on the fringes of superposition. These values have been used in formulas developed by Edlen [1966] and by Jones [1981] which can be applied to estimate the refractive index correction, provided the air is free from contamination. More recent work on the refractive index of air has been reviewed by Ciddor [1996] and Ciddor and Hill [1999], who have combined these results to obtain a set of equations that can be used over a wide range of atmospheric parameters and wavelengths.

Interferometry has also been widely used to measure the refractive indices of other gases, as well as mixtures of gases, since they are all so close to unity that measurements by conventional methods of refractometry are not possible. Two interferometers that have been used for this purpose are the Rayleigh interferometer and the Jamin interferometer. As shown in Fig. 7.2, the two beams pass through two cells of the same length containing the gases whose refractive indices are to be compared. For absolute measurements of the refractive index, one cell is evacuated and the number of fringes crossing a given point in the field is counted as the gas is admitted to the cell. Since this procedure can be rather tedious, a compensator is frequently used. This consists of a fixed glass plate in one beam, and an identical plate in the other beam which can be tilted to introduce a known change in the optical path. However, it should be noted that, as discussed in Section 3.3, such measurements only give the phase index when the actual interference order is measured with a quasi-monochromatic

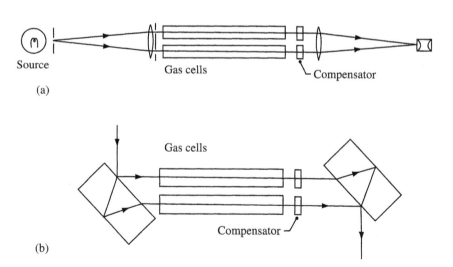

Figure 7.2. (a) Rayleigh and (b) Jamin interferometers.

source. Measurements which involve judgments of equality of the two optical paths using white-light fringes give values of the group refractive index.

7.6 The International Prototype Metre

The first measurement of the length of the International Prototype Metre, in wavelengths of the red cadmium line ($\lambda \approx 644$ nm), was carried out by Michelson in 1892, using a set of nine transfer standards, each consisting, as shown in Fig. 7.3, of two mirrors M_1, M_1' set parallel to each other on a metal support. The distance d separating the mirrors increased by a factor of 2 from approximately 0.39 mm in the shortest (standard 1) to 100 mm in the longest (standard 9).

The first step was to measure the separation of the mirrors M_1, M_1' in standard 1, in wavelengths. For this, standard 1 was set up in an interferometer, as shown in Fig. 7.4a. The half of the field covered by M_1 and M_1' was illuminated with white light, while the other half was illuminated with monochromatic light from the cadmium lamp.

To start with, the end mirror M_B of the interferometer was set so that its image M_B' coincided with M_1, by locating the zero-order fringe at the center of M_1. The end mirror M_B was then moved slowly until M_B' coincided with M_1', and the number of fringes passing a given point in the other half of the field was counted.

To compare standard 1 with standard 2, the two standards $M_1 M_1'$ and $M_2 M_2'$ were set up, side by side, in one arm of the interferometer, as shown in Fig. 7.4b. Observations of the zero-order fringe were then used to move standard 1 through a distance exactly equal to its own length, and the residual difference was measured by fringe counting. This process was repeated with successive standards.

In the final step, a microscope was used to bring a line on standard 9 into coincidence with a line on the prototype metre, which was placed next to it. Standard 9 was then moved in steps, as described above, through a distance equal to 10 times its length. The residual distance separating the line on standard 9 and the line at the other end of the prototype metre was then read off, using another microscope.

Another series of measurements was carried out by Benoit, Fabry, and Perot in 1913 using multiple-beam fringes formed between pairs of plane, semitransparent

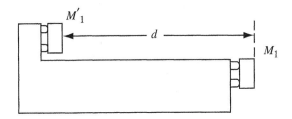

Figure 7.3. Transfer standard used by Michelson.

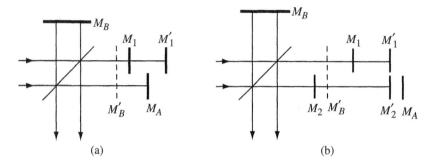

Figure 7.4. (a) Measurement of transfer standard 1 in wavelengths, and (b) comparison of transfer standard 1 with transfer standard 2.

mirrors (etalons). Five etalons were used with U-shaped invar separators, their lengths being 0.0625 m, 0.125 m, 0.25 m, 0.5 m, and 1 m. The last etalon had two fine lines engraved on the upper edges of the mirrors, parallel to their surfaces.

In the first step, the distance between the two marks on the 1-m etalon was determined with reference to the prototype metre, using an optical comparator. In the next step, N_0, the sum of the distances between the lines and the faces of the mirrors (expressed in wavelengths) was measured by mounting the plates to form, in turn, two short etalons, in one of which the distance between the lines was twice that in the other. If N_1 and N_2 are the orders of interference with these two short etalons, $N_0 = 2N_2 - N_1$.

Finally, the separation of the plates in the 0.0625-m etalon was measured by the method of exact fractions. This etalon was then compared with the next longer one using fringes of superposition, the residual difference being measured, as shown in Fig. 7.5, by means of a third, slightly wedged air film, which could be moved across the field to introduce an additional optical path. This process was repeated to obtain, in wavelengths, the distance between the faces of the plates in the 1-m etalon.

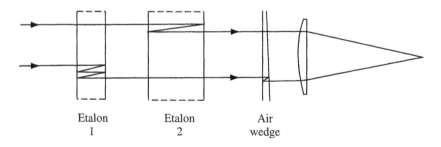

Etalon 1   Etalon 2   Air wedge

Figure 7.5. Comparison of two etalons using fringes of superposition.

7.7 The ^{86}Kr Standard

The measurements described in Section 7.6, as well as subsequent measurements at a number of national laboratories and the *Bureau International des Poids et Mésures* (BIPM), showed that while the definition of the metre based on the International Prototype Metre was significant to about 0.2 μm, or 2 parts in 10^7, lengths could be measured to a much higher degree of precision in terms of the wavelength of a suitable spectral line. This resulted in a considerable body of opinion that favored replacement of the International Prototype Metre by a spectral line. Finally, after detailed studies of the characteristics of the spectral lines of ^{114}Cd, ^{198}Hg, and ^{86}Kr, the metre was redefined in 1960 in terms of the wavelength of the orange line of ^{86}Kr ($\lambda = 606$ nm). Because ^{86}Kr has no nuclear magnetic moment, this line exhibited no hyperfine structure and had a symmetrical profile. In addition, the broadening of the line due to the Doppler effect could be minimized by operating the source in a bath of nitrogen at its triple point (64 K). Measurements with this source were reproducible to better than 1 part in 10^8 [Baird and Howlett, 1963].

7.8 Frequency Measurements

With the development of frequency-stabilized lasers (see Section 5.4) sources of monochromatic light became available with much narrower line widths and much better reproducibility than the ^{86}Kr lamp. The wavelength of the 3.39-μm line from the He-Ne laser stabilized by saturated absorption in CH_4 was the first to be measured with high accuracy relative to the ^{86}Kr standard by several national laboratories and the BIPM. These measurements had a spread of only 6 parts in 10^9, which was largely attributable to the uncertainty associated with the ^{86}Kr standard itself.

Simultaneously, the major national laboratories undertook measurements to compare the frequency of the same laser with the frequency of the ^{133}Cs clock, which is the primary standard of time. This comparison, necessarily, had to be carried out in stages. In one experiment [Evenson *et al.*, 1972, 1973], the first stage involved a klystron oscillating at 74 GHz whose frequency could be compared directly with the ^{133}Cs standard ($\nu \approx 9.19$ GHz). The 12th harmonic of this klystron gave a beat frequency of about 2 GHz with an HCN laser ($\nu \approx 890$ GHz), which could be measured precisely using a diode detector consisting of a tungsten wire, a few micrometres in diameter with a sharpened end, contacting a polished nickel surface. Measurements were made, in a similar manner, of the beat frequency between the 12th harmonic of the HCN laser and a H_2O laser ($\nu \approx 10.72$ THz). The third harmonic of this laser gave a measurable beat with a CO_2 laser ($\nu \approx 32.13$ THz). This laser was, in turn, compared with a CO_2 laser oscillating on another transition ($\nu \approx 29.44$ THz) whose third harmonic gave a beat with the He-Ne laser ($\nu \approx 88.38$ THz). It was possible, in this fashion, to measure the frequency of the He-Ne laser with an uncertainty of 6 parts in 10^{10}.

A simpler setup for such a frequency chain using beats between five CO_2 lasers, with different isotope fillings, operating on slightly different wavelengths was described by Whitford [1979], while the extension of frequency measurements to the visible region was discussed by Evenson, Jennings, and Petersen [1981] and Baird [1981, 1983a].

Since the speed of light is given by the product of the wavelength of a source and its frequency, these results gave a value for the speed of light which brought into question the value

$$c = 299\ 792\ 458 \text{ m s}^{-1} \tag{7.4}$$

which had been formally adopted in 1975. The accuracy of this value was limited essentially by the uncertainties relating to the standard of length at the time, that is to say, the wavelength of the ^{86}Kr lamp.

7.9   The Definition of the Metre

It is apparent from Section 7.8 that a point had been reached at which the ^{86}Kr standard was no longer adequate for absolute measurements of the wavelengths of stabilized lasers with the degree of precision justified by their characteristics, and an improved standard of length was necessary. An obvious possibility was to replace the ^{86}Kr lamp by a stabilized laser. However, this could lead to the need for further revisions, each time a better laser source was developed. After careful consideration of various alternatives, the *Comité Consultatif pour la Définition du Métre* (CCDM) therefore proposed a different solution which offered many advantages in the long term [Terrien, 1976; see also the review by Petley, 1983]. This was to assign a fixed value, based on the best results obtained so far, to the speed of light, which is a physical constant. Specified laser radiations would then be involved in the methods of realization of the metre, rather than in its definition.

Accordingly, from October 1983, the metre was redefined as follows:

The metre is the length of the path traveled by light in vacuum during a time interval of 1/229 792 458 of a second.

Three distinct methods of realizing the metre were recommended by the CCDM. The first was by direct measurements of the travel time of an electromagnetic wave. The second was by measurements in terms of the wavelength in vacuum of an electromagnetic wave of known frequency. The third was by means of the vacuum wavelengths of a number of specified sources; these included spectral lamps using ^{86}Kr, ^{198}Hg, and ^{114}Cd as well as lasers stabilized by saturated absorption (see Sections 5.4.5 and 10.11).

7.10 Length Measurements with Lasers

Problems due to the limited coherence length of light from thermal sources, such as discharge lamps, have been virtually eliminated with laser sources, making possible direct measurements of quite large lengths. Lasers have also opened up new applications of length interferometry, such as the measurement of earth strains, for which interferometers with a length of more than a kilometre have been used [Vali and Bostrom, 1968; Berger and Lovberg, 1969]. These interferometers provide a strain sensitivity of a few parts in 10^{11} for strain rates up to a maximum value of 10^{-4} s^{-1}, with a long-term stability of the same order.

More recently, an absolute determination of the length (≈ 300 m) of a Fabry–Perot cavity in the TAMA gravitational wave detector (see Section 15.4) has been carried out by measuring the frequency difference between a carrier and the sidebands produced by phase modulation, both of which resonate in the cavity [Araya *et al.*, 1999].

The range of wavelengths available with lasers also makes it possible to apply the method of exact fractions very effectively to absolute measurements of large lengths. The CO_2 laser is very well suited to such measurements, since it can be tuned rapidly, under computer control, to a number of wavelengths that are known to a high degree of accuracy [Bourdet and Orszag, 1979; Gillard and Buholz, 1983; Walsh and Brown, 1985].

7.10.1 Two-Wavelength Interferometry

Lasers have also made possible new techniques in length interferometry.

In one method, the interferometer is illuminated simultaneously with two wavelengths, λ_1 and λ_2. The envelope of the fringes then yields the interference pattern that would be obtained with a synthetic wavelength

$$\lambda_s = \frac{\lambda_1 \lambda_2}{|\lambda_1 - \lambda_2|}. \tag{7.5}$$

This method has been implemented by switching a CO_2 laser rapidly between two wavelengths as one of the interferometer mirrors is moved over the distance to be measured. The output signal from the detector is squared, low-pass filtered, and processed in a computer to obtain the phase difference. Distances up to 100 m can be measured with an accuracy of 1 part in 10^7 using synthetic wavelengths of 300 μm and 3.5 mm [Matsumoto, 1986].

Measurements with a large ambiguity-free range can be made using a stable and accurate synthetic wavelength generated by using two CO_2 lasers locked to two different lines of the CO_2 molecule [Walsh, 1987] or two diode-pumped Nd:YAG lasers locked to sub-Doppler transitions of $^{133}Cs_2$ or $^{127}I_2$ [Mahal and Arie, 1996]. The dense spectra of the molecular absorbers allow selection of the synthetic wavelength over a wide range. With a synthetic wavelength of ≈ 19 mm, an accuracy of 70 μm can be achieved by phase shifting.

High accuracy can also be obtained with diode lasers using dual-wavelength phase-shifting interferometry [Ishii and Onodera, 1991], dual-wavelength sinusoidal phase-modulating interferometry [Sasaki, Sasazaki, and Suzuki, 1991], dual-wavelength phase-locked interferometry [Suzuki, Sasaki, and Maruyama, 1996; Suzuki *et al.*, 1997a], or dual-wavelength heterodyne interferometry [Tiziani, Rothe, and Maier, 1996; Onodera and Ishii, 1999a; Yokoyama *et al.*, 1999; Zhao, Zhou, and Li, 1999]. Onodera and Ishii [1994] have described a two-wavelength phase-shifting interferometer that uses a phase-extraction algorithm that is insensitive to the intensity variations associated with frequency shifting as well as a two-wavelength interferometer using a fractional-fringe technique with two frequency-ramped diode lasers [Onodera and Ishii, 1995].

Some unusual techniques for absolute distance measurements have been reviewed by de Groot [2001].

7.10.2   Wavelength-Scanning Interferometry

The earliest implementation of this technique involved measurements, at a mean wavelength λ, of $\Delta\lambda$, the change in the wavelength required to change the phase difference between the two interfering beams by an integral multiple (say, N) of 2π [Olsson and Tang, 1981]. The optical path difference d can then be obtained from the relation

$$N = d\Delta\lambda/\lambda^2. \tag{7.6}$$

Several later implementations have used the ability of diode lasers to be tuned electrically over a range of wavelengths to make absolute measurements of distance.

In a system described by Kubota, Nara, and Yoshino [1987], shown in Fig. 7.6, interference takes place between a reference beam reflected from the front surface of the fixed $\lambda/8$ plate and the signal beam reflected from the movable end mirror. The signal beam returns as a circularly polarized beam, since it has traversed the $\lambda/8$ plate twice, and its two orthogonally polarized components are divided at a polarizing beamsplitter. As a result the two photodetectors see fringe patterns whose intensities vary in quadrature. Fringe counting techniques can then be used to determine the magnitude and sign of a displacement of the mirror.

Absolute measurements of distance are made by sweeping the frequency of the laser linearly with time by using a function generator to ramp the injection current of the laser. If the optical path difference between the two beams in the interferometer is L, they reach the detector with a time delay L/c, where c is the speed of light, and interfere to yield a beat signal with a frequency

$$F = (L/c)(\mathrm{d}f/\mathrm{d}t), \tag{7.7}$$

where $\mathrm{d}f/\mathrm{d}t$ is the rate at which the laser frequency varies with time.

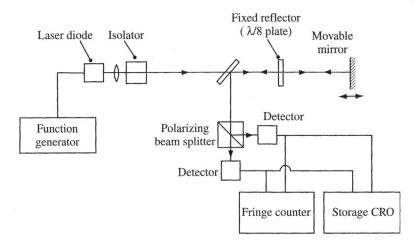

Figure 7.6. Laser interferometer for measurements of displacements and absolute distances [Kubota, Nara, and Yoshino, 1987].

Absolute measurements of length can also be made by counting fringes as the laser wavelength is swept. The fringe count is proportional to the product of the amount by which the frequency of the laser source is shifted and the difference in the lengths of the two arms of the interferometer [Stone, Stejskal, and Howard, 1999]. Higher accuracy can be obtained by using wavelength scanning to determine the integral fringe order and heterodyne interferometry to determine the fractional fringe order [Dai and Seta, 1998].

Wavelength scanning can also be used for surface profilometry by changing the wavelength in a stepwise manner and storing the interference patterns obtained at each step. The image data obtained at a sequence of steps (typically 256–1024 steps) are then analyzed to obtain the period and the phase at each point [Kuwamura and Yamaguchi, 1997; Yamaguchi, Yamamoto, and Yano, 2000].

7.10.3 Sinusoidal Wavelength Scanning

In an arrangement described by Kikuta, Iwata, and Nagata [1986], the input wavelength to a heterodyne interferometer is varied sinusoidally by varying the injection current of a diode laser. If the wavelength variation is known, the optical path difference can be obtained from the resulting variation in the phase difference between the interfering beams. Real-time measurements of displacements can be made by sampling the interference signal at times that satisfy certain conditions [Suzuki *et al.*, 1989].

Sasaki, Yoshida, and Suzuki [1991] have also described a double sinusoidal phase-modulating interferometer using a diode laser as the source, in which a carrier signal is provided by a vibrating mirror, and the signal used for measurements of the optical path difference is produced by modulating the injection current of the diode laser (see Section 6.8). Phase fluctuations caused by mechanical vibrations are eliminated by a feedback control system using the carrier signal.

Higher accuracy can be obtained by filtering the spectrum of a superluminescent diode by a sinusoidally moving slit, to scan over a larger range of wavelengths [Sasaki, Murata, and Suzuki, 2000].

Steps whose height is greater than one-half of one wavelength can be measured by two-wavelength interferometry using a technique of time-shared sinusoidal phase modulation in which two diode lasers are alternately modulated with a sinusoidal signal, and the overlapping interference images detected by a CCD camera are separated by an integrating-bucket method [Sasaki, Okazaki, and Sakai, 1987; Suzuki, Yazawa, and Sasaki, 2002].

7.11   Changes in Length

An important application of heterodyne interferometry is for precise measurements of changes in length, such as in measurements of coefficients of thermal expansion [White, 1967].

Figure 7.7 is a schematic of an experimental arrangement used for such measurements on low-expansion materials such as fused silica [Jacobs and Shough, 1981]. In this arrangement, mirrors are attached to the two ends of the sample to form a Fabry–Perot interferometer, and the frequency of a slave laser is locked to a transmission peak of this interferometer, so that the wavelength of the slave laser is an integral submultiple of the optical path difference in the interferometer. Any change in the separation of the mirrors then results in a corresponding change in the wavelength of the slave laser and, hence, in its frequency. These changes are measured by mixing the beam from the slave laser with the beam from a frequency-stabilized reference laser at a fast photodiode and measuring the beat frequency.

A problem with such a setup is that measurements can only be made over a very limited range, beyond which the transmission peak of the Fabry–Perot interferometer moves outside the gain profile of the laser. Accordingly, after a preassigned frequency excursion, the slave laser is locked to an adjacent transmission peak of the Fabry–Perot interferometer that lies well within the gain profile of the laser.

Measurements have been made by this method, using a 100-mm sample, with a precision better than 1 part in 10^8. A modified version of this method has also been described for comparing the coefficients of thermal expansion of samples taken from different parts of a large mirror blank of low-expansion material [Jacobs, Shough, and Connors, 1984].

7.12   Displacements

A simple arrangement for measuring small displacements with high accuracy uses two diode lasers with external cavities. A displacement of the reflecting mirror of one

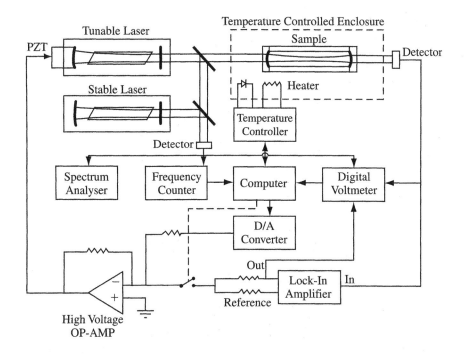

Figure 7.7. Interferometer for measurements of thermal expansion [Jacobs and Shough, 1981].

cavity causes a shift in the laser frequency which can be evaluated from measurements of the heterodyne beat frequency [Takahashi, Kakuma, and Ohba, 1996].

Real-time measurements of large, high-velocity displacements present a problem with conventional techniques of heterodyne interferometry, since the precision of measurements decreases as the bandwidth of phase detection is increased.

One approach to this problem is by combining two complementary techniques: fringe counting at a high beat frequency (20 MHz) for initial measurements of displacements at high velocity, and synchronous phase demodulation with a 2-kHz beat frequency for final high-resolution measurements [Yim, Eom, and Kim, 2000]. The two techniques are operated in a switching mode, in accordance with the speed of movement of the object.

Another approach uses two-wavelength interferometry. However, the usual method of estimating a displacement by two-wavelength interferometry involves evaluating the change in the difference of the phases obtained for the two wavelengths used. Since each of the phases is measured separately, real-time measurements present difficulties.

This problem can be solved by using two diode lasers whose output wavelengths are modulated by two separate sinusoidal currents at the same frequency [Suzuki,

Kobayashi, and Sasaki, 2000]. Changes in the phase difference between the two beams can then be monitored in real time.

7.12.1 Vibrations

Real-time measurements of low-frequency ($\approx$50 Hz) vibrations can be made with sinusoidal phase-modulating interferometry by sampling the interference signal in synchronism with the sinusoidal phase-modulating signal at specified times [Suzuki *et al.*, 1989].

   An alternative method is based on phase-locked interferometry [Suzuki, Sasaki, and Maruyama, 1989], where the fluctuations in the feedback signal are a measure of the amplitude of the vibration [Suzuki *et al.*, 1997b].

7.13 Dynamic Angle Measurements

In a simple technique for dynamic angle measurements described by Malacara and Harris [1970], two beams reflected, respectively, from the front and rear surfaces of a plane-parallel glass plate are made to interfere. A rotation of the glass plate results in a change in the optical path difference between the beams. This technique can cover a large range of angles, but the variation of the optical path difference has a significant nonlinear component.

   This technique was extended by Shi and Stijns [1993], who used a right-angle prism in the probe beam and calculated its optimum position by a parametric compensation method to obtain an improvement in linearity. More recently, Ikram and Hussain [1999] were able to reduce the error by up to three orders of magnitude, for a rotation angle of $\pm$20°, by using a glass plate and a prism, or two plates and two prisms, at an angle to the input beam.

Chapter 8

Optical Testing

Interferometry can be used to study phase variations across an optical wavefront. As mentioned in Chapter 2, one field of applications is in studies of gas flows, combustion, and diffusion, where local changes in the refractive index can be related to changes in the pressure, the temperature or the relative concentration of different components. Another important application is in testing optical components and optical systems.

8.1   The Fizeau Interferometer

Fringes of equal thickness (see Section 2.4.2) can be used to compare a nominally flat surface with a standard flat surface. The simplest way to do this is to bring the two surfaces together and view, at near normal incidence, the fringes formed in the thin air film separating them. If the deviations of the test surface from flatness are small, the top plate is tilted slightly to obtain a wedged air film. If, then, the fringe spacing is a, and the peak deviation of the diametral fringe from straightness is δ, the peak error of the surface is $(\delta/a)(\lambda/2)$.

However, as shown in Section 2.4.3, fringes of equal thickness are obtained with an extended source only when the air gap between the surfaces is less than a few micrometres. To obtain fringes of equal thickness with a larger air gap, it is necessary to use collimated light. Figure 8.1 is a schematic diagram of a Fizeau interferometer using a lens for collimation; alternatively, a concave mirror can be used for this purpose. Higher accuracy can be obtained if reflecting coatings are applied to the reference flat and the surface under test, so that multiple-beam fringes (see Section 4.3) are formed [Koppelmann and Krebs, 1961a, 1961b].

Fairman *et al.* [1999] have described a 300-mm aperture digital phase-shifting Fizeau interferometer for precision metrology of optical components; the calibration of this instrument has been discussed by Oreb *et al.* [2000].

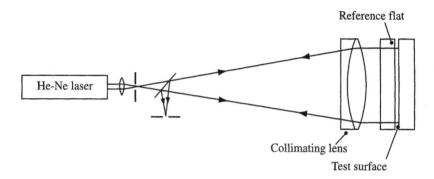

Figure 8.1. Optical system of a Fizeau interferometer.

With a Fizeau interferometer it is possible, provided care is taken to exclude vibrations, to make absolute measurements of deviations from flatness by using a liquid surface as a reference flat surface [Bünnagel, 1956, 1965]. Alternatively, three surfaces can be tested in a series of combinations to evaluate their individual errors [Schulz and Schwider, 1967, 1976] (see Section 8.10.1).

One common application of the Fizeau interferometer is to check the parallelism of the faces of a transparent plate. With a laser source, problems of coherence are eliminated and quite thick plates can be tested.

Modified forms of the Fizeau interferometer have also been described using convergent or divergent light to test curved surfaces [Heintze, 1967; Biddles, 1969; Shack and Hopkins, 1979]. The best arrangement is one in which, as shown in Fig. 8.2,

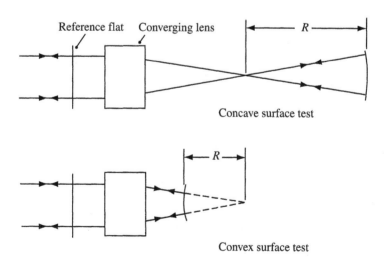

Figure 8.2. Fizeau interferometer used to test concave and convex surfaces.

the collimated beam leaving the plane reference surface is brought to a focus by a well-corrected converging lens, so that a plane wavefront is reflected back.

A simpler system is obtained if the last surface of the lens is used as the reference surface [Mantravadi, 1992a]. However, this arrangement can produce errors with high numerical-aperture surfaces in a phase-shifting interferometer [Creath and Hariharan, 1994]. These errors can be minimized by a proper choice of the algorithm used to calculate the phase.

A problem, commonly encountered in flatness testing of transparent plates, is unwanted interference effects (ghost fringes) arising from reflections from the back surface. This problem can be overcome by using a multimode laser. The coherence function of such a laser is periodic, with very narrow maxima, and ghost fringes can be eliminated by placing the test surface at a distance from the reference surface equal to a multiple of the length of the laser cavity [Ai, 1997]. It is also possible to use a tunable source and a phase-shifting algorithm that suppresses interference modulations due to spurious reflections [Okada *et al.*, 1990; de Groot, 2000b].

8.2 The Twyman–Green Interferometer

A wider range of optical components can be tested with the Twyman–Green interferometer [see Twyman, 1957]. This is a Michelson interferometer modified to use collimated light, so that fringes of equal thickness are obtained. Detailed descriptions of a number of test setups for various optical elements have been given by Briers [1972] and by Malacara [1992a]. The optical arrangement used to test a prism is shown in Fig. 8.3a, while that used to test a lens is shown in Fig. 8.3b. With the addition of a nodal slide, lenses can also be tested off-axis.

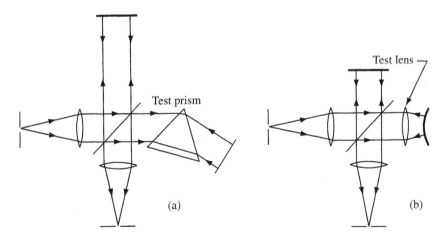

Figure 8.3. Twyman–Green interferometer used to test (a) a prism and (b) a lens.

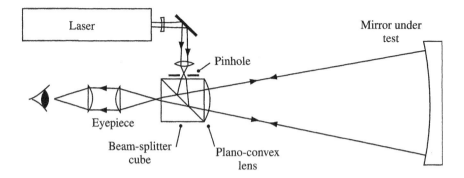

Figure 8.4. Laser unequal-path interferometer used for testing a large concave mirror [Shack and Hopkins, 1979].

8.3   Unequal-Path Interferometers

To obtain fringes of good contrast with a thermal source, its size and spectral bandwidth must satisfy the conditions for spatial and temporal coherence [Hansen, 1955; Hansen and Kinder, 1958] (see Sections 3.10 and 3.11). However, with a laser source, these problems disappear, and fringes can be obtained even with large optical path differences between the beams [Houston, Buccini, and O'Neill, 1967].

Figure 8.4 shows a compact unequal path interferometer which can be used for testing large concave mirrors [Shack and Hopkins, 1979]. This interferometer contains only one precision optical component, a beamsplitter cube with a plano-convex lens cemented to one surface. The image of the center of curvature of the convex surface of this lens lies just outside the input face of the cube. A pinhole spatial filter is placed at this point in the beam from the laser, which is brought to a focus by a microscope objective. Interference fringes are formed by the beams reflected from the surface under test and the spherical surface of the lens cemented to the beamsplitter cube.

8.4   Phase Unwrapping

A problem in analyzing an interferogram is that values of the phase difference obtained at any point are known only to modulo 2π. The process of determining the number of 2π steps to be added to these raw values to obtain the actual phase difference is called phase unwrapping [see Malacara, Servin, and Malacara, 1998].

Phase unwrapping requires a knowledge of the sign as well as the magnitude of the raw phase [Bone, 1991]. It is then possible, by starting from a point at which the phase difference is known to be zero, and checking the values of the raw phase at successive pixels along a line, to decide whether to add or subtract 2π when crossing each fringe. This procedure is extended to two dimensions by taking the values of the unwrapped phase at successive pixels along this line as new starting points. Phase unwrapping can also be carried out along an edge [Stetson, 1992], or along lines chosen so that the change in phase between adjacent data points is a minimum [Judge, Quan, and Bryanston-Cross, 1992].

In another method for phase unwrapping, a simple operation for local phase unwrapping is repeated many times over a given area to generate a consistent phase distribution [Ghiglia, Mastin, and Romero, 1987]. Subsequently, a major step toward a path-independent phase unwrapping procedure was taken by Ghiglia and Romero [1994], who used the wrapped phase differences along the x and y directions to obtain the phase gradients along these directions. These wrapped phase gradients were then least-squares integrated to obtain the unwrapped phase.

A basic problem in two-dimensional phase unwrapping arises from the assumption that the phase difference between adjacent pixels is less than π. This assumption can result in errors with noisy phase maps, since a single error along any of the paths is propagated up to the end of that path. A number of image-processing schemes have been suggested to handle this problem, including the use of adaptive filters [Ghiglia and Pritt, 1998; Qian *et al.*, 2001].

An alternative way of obtaining the phase function without phase unwrapping is by line integration of the phase gradients [Paez and Strojnik, 1997, 1999]. This technique is insensitive to spatial nonuniformity of the illuminating beam and the shape of the boundary of the interferogram. It can also be applied to undersampled interferograms to obtain a first approximation to the phase function. An iterative method of successive approximations (the method of synthetic interferograms) is then used to determine the unknown phase with the required degree of precision [Paez and Strojnik, 1998, 2000]. A comparison of global and local phase-unwrapping techniques has been presented by Fornaro *et al.* [1997].

The problem of 2π ambiguities is particularly serious in wavefront sensing with segmented mirrors, where normal phase unwrapping methods, which work within segment boundaries, break down between neighboring segments. One way of solving this problem is by using multiple wavelengths to extend the capture range [Löfdahl and Eriksson, 2001].

In addition, the application of wavelength-scanning interferometry to surface profiling (see Section 7.10.2) has led to the development of a technique known as temporal phase unwrapping for such data [Saldner and Huntley, 1997]. Paulsson *et al.* [2000] have identified two strategies that have advantages; one minimizes the number of frames needed, while the other has superior reliability.

8.5 Analysis of Wavefront Aberrations

For a lens used off-axis, the deviation of the transmitted wavefront from a sphere centered on the Gaussian image point can be written as

$$W(x, y) = A(x^2 + y^2)^2 + By(x^2 + y^2) + C(x^2 + 3y^2)$$
$$+ D(x^2 + y^2) + Ey + Fx, \tag{8.1}$$

where A is the coefficient of spherical aberration, B the coma coefficient, C the astigmatism coefficient, D the defocusing coefficient and E and F are measures of tilt about the x and y axes, respectively [Kingslake, 1925–26].

Pictures of interferograms corresponding to the various primary aberrations have been presented by Kingslake [1925–26] and by Malacara [1992a]. An interferogram containing a mixture of aberrations can be evaluated by measuring the optical path difference at several points on the x and y axes and then calculating the coefficients in Eq. (8.1) from a set of linear equations.

Equation (8.1) can be generalized and written in polar coordinates in the form

$$W(\rho, \theta) = \sum_{n=0}^{k} \sum_{l=0}^{n} \rho^n (a_{nl} \cos^l \theta + b_{nl} \sin^l \theta), \tag{8.2}$$

where ρ and θ are polar coordinates over a circle of unit radius defining the pupil.

8.5.1 Zernike Polynomials

The Zernike polynomials are a convenient way of representing wavefront aberrations over a circular pupil. Any wavefront $W(\rho, \theta)$ of degree k can be expressed as a linear combination of Zernike circular polynomials as

$$W(\rho, \theta) = \sum_{n=0}^{k} \sum_{l=-n}^{n} C_{nl} R_n^{|l|} e^{il\theta}. \tag{8.3}$$

The orthogonality of the polynomials simplifies the process of fitting the polynomials to the interferometric data. They are also easily related to the classical aberrations [Malacara and DeVore, 1992].

8.5.2 Wavefront Fitting

The data obtained from measurements is at discrete points, while the polynomials represent continuous functions. It is, therefore, necessary to fit the chosen functions to the data; this can be done using a least-squares program [Rimmer, King, and Fox, 1972; Freniere, Toler, and Race, 1981; Malacara, Carpio-Valadez, and Sanchez-Mondragon, 1987, 1990]. Swantner and Chow [1994] have described the use of the Gram–Schmidt method for fitting polynomials with general aperture shapes.

8.6 Shearing Interferometers

In a shearing interferometer, both wavefronts are derived from the system under test, and the interference pattern is produced by shearing one wavefront with respect to the other. Shearing interferometers have the advantage that they do not require a reference surface of the same dimensions as the system under test. In addition, since both the interfering beams traverse very nearly the same optical path, the fringe pattern is less affected by mechanical disturbances.

To produce fringes of good visibility with a shearing interferometer, the beams must have adequate spatial coherence. With thermal radiation, this requires a very small illuminated pinhole and, hence, an intense source, usually generating radiation with limited temporal coherence. As a result, much effort went into the design of shearing interferometers which were compensated for white light (see Bryngdahl [1965]). With the availability of lasers, which give light with a high degree of spatial and temporal coherence, many simpler arrangements have come into use.

8.6.1 Lateral Shear

The simplest type of shear is a lateral shear (see Section 3.10). With a nearly plane wavefront, this involves producing two images of the wavefront with a mutual lateral displacement, whereas with a nearly spherical wavefront it requires a similar displacement of the two images of the wavefront over the surface of the reference sphere. Figure 8.5 shows three typical optical systems which can be used with converging wavefronts. These systems, which are based on the Michelson, Mach–Zehnder, and Sagnac interferometers, were described, respectively, by Lenouvel and Lenouvel [1938], Bates [1947], and Hariharan and Sen [1960d]. A number of modifications of these systems for use with spherical and plane wavefronts have been reviewed by Mantravadi [1992b].

A particularly simple arrangement devised by Murty [1964], which can be used with a laser source, consists of a plane-parallel plate. As shown in Fig. 8.6, the light from the laser is focused by a microscope objective on a pinhole located at the focus of the lens under test. The beam emerging from this lens gives rise to two wavefronts reflected from the front and back surfaces of the plate. The lateral shear between these wavefronts can be varied by tilting the plate and is given by the relation

$$s = (n^2 - \sin^2 \theta)^{-1/2} d \sin 2\theta, \tag{8.4}$$

where d is the thickness of the plate and θ is the angle of incidence. A modification of this arrangement [Hariharan, 1975b] uses two separate plates with a variable air gap; this has the advantage that a tilt can be introduced between the two sheared wavefronts to make the interpretation of the fringes easier. The use of a liquid-crystal phase retarder for phase-shifting has been described by Griffin [2001].

A vectorial shearing interferometer based on the Mach–Zehnder configuration, incorporating a pair of wedge prisms, has been described by Paez, Strojnik, and Torales [2000]. Variable shear and tilt can be implemented along any direction, permitting the number of fringes and their direction to be controlled.

8.6.2 Interpretation of Interferograms

The interpretation of a lateral shearing interferogram is more difficult than that of an interferogram obtained with a Twyman–Green interferometer, since interference

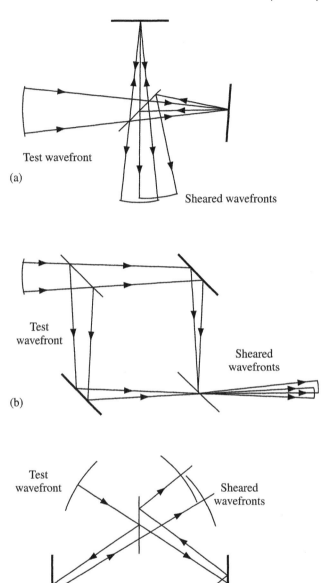

Figure 8.5. Lateral shearing interferometers based on (a) the Michelson, (b) the Mach–Zehnder, and (c) the Sagnac interferometers.

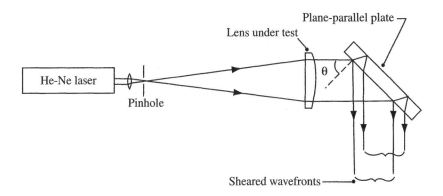

Figure 8.6. Lateral shearing interferometer using a laser source and a tilted plane-parallel plate [Murty, 1964].

takes place between two aberrated wavefronts instead of between the aberrated wavefront and a perfect reference wavefront. The analysis of such an interferogram has been discussed by Malacara and Mendez [1968] and by Rimmer and Wyant [1975]. A polynomial $\Delta W(x, y)$ is fitted to the measured values of the optical path difference in two interferograms with mutually perpendicular directions of shear; the coefficients of $W(x, y)$ the polynomial representing the errors of the test wavefront, can then be derived from the coefficients of $\Delta W(x, y)$ [Servin, Malacara, and Marroquin, 1996; Harbers, Kunst, and Leibbrandt, 1996; Shen, Chang, and Wan, 1997; Okuda *et al.*, 2000]. Other approaches use Fourier filtering [Loheide, 1997] or a least-squares method [Elster and Weingärtner, 1999a, 1999b; Elster, 2000]. Alternatively, the differences obtained by shearing the test wavefront in a number of directions can be analyzed [Nomura *et al.*, 2002].

A possibility that is being explored is the use of a neural network for rapidly identifying and evaluating the primary aberrations in an interferogram [Yang and Oh, 2001].

8.6.3 Rotational and Radial Shearing

As can be seen from Fig. 3.5, other forms of shear besides a lateral shear are possible. One is rotational shear in which interference takes place between two images of the test wavefront, one of which is rotated with respect to the other [Armitage and Lohmann, 1965; Murty and Hagerott, 1966]. Another is reversal shear [Gates, 1955; Saunders, 1955]. Perhaps the most useful is radial shear, in which one of the images of the wavefront is contracted or expanded with respect to the other [Brown, 1959; Hariharan and Sen, 1961c].

A number of optical arrangements for radial shearing interferometers are available, which have been described by Malacara [1992b]. With a thermal light source, a convenient arrangement uses a triangular path (Sagnac) interferometer in which two

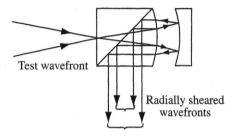

Test wavefront

Radially sheared
wavefronts

Figure 8.7. Radial shear interferometer for use with a laser source [Zhou, 1985].

images of the test wavefront of different sizes are produced by a lens system which is traversed in opposite directions by the two beams [Hariharan and Sen, 1961c]. With a laser source, a simple system can be used, consisting of a thick lens [Steel, 1975]. In this arrangement, interference takes place between the directly transmitted wavefront and the wavefront which has undergone one reflection at each surface. An even simpler system is shown in Fig. 8.7, in which interference takes place between the wavefronts reflected from two spherical surfaces [Zhou, 1984, 1985]. A compact, in-line, radial shearing interferometer using a beamsplitting cube has been described by Shukla, Moghbel, and Venkateswarlu [1992]; one can also be set up with two zone plates [Smartt and Hariharan, 1985; Kohno et al., 2000].

To analyze a radial shearing interferogram it is convenient to express the deviations of the wavefront under test from a reference sphere in the form presented in Eq. (8.2), namely,

$$W(\rho, \theta) = \sum_{n=0}^{k} \sum_{l=0}^{n} \rho^n (a_{nl} \cos^l \theta + b_{nl} \sin^l \theta),\qquad(8.5)$$

where ρ and θ are polar coordinates over a circle of unit radius defining the pupil.

If the wavefront with which the test wavefront is compared is magnified by a factor $(1/S)$, where $S(< 1)$ is known as the shear ratio, the deviations of this magnified wavefront from the same reference sphere are

$$W'(\rho, \theta) = \sum_{n=0}^{k} \sum_{l=0}^{n} S^n \rho^n (a_{nl} \cos^l \theta + b_{nl} \sin^l \theta).\qquad(8.6)$$

Accordingly, the optical path differences in the interferogram are given by the relation

$$p(\rho, \theta) = W(\rho, \theta) - W'(\rho, \theta)$$

$$= \sum_{n=0}^{k} \sum_{l=0}^{n} (1 - S^n) \rho^n (a_{nl} \cos^l \theta + b_{nl} \sin^l \theta).\qquad(8.7)$$

It is apparent from a comparison of Eq. (8.7) with Eq. (8.5) that, with a reasonably small value of the shear ratio ($S < 0.5$), the radial shear interferogram is very similar to the interferogram that would be obtained with a Twyman–Green interferometer, and the wavefront aberrations can be computed in a very similar fashion [Hariharan and Sen, 1961c; Malacara, 1974].

8.7 Grating Interferometers

Gratings can be used, instead of thin-film beamsplitters, in any of the conventional two-beam interferometers such as the Michelson, Mach–Zehnder, or Sagnac interferometers. Such an arrangement is very stable, and the angle between the two beams is affected only to a small extent by the orientation of the grating. With reflecting gratings, it is possible to build an interferometer for use in the infrared or far ultraviolet regions, where good-quality thin-film beamsplitters are not available [Munnerlyn, 1969].

A very simple type of grating interferometer is based on the Talbot effect. It can be shown that when a grating is illuminated with collimated monochromatic light, a series of images of the grating are formed at distances from it given by the relation

$$z_m = 2m\Lambda^2/\lambda, \tag{8.8}$$

where Λ is the period of the grating and m is an integer. This phenomenon is also known as Fourier or self-imaging. If another identical grating is placed in the plane of one of these images and rotated slightly, high-contrast moire fringes are obtained. If, then, a phase object is introduced between the gratings, close to the first grating, the moire fringes are deformed and correspond to lines of equal deflection of the rays [Lohmann and Silva, 1971; Yokozeki and Suzuki, 1971]. The smallest detectable deflection is

$$\alpha_n = \Lambda/2z_m,$$
$$= \lambda/m\Lambda. \tag{8.9}$$

The sensitivity, therefore, increases with the separation of the gratings.

Similar interference phenomena can also be seen even with an extended white light source (the Lau effect). In this case, colored moire fringes are formed. The theory of these fringes has been discussed by Jahns and Lohmann [1979], who have shown their similarity to Talbot fringes, and by Sudol and Thompson [1979]. They can also be used to study phase objects with a modified system in which one grating is imaged on the other by a telecentric optical system [Bartelt and Jahns, 1979].

A simple type of grating interferometer is the Ronchi interferometer (see Ronchi [1964]). As shown in Fig. 8.8a, a coarse grating (≈ 10 lines/mm) is placed at the focus of a converging wavefront from the system under test, and the resulting interference fringes are observed from behind the grating. This arrangement can be regarded as a shearing interferometer in which interference takes place between the directly

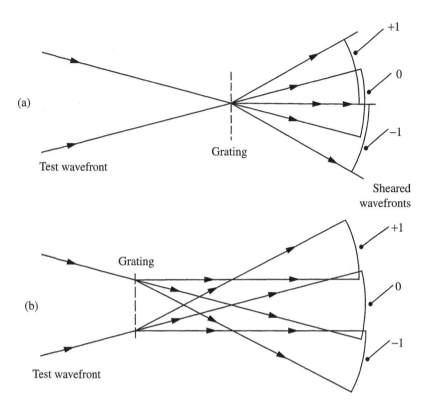

Figure 8.8. Production of (a) shear and (b) shear with tilt in the Ronchi interferometer.

transmitted wave and the diffracted orders which overlap it. The shear is determined by the period of the grating, while the tilt introduced between the interfering wavefronts can be increased by moving the grating away from the focus, as shown in Fig. 8.8b. Additional phase shifts can be introduced, by moving the grating sideways, for synchronous phase detection [Omura and Yatagai, 1988].

The Ronchi interferometer is easy to set up, but has the drawback that, for small shears, different orders overlap and interfere with each other. One solution to this problem is the use of a double-frequency grating [Wyant, 1973]. The lower of the two grating frequencies is chosen so that the diffracted orders from both the gratings are separated from the directly transmitted beam, while the difference in the grating frequencies is chosen to give the required shear. A better solution is, as shown in Fig. 8.9a, to use two gratings with the same spacing [Hariharan, Steel, and Wyant, 1974; Rimmer and Wyant, 1975]. In this case, as shown in Fig. 8.9b, the shear can be varied by rotating one grating in its own plane. In addition, the amount and direction of the tilt introduced can be controlled by varying the separation of the gratings and their position with respect to the focus.

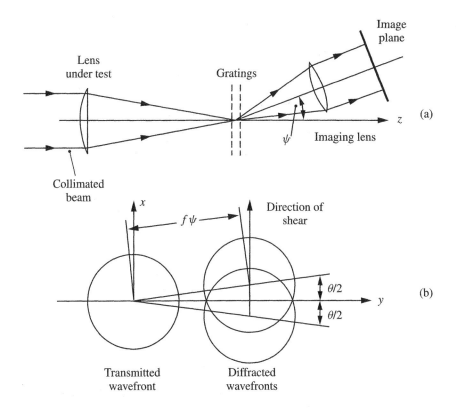

Figure 8.9. Double-grating interferometer: (a) optical system, and (b) sheared wavefronts [Hariharan, Steel, and Wyant, 1974].

Modified two-grating interferometers which can be used with collimated beams have been described by Hariharan and Hegedus [1974], Leibbrandt, Harbers, and Kunst [1996], and Schreiber and Schwider [1997]. The shear can be varied by varying the separation of the gratings, and phase shifts can be introduced by a lateral movement of one grating. The use of such a two-grating interferometer for testing extreme-UV optics at the wavelength at which they are to be used has been described by Hegeman *et al.* [2001].

8.8 The Scatter-Plate Interferometer

The scatter-plate interferometer [Burch, 1953, 1972] also makes use of diffraction, but gives a fringe pattern which can be interpreted directly. A detailed description of its use has been given by Scott [1969]. To understand its operation, consider the optical system shown in Fig. 8.10, which can be used to test a concave mirror. In

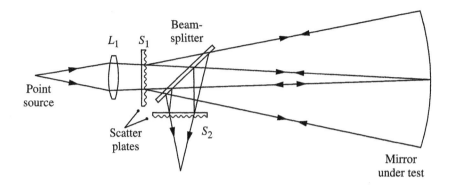

Figure 8.10. Scatter-plate interferometer used to test a concave mirror.

this arrangement, a lens L_1 forms an image of a point source on the mirror under test through a scatter plate S_1. This is a weak diffuser which transmits part of the light and scatters the rest to fill the aperture of the mirror. The scattered light is brought back to a focus, by means of a small beamsplitter, at another scatter plate S_2 which is identical to S_1 and is positioned so that it coincides with its image. Interference takes place between the wavefront transmitted by S_1 and scattered by S_2 and the wavefront scattered by S_1 and transmitted by S_2. The interference pattern is similar to that produced in a Twyman–Green interferometer, since the wavefront from the mirror interferes with a virtually perfect reference wavefront reflected from its central zone. A displacement of S_1 or S_2 along the axis changes the radius of curvature of the reference wavefront, while tilt can be introduced between the wavefronts by moving S_1 or S_2 in its own plane.

A very similar arrangement is possible using Fresnel zone plates [Murty, 1963; Smartt, 1974], which behave like lenses with multiple foci.

A computer-generated scatter plate with a large scattering angle and negligible symmetry error can be produced using an electron beam writer [Räsänen *et al.*, 1997]. A birefringent scatter plate can also be used to separate the test and reference beams and permit measurements by phase shifting [North-Morris, VanDelden, and Wyant, 2002].

8.9   The Point-Diffraction Interferometer

Another, very simple, common-path interferometer is the point-diffraction interferometer (PDI) [Smartt and Steel, 1975]. In this interferometer, as shown in Fig. 8.11, a pinhole in a partially transmitting film is placed at the focus of a converging wavefront from the system under test. Interference takes place between the test wavefront, which is transmitted through the film, and a reference wavefront produced by diffraction at the pinhole. With a film having a transmittance of about 0.01, the amplitudes of the two wavefronts are very nearly equal, and fringes with good contrast can be obtained.

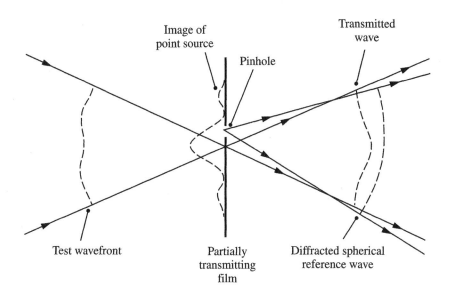

Figure 8.11. Point-diffraction interferometer [Smartt and Steel, 1975].

If the pinhole is smaller than the Airy disc for an aberration-free wavefront, the reference wave is a spherical wave, and the fringe pattern resembles a Twyman–Green interferogram. A major advantage of the PDI is that it can be used to test a telescope objective, under actual working conditions, with light from a bright star. The PDI has also been used to measure temperature profiles in boundary layers and flames [Goldmeer, Urban, and Yuan, 2001].

With the PDI, a nematic liquid-crystal layer can be used to introduce phase shifts between the object beam and the reference beam generated by a microsphere embedded within the liquid-crystal layer [Mercer and Creath, 1996]. Guardalben *et al.* [2002] have shown that the main sources of errors in such an interferometer are frame-to-frame intensity changes caused by alignment distortions of the host liquid crystal and dichroism of the dye incorporated in it. Accurate measurements can be made if these effects are minimized.

A modified PDI in which phase-shifting is implemented by translating a grating has been described by Medecki *et al.* [1996]. An improved version of this interferometer [Naulleau and Goldberg, 1999] has been used for measurements on extreme uv optics at a wavelength of 193 nm [Lee *et al.*, 2000].

8.10   Computerized Test Methods

Computer processing of the test data has opened the way to new test methods for specific applications.

8.10.1 Absolute Tests for Flatness

Absolute measurements of the deviations from flatness of optical surfaces can be made with a Fizeau interferometer. The conventional procedure involves an intercomparison of three surfaces, taken two at a time. However, a problem with this simple procedure is that, in any such comparison, the locations of all the points in one of the surfaces are mirror-reflected about a diameter. As a result, unambiguous values of the surface errors can be obtained only along this diameter, and there are difficulties in extending these measurements to cover the whole surface [Schulz and Schwider, 1967, 1976].

Modifications of this procedure, which involve making additional measurements with one of the surfaces rotated through known angles, have been described by Fritz [1984] and by Grzanna and Schulz [1990, 1992]. These methods involve fitting orthogonal polynomials to the data, with the result that information on local irregularities is lost. In another method described by Ai and Wyant [1993], four additional measurements are made with one of the surfaces rotated in its own plane through angles of $0°$, $45°$, $90°$, and $135°$, yielding a total of six measurements. From these measurements, it is possible to obtain exact profiles along diameters oriented at $0°, 22.5°, 45°, \ldots, 157.5°$, as well as an approximation to the surface deviations between these diameters. A rotation method for absolute testing of flats, using four or five positional combinations, has been described by Schulz and Grzanna [1992], Schulz [1993], Elssner *et al.* [1994], and Grzanna [1994].

An alternative approach is based on the fact that averaging a series of measurements made with one surface rotated, in steps, in its own plane, smears out the errors of that surface which are not rotationally invariant.

In one method, the test data obtained as one of the surfaces is rotated in its own plane to n equally spaced positions are rotated back to a common orientation by using appropriate software and are then averaged. All rotationally variant errors of this surface, except those of angular order kn, where k is an integer, are eliminated [Evans and Kestner, 1996]. However, a problem is that the original data are acquired on a square grid of points. With increments in the rotation angle other than $90°$, interpolation is necessary to obtain data on points that are not on the nodes of the grid, resulting in a loss of information on local irregularities.

One way to solve this problem is by averaging the results of intercomparisons of a set of three surfaces, taken two at a time, with both surfaces rotated in their own plane in a series of steps [Hariharan, 1997b]. It is then possible to separate the rotationally invariant and rotationally variant components of the surface errors. The absolute values of the errors of each of the surfaces are then obtained directly from these measurements, with a spatial resolution corresponding to the original sampling grid. Alternative procedures which require fewer measurements have been discussed by Evans [1998], Freischlad [2001], and Küchel [2001].

8.10.2 Small-Scale Irregularities

Some applications place very stringent limits on the amplitude of small-scale surface irregularities (ripples). Direct measurements of such small-scale irregularities can be made with a Fizeau interferometer by translating the reference surface laterally, in a series of steps in the x and y directions, and recording interferograms with the reference surface moved to an $n \times n$ matrix of positions distributed at random over a square of dimensions $na \times na$, where a is the average displacement of the reference surface between measurements. If, then, we average the data obtained at each point on the test surface, small-scale irregularities on the reference surface, with spatial wavelengths less than a, are smeared out, and their apparent amplitude is reduced by a factor of n in the resultant final interferogram. By averaging a sufficiently large number of data sets (say, $n = 10$), small-scale irregularities on the test surface can be mapped with an error less than 10% [Hariharan, 1996b, 1998].

The spatial frequency spectrum of these small-scale irregularities can also be obtained directly by recording a reference set of values along a diameter and taking the differences of these values and the values obtained in a series of measurements with the test surface shifted laterally along this diameter in a series of steps. The envelope of the values obtained by taking the Fourier transform of these differences then corresponds to the spatial frequency spectrum of the deviations of the test surface from a reference plane [Hariharan, 1996a].

8.10.3 Sources of Error

Any aberrated wavefront changes shape as it travels. Accordingly, to avoid errors, the interference pattern must be imaged on the detector array. To satisfy this requirement, the separation of the reference and test surfaces must be as small as possible. When testing lenses, the pupil of the lens must be imaged back on itself [Malacara, Servin, and Malacara, 1998]. In addition, when making measurements of small-scale surface irregularities, errors can arise due to scattered light and the transfer function of the imaging system and the CCD array camera [Hariharan, 1997a].

Finally, with a phase-shifting interferometer, systematic errors can arise due to errors in the phase shifts and deviations in the fringe profile from a sinusoid. One way to minimize these errors is by using more sophisticated phase-calculation algorithms (see Section 6.6.1). A simpler way is to average a series of m measurements made with the initial optical path difference increased in steps, so as to introduce additional phase offsets of $2\pi/m$ [Hariharan, 2000]. This procedure has the advantage that it also reduces additive noise and quantization errors [Surrel, 1997a; Zhao and Surrel, 1997].

8.10.4 Subaperture Testing

Another technique that has become feasible with computer processing of the test data is subaperture testing. In this method, measurements are made with a small reference surface at a number of overlapping positions, so as to cover a larger aperture. Computational techniques are used to link these observations and minimize errors due to misalignment of the reference surface in the individual measurements [Jensen, Chow, and Lawrence, 1984; Negro, 1984].

8.11   Aspheric Surfaces

Another application of computer processing of the test data has been in testing aspheric surfaces, where it is possible to calculate the theoretical deviations of the wavefront from the best-fit sphere and subtract these from the measured values. However, this method breaks down if the phase difference changes by more than 90° between adjacent elements in the detector array. This problem can be overcome, if some *a priori* information is available regarding the shape of the surface, by using a fringe-analysis technique known as sub-Nyquist interferometry [Greivenkamp, 1987]. Surfaces with departures of several hundreds of waves from a reference sphere have been measured using sub-Nyquist interferometry [Greivenkamp, Lowman, and Palum, 1996].

A method of demodulating such undersampled interferograms using a computer-stored reference wavefront has been described by Servin *et al.* [1994].

Another approach to this problem is by making measurements of the phase differences ϕ_1 and ϕ_2 at two wavelengths λ_1 and λ_2 (see Section 7.10.1). The difference of these values $(\phi_1 - \phi_2)$ corresponds to the phase difference that would have been obtained with a longer synthetic wavelength [Cheng and Wyant, 1984, 1985a],

$$\lambda_S = \frac{\lambda_1 \lambda_2}{|\lambda_1 - \lambda_2|}. \tag{8.10}$$

Yet another approach is the use of a shearing interferometer, in which case the sensitivity can be varied by adjusting the amount of shear. With a small lateral shear, the interferogram yields the derivative of the wavefront errors along the direction of shear. Integration of the data from two interferograms with orthogonal directions of shear then gives the wavefront aberrations [Yatagai and Kanou, 1984]. It is also possible to use a radial shear interferometer in which the shear ratio (the ratio of the diameters of the two images of the pupil) is close to unity to give a desensitized interferogram [Hariharan, Oreb, and Wanzhi, 1984]. A polynomial is fitted to the data; the errors of the test wavefront can then be calculated from the coefficients of this polynomial and the shear ratio.

A problem in non-null tests of aspheres is systematic errors introduced by the imaging optics. These errors need to be evaluated in each case and removed [Murphy, Brown, and Moore, 2000].

8.11.1   Computer-Generated Holograms

In large-scale production, aspheric optical surfaces are often tested by using an aspheric reference surface or an additional element, called a null lens, which converts the wavefront leaving the surface under test into a spherical or plane wavefront. An alternative is to use a computer-generated hologram (CGH) [MacGovern and Wyant, 1971] (see Appendix F.2).

A modified Twyman–Green interferometer using a CGH to test an aspherical mirror is shown in Fig. 8.12 [Wyant and Bennett, 1972]. The CGH is located in the

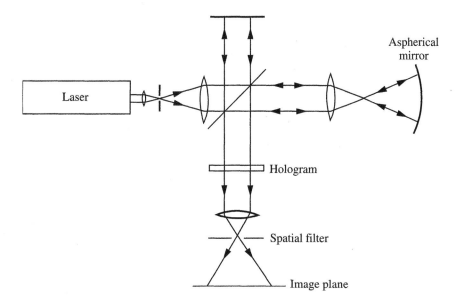

Figure 8.12. Modified Twyman–Green interferometer using a computer-generated hologram to test an aspherical mirror [Wyant and Bennett, 1972].

plane in which the mirror under test is imaged and is equivalent to the interferogram formed by the wavefront from an aspheric surface with the desired profile and a tilted plane wavefront. The moire pattern formed by the superposition of the actual interference fringes and the CGH gives the deviation of the surface under test from the ideal surface.

The contrast of the moire pattern is improved by reimaging the hologram. A small aperture placed in the focal plane of the imaging lens acts as a spatial filter, passing only the wavefront from the mirror under test which has been transmitted by the hologram and the wavefront reconstructed by the hologram when it is illuminated with a plane wavefront. Typical fringe patterns obtained with such an arrangement, without and with the CGH, are shown in Fig. 8.13.

A limit is set to the deviations from a sphere which can be handled by a CGH by the requirement that the spatial carrier frequency of the CGH must be at least three times the highest spatial frequency in the uncorrected interference pattern. This restriction can be overcome by using a combination of a simple null lens, which reduces the residual aberrations to an acceptable level, and a CGH [Faulde *et al.*, 1973; Wyant and O'Neill, 1974]. In addition, with a binary CGH, it is necessary to ensure that unwanted diffraction orders do not produce disturbing interference effects [Lindlein, 2001].

Computer-generated holograms are now used widely to test aspheric surfaces (see Loomis [1980]). With the development of improved plotting routines and the use of techniques such as electron-beam recording on layers of photoresist coated on

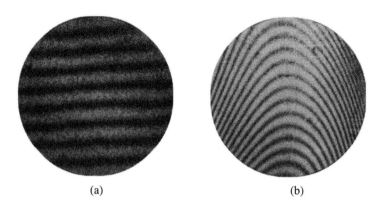

(a)   (b)

Figure 8.13. Interferograms of an aspherical surface (a) with and (b) without a compensating computer-generated hologram [Wyant and Bennett, 1972].

optically worked substrates, computer-generated holograms of very high quality can now be produced for such work [Arnold, 1985, 1988, 1989; Dörband and Tiziani, 1985; Urquhart *et al.*, 1989].

In addition, it is possible to use a CGH as a diffractive optical element (DOE). The use of such a DOE as a beamsplitter to test surfaces with complex forms has been discussed by Schwider [1998].

8.12   Rough Surfaces

The natural contour interval of half a wavelength (about 0.25 μm with visible light) with conventional interferometers makes it difficult to use them for measurements on surfaces whose deviations from flatness, or a reference sphere, exceed about 10 μm. It is also not possible to obtain fringes with surfaces that are rough on an optical scale and do not give a specular reflection. A typical example is large aspheric surfaces, which often need to be checked before polishing.

One way to solve these problems is to use a longer wavelength which is specularly reflected even from a fine-ground surface. Infrared interferometry with a CO_2 laser at a wavelength of 10.6 μm has been used successfully to test ground surfaces, as well as surfaces with relatively large deviations from a reference sphere [Munnerlyn and Latta, 1968; Kwon, Wyant, and Hayslett, 1980]. Accurate measurements have been made with a pyroelectric vidicon using phase-shifting techniques [Stahl and Koliopoulos, 1987]. A rigorous analysis of the intensity distribution in the fringe patterns, taking into account the effects of surface roughness, has been presented by Verma and Han [2001].

A simpler alternative with nominally flat surfaces is to reduce the sensitivity of the interferometer by using collimated light incident obliquely, instead of normally,

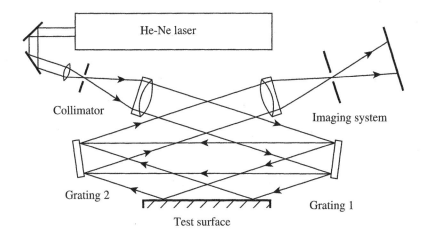

Figure 8.14. Grazing incidence interferometer using reflection gratings [Hariharan, 1975a].

on the surface [Linnik, 1942]. The contour interval is then $\lambda/2 \cos \theta$, where θ is the angle of incidence, and interference fringes can be obtained with surfaces such as fine-ground glass and metal [Abramson, 1969; Birch, 1973]. The low reflectivity of the test surface can be compensated for by means of a system, such as that shown in Fig. 8.14, using blazed reflection gratings to divide and recombine the beams [Hariharan, 1975a], or by using a pair of wedged beamsplitters, coated on only one side [Murty and Shukla, 1976]. A symmetric geometry, using a reference mirror to correct for relative inversion of the measurement and reference wavefronts, has been described by de Groot [2000a].

The use of multiple wavelengths to increase the range of measurements and to solve ambiguities in the height of steps has been described by Franze and Tiziani [1998]. An alternative is to use a pair of phase diffraction gratings to illuminate the object simultaneously at two different angles. The resulting vernier effect provides an equivalent wavelength of 12.5 μm at angles of incidence less than 25° [de Groot, de Lega, and Stephenson, 2000].

8.13 The Optical Transfer Function

The intensity distribution in the image of a point yields the point-spread function (PSF) of an optical system. A useful way of specifying the performance of an optical system is by the Fourier transform of its PSF, which is known as the optical transfer function (OTF). This is a measure of the transmittance of the system for different spatial frequencies in the object and has the advantage that, for a system built up of a number of linear elements, the OTF of the system is given by the product of the OTFs of the individual elements.

It can be shown (see Appendix B) that with coherent illumination $g(\xi, \eta)$, the pupil function of the system, and $G(x, y)$, the amplitude distribution in the image of an infinitely distant point, are related by a Fourier transform, so that we can write (see Appendix A.1)

$$G(x, y) \leftrightarrow g(\xi, \eta). \tag{8.11}$$

With incoherent illumination, the intensity distribution in the image is given by $|G(x, y)|^2$, so that the OTF is

$$\Omega(\xi, \eta) = \mathcal{F}\{|G(x, y)|^2\},$$
$$= g(\xi, \eta) \star g(\xi, \eta). \tag{8.12}$$

The right-hand side of Eq. (8.12) is usually normalized so that it has a maximum value of unity.

The OTF of an optical system can be measured directly by a shearing interferometer [Hopkins, 1955]. To see how this is done, consider two images of the wavefront leaving the pupil of the system under test, which have been laterally sheared by an amount η_s along (say) the η axis, and which have an additional phase difference ϕ introduced between them. The complex amplitudes in these two wavefronts at a point (ξ, η) are $g(\xi, \eta)$ and $g(\xi, \eta - \eta_s) \exp(i\phi)$, respectively. The total flux leaving the interferometer, which is also the total flux in the image of a point source, is then

$$I(\phi) = \iint |g(\xi, \eta) + g(\xi, \eta - \eta_s) \exp(i\phi)|^2 d\xi \, d\eta$$

$$= \iint |g(\xi, \eta)|^2 d\xi \, d\eta + \iint |g(\xi, \eta - \eta_s)|^2 d\xi \, d\eta$$

$$+ 2\mathrm{Re}\left[\exp(i\phi) \iint g(\xi, \eta)g^*(\xi, \eta - \eta_s)d\xi \, d\eta\right]$$

$$= 2\bar{I}[1 + |\Omega(s)| \cos(\psi + \phi)], \tag{8.13}$$

where

$$\bar{I} = \iint |g(\xi, \eta)|^2 d\xi \, d\eta$$

$$= \iint |g(\xi, \eta - \eta_s)|^2 d\xi \, d\eta, \tag{8.14}$$

and $\Omega(s) = |\Omega(s)| \exp(i\psi)$, where s is a spatial frequency defined by the relation $s = \eta_s / \lambda f$.

Equation (8.13) shows that if the additional phase shift ϕ between the two sheared wavefronts is varied linearly, the transmitted flux varies sinusoidally. The difference

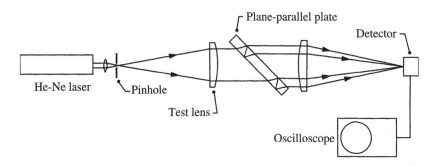

Figure 8.15. Parallel-plate shearing interferometer for measurements of the MTF of a lens [Wyant, 1976].

between the maximum and minimum values of $I(\phi)$ divided by their average $\bar{I}$ gives the modulus $|\Omega(s)|$ of the OTF (known as the modulation transfer function, or the MTF) for the spatial frequency s, while its phase ψ can be evaluated from the values of ϕ corresponding to these maxima and minima. These measurements are repeated at increasing values of the shear to trace out the OTF curve.

To carry out such measurements, we require an interferometer in which the shear can be set at any desired value without introducing tilt between the beams, while the path difference is varied over a convenient range. Early systems used modified Twyman–Green [Kelsall, 1959], Sagnac [Hariharan and Sen, 1960e], polarizing [Tsuruta, 1963], and Mach–Zehnder [Montgomery, 1964] interferometers. Perhaps the simplest system is that due to Wyant [1976], shown in Fig. 8.15.

In this arrangement, the beam transmitted through a tilted plane-parallel plate and the beam reflected off its two surfaces interfere to give a lateral-shear interferogram. The shear is given by the relation

$$s = (n^2 - \sin^2 \theta)^{-1/2} d \sin 2\theta, \tag{8.15}$$

while the optical path difference is

$$p = 2nd\{[(n^2 - \sin^2 \theta)/n^2]^{1/2} - 1\}, \tag{8.16}$$

where n is the refractive index and d the thickness of the plate, and θ is the angle of incidence. The two beams are brought to a focus on a detector whose output is displayed on an oscilloscope, as a function of θ, as the plate is rotated. The envelope of the time-varying signal of the oscilloscope then gives a direct display of the MTF of the system under test.

Chapter 9

Interference Microscopy

Two-beam interference microscopes are widely used to study transparent living objects, which cannot be stained without damaging them, as well as to study the structure of surfaces. Early instruments used for such studies were based on the Michelson and Mach–Zehnder interferometers. For high magnifications, a suitable configuration is that described by Linnik [1933] in which a beamsplitter directs the light onto two identical objectives, one viewing the test surface, and the other a reference mirror. However, the arrangement most commonly used now is the Mirau interferometer.

9.1 The Mirau Interferometer

The Mirau interferometer [Delaunay, 1953] uses a very compact optical system which can be incorporated in a microscope objective. As shown in Fig. 9.1, light from an illuminator is incident through the microscope objective on a beamsplitter. The transmitted beam goes to the test surface, while the reflected beam is directed to an aluminized spot on the front surface of the microscope objective. The two reflected beams are recombined at the same beamsplitter and return through the objective. The interference pattern formed in the image plane contours the deviations from flatness of the test surface.

9.2 Common Path Interference Microscopes

Common-path interference microscopes with double-focus or lateral-shearing systems, using a Sagnac interferometer or birefringent elements, were developed at a

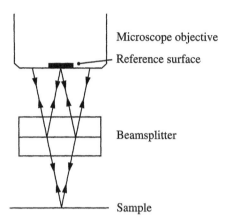

Figure 9.1. The Mirau interferometer.

fairly early stage. Details of these instruments are to be found in books by Françon [1961] and Krug, Rienitz, and Schulz [1964], as well as in a review by Bryngdahl [1965]. With a relatively large shear, it is possible to have one beam passing through a uniform area on the slide, so that an interference pattern corresponding to the actual variations of the optical thickness of the specimen is obtained. Alternatively, the shear may be made very small, to give the gradient of the optical thickness. A very simple interference microscope based on the point-diffraction interferometer has also been described by Ross and Singh [1983].

9.3   Polarization Interferometers

Polarization interferometers use birefringent elements as beamsplitters (see Françon and Mallick [1971]). They can be classified under three broad categories: systems that produce lateral shear with plane wavefronts, systems that produce lateral shear with spherical wavefronts, and systems that produce radial shear.

9.3.1   Lateral Shear: Plane Wavefronts

With a plane wavefront, lateral shear can be produced most conveniently with a Savart polariscope. This consists, as shown in Fig. 9.2, of two identical birefringent plates cut from a uniaxial crystal (usually calcite or quartz) with the axis at approximately 45° to the entrance and exit faces and put together with their principal sections crossed.

A ray, linearly polarized at 45° by means of a polarizer P_1, incident on the first plate, is split into two rays, the ordinary (O) and the extraordinary (E). Because the second plate is rotated by 90° with respect to the first plate, the ordinary ray in the first plate becomes the extraordinary (OE) in the second, while the extraordinary

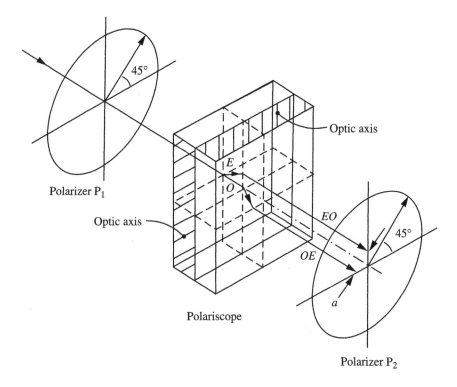

Figure 9.2. The Savart polariscope.

ray in the first becomes the ordinary (EO) in the second. Both rays therefore emerge parallel to each other, but with a mutual displacement

$$a = 2^{1/2}d(n_e^2 - n_o^2)/(n_e^2 + n_o^2), \qquad (9.1)$$

where d is the thickness of the plates and n_o and n_e are the ordinary and extraordinary indices of refraction of the crystal. Since these rays are orthogonally polarized, a second polarizer P_2 is used to make them interfere. For small angles of incidence, the interference pattern consists of straight, equally spaced fringes localized at infinity, with an angular separation

$$\alpha = \lambda/a. \qquad (9.2)$$

Straight fringes can be obtained over a wider field with a modified Savart polariscope [Françon, 1957]. This consists, as shown in Fig. 9.3, of two identical plates cut, as before, at 45° to the optic axis, but with one plate rotated through 180° so that, while the principal sections of the two plates are parallel, the optic axes are perpendicular. A half-wave plate is inserted between the two plates with its axis at

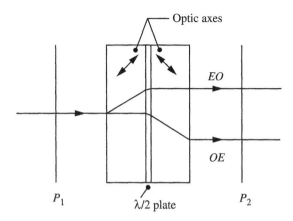

Figure 9.3. Savart polariscope modified to give a wider field [Françon, 1957].

45° to the principal sections. In this case, the rays stay in the same plane and their separation is

$$a = 2d(n_e^2 - n_o^2)/(n_e^2 + n_o^2). \qquad (9.3)$$

A variable shear can be obtained with this polariscope if the polarizers P_1, P_2 are fixed to the plates, and the two components are rotated in opposite directions through the same angle [Steel, 1964a]; if this angle is θ, the lateral shear is

$$a = 2d[(n_e^2 - n_o^2)/(n_e^2 + n_o^2)] \cos \theta. \qquad (9.4)$$

Two identical Savart polariscopes of the type shown in Fig. 9.3 can be used in a system, such as that shown in Fig. 9.4, for the examination of phase objects. A half-wave plate with its optic axis at 45° to the principal sections of the two polariscopes Q_1, Q_2 is interposed between them. As a result, the OE ray emerging from Q_1 becomes the EO ray in Q_2 and vice versa, so that the shear produced by Q_1 is compensated by Q_2. It is, therefore, possible to use an extended white-light source with this system.

9.3.2   Lateral Shear: Spherical Wavefronts

With a spherical wavefront, lateral shear can be produced by a Wollaston prism. This is made up, as shown in Fig. 9.5a, of two prisms having the same angle θ and cut from a uniaxial crystal in such a way that the optic axes are parallel to the outer faces, but are at right angles to each other. An incident ray is split into two rays making an angle α with each other which, when θ is small, is given by the relation

$$\alpha = 2(n_e - n_o) \sin \theta. \qquad (9.5)$$

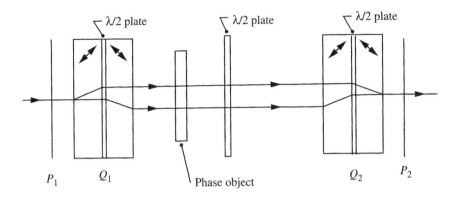

Figure 9.4. Shearing interferometer using two Savart polariscopes.

A wider useful field can be obtained with the modified form of the Wollaston prism shown in Fig. 9.5b, in which the optic axes in the two elements are parallel to each other, and a half-wave plate is inserted between them.

Lateral shearing interferometers using a Wollaston prism have been described by Philbert [1958] and by Dyson [1963].

9.3.3 Radial Shear

A lens made from a birefringent crystal produces two images at different points along the axis and can be used to obtain radial shear. Radial shear interferometers using such birefringent lenses have been described by Dyson [1957] and by Steel [1965].

9.4 The Nomarski Interferometer

The Nomarski interferometer is used widely for interference microscopy. As shown in Fig. 9.6, it uses two modified Wollaston prisms to split and recombine the beams.

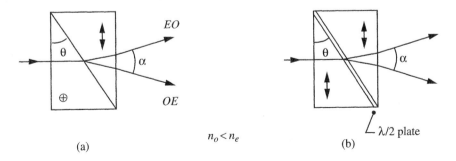

Figure 9.5. (a) The Wollaston prism, and (b) modification to give a wider field.

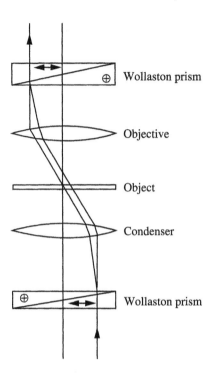

Figure 9.6. The Nomarski interferometer (transmission version).

Two modes of observation can be used with this system. With small isolated objects, it is convenient to use a lateral shear that is larger than the dimensions of the object. Two images of the object are then seen, covered with fringes that contour the phase changes due to the object. More commonly, the shear is made much smaller than the dimensions of the object. The interference pattern then shows the phase gradients and edges are enhanced (differential interference contrast, or DIC microscopy). As shown in Fig. 9.7, this makes it very easy to detect grain structure and local defects, such as scratches.

It is possible to use a white-light source with the Nomarski interferometer. Very small changes in phase are then revealed by changes in color. In addition, ambiguities arising at steps can be resolved, since corresponding fringes on either side of the step can be identified easily.

9.5   Electronic Phase Measurements

Extremely high sensitivity can be attained with polarization interferometers by using heterodyne techniques. One arrangement uses two orthogonally polarized beams with

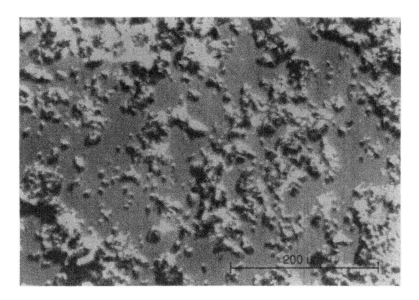

Figure 9.7. Nomarski micrograph of a partially polished glass surface, showing residual grinding pits [Bennett and Mattson, 1989].

a small frequency difference. One of these beams is focused at a fixed reference point on the sample, while the other scans the surface along a circle around this point [Sommargren, 1981]. The phase difference between the two beams is measured by a heterodyne technique. A better arrangement, which eliminates errors due to vibration almost completely, is a scanning system using two coaxial beams. One of the beams is focused to a 2-μm spot, while the other illuminates a concentric area with a diameter of about 50 μm [Huang, 1984].

Polarimetric techniques can also be used to measure the phase difference [Downs, McGivern, and Ferguson, 1985]. In this case, the two orthogonally polarized beams, which have the same frequency, pass through a λ/4 plate after leaving the interferometer, yielding a linear vibration whose azimuth depends on their phase difference. Shifts in the plane of polarization, which correspond to changes in the surface height, are measured with a Faraday-effect polarimeter. Surface irregularities as small as a few hundredths of a nanometre rms can be detected with such a system.

9.5.1 Phase-Shifting Techniques

If the beamsplitter in a Mirau interferometer is mounted on a PZT, very accurate measurements of surface profiles can be made using digital phase-shifting techniques. A resolution in depth of 0.5 nm can be attained by averaging the results of several runs [Bhushan, Wyant, and Koliopoulos, 1985]. Figure 9.8 shows typical results obtained with a hard-disk head [Wyant and Creath, 1985]. In the case of measurements on rough

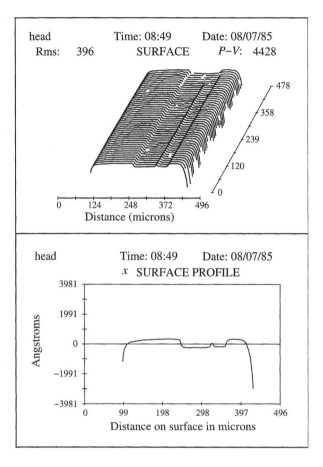

Figure 9.8. (a) Three-dimensional map and (b) profile of a hard-disk head, obtained with a phase-shifting interference microscope [Wyant and Creath, 1985].

surfaces, the data can be processed to plot a histogram of the surface deviations, or to obtain the rms surface roughness and the autocovariance function of the surface irregularities.

Phase-shifting techniques can also be implemented with polarization interference microscopes [Hariharan, 1996c]. A simple arrangement uses a quarter-wave retarder, oriented at 45°, inserted in the two orthogonally polarized beams emerging from the interferometer. This retarder converts one beam into right-handed and the other into left-handed circularly polarized light. If these two circularly polarized beams are made to interfere by a linear polarizer, rotation of the polarizer through an angle θ introduces an additional phase difference between the beams (see Section 6.7.2)

$$\Delta\phi = 2\theta. \tag{9.6}$$

Four sets of intensity values recorded with phase shifts of 0°, 90°, 180°, and 270° can then be used to calculate the original phase distribution in the image [Cogswell *et al.*, 1997].

It should be noted that, with all these systems, the fringe spacing increases as the numerical aperture is increased. A correction must be applied for this effect [Dubois *et al.*, 2000]. In addition, the fringe visibility decreases with the optical path difference, limiting the range over which measurements can be made.

9.6 Phase Ambiguities

While interferometric profilers using monochromatic light offer excellent vertical resolution, a serious limitation to their use is 2π phase ambiguities which arise if the measurement range involves a change in the optical path difference greater than a wavelength. Phase unwrapping techniques can be used only with smooth continuous surfaces and break down where the test surface exhibits a sharp step, or a discontinuity.

One way to extend the measurement range is by making measurements of the phase differences at two wavelengths (see Sections 7.10.1 and 8.11), in which case the repeat distance corresponds to the longer synthetic wavelength [Cheng and Wyant, 1984; Creath, 1987; Hariharan and Roy, 1996a].

A technique which can be used for large step heights is wavelength scanning using a tunable diode laser (see Section 7.10.2). A wider frequency-scan range can be obtained by using a superluminescent diode (SLD) as a light source, in conjunction with a liquid-crystal Fabry–Perot interferometer as a frequency scanning device. A constant input to the interferometer is obtained over the entire frequency-scan range by adjusting the injection current to the SLD. The height distribution of the object is obtained from the phase information recovered, using a Fourier-transform technique [Mehta *et al.*, 2002].

9.7 White-Light Interferometry

The problem of 2π phase ambiguities can be overcome by using white light and scanning the object in depth [Kino and Chim, 1990; Lee and Strand, 1990]. We assume that the origin of coordinates is taken at the point on the z axis at which the two optical paths are equal, and that the test surface is moved along the z axis in a series of steps of size Δz. With a broad-band source, the intensity at any point in the image plane, corresponding to a point on the object whose height is h, can then be written as

$$I(z) = I_1 + I_2 + 2(I_1 I_2)^{1/2} g(p) \cos[(2\pi/\bar{\lambda})p + \phi_0], \qquad (9.7)$$

where I_1 and I_2 are the intensities of the two beams acting independently, $g(p)$ is the fringe visibility, or coherence function (which corresponds to the envelope of

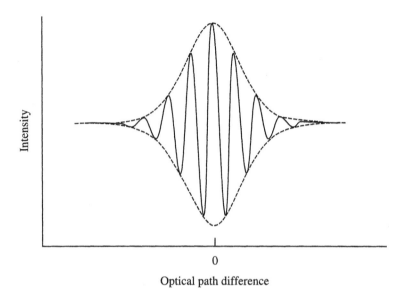

0

Optical path difference

Figure 9.9. Output from a white-light interferometer as a function of the position of one of the end mirrors along the z axis.

the interference fringes), and $\cos[(2\pi/\bar{\lambda})p + \phi_0]$ is a cosinusoidal modulation. In Eq. (9.7), $\bar{\lambda}$ corresponds to the mean wavelength of the source, $p = 2(z - h)$ is the difference in the lengths of the optical paths traversed by the two beams, and ϕ_0 is the difference in the phase shifts on reflection at the beamsplitter and the mirrors.

Figure 9.9 shows the variations in intensity at a given point in the image as the object is scanned along the height (z) axis. Each of these interference patterns can be processed to obtain the envelope of the intensity variations (the fringe visibility function) and determine the peak amplitude of the intensity variations as well as the location of this peak along the scanning axis. The values of the peak amplitude correspond to an image of the test object, while the values of the location of this peak along the scanning axis yield the height of the surface at the corresponding points.

The method most widely used to recover the fringe visibility function from the sampled data has been by digital filtering in the frequency domain [Lee and Strand, 1990]. This process involves two Fourier transforms (forward and inverse) along the z direction for each pixel in the sample. It is necessary, therefore, for the step size in the z direction to correspond to a change in the optical path difference of less than a fourth of the shortest wavelength; typically, the step Δz is around 50 nm. Consequently, this procedure requires a large amount of memory and processing time, though these requirements can be reduced to some extent, and good accuracy obtained, by modified sampling and processing techniques [Chim and Kino, 1992; Caber, 1993; de Groot and Deck, 1993; de Groot et al., 2002]. An alternative method of analysis uses a wavelet filter and has been shown to have advantages with sub-Nyquist sampling [Recknagel and Notni, 1998]. A two-step procedure for locating the true peak of the fringe-visibility function has been described by Park and Kim [2000].

9.7.1 Phase Shifting

A more direct approach, which is also computationally much less intensive, involves shifting the phase of the reference wave by three or more known amounts at each step along the z axis and recording the corresponding values of the intensity; these intensity values can then be used to evaluate the fringe visibility directly at that step [Dresel, Hausler, and Venzke, 1992]. If this is done by changing the optical path difference, two problems arise. The first is that the dynamic phase introduced by a change in the optical path varies inversely with the wavelength; the second is that, since the value of p is changing, the value of the envelope function $g(p)$ is not the same for all the intensity values obtained at a given position of the object. Both these factors can lead to systematic errors in the values of the fringe visibility.

A way to solve these problems is by using a modified five-step phase-shifting algorithm to calculate values of the visibility [Larkin, 1996]. The location of the peak is then found by a simple least-squares fit, after which the phase at this position can be evaluated. In an alternative approach, the coherence envelope is considered as locally linear, and a seven-point algorithm is used to calculate the relative phase and the fringe contrast at each point. The absolute phase is then obtained by combining the relative phase with the fringe order derived from the value of the contrast [Sandoz, 1996; Sandoz, Devillers, and Plata, 1997].

Another possibility is the analysis of white-light interferograms in the spatial frequency domain [de Groot and Deck, 1995]. White-light interferograms generated by scanning the object in a direction perpendicular to its surface may be considered as the incoherent sum of fringe patterns formed by different wavelengths. Fourier analysis of these interferograms can recover these virtual single-color fringe patterns to determine their relative strengths and phases as a function of wavenumber. It is then possible to measure distances precisely by observing how the phases vary as a function of wavenumber.

Yet another possibility is a wavelength sampling approach [Sandoz, Calatroni, and Tribillon, 1999] in which the phase shift is introduced through a variation of the wavelength. Since the necessary wavelengths can illuminate the object simultaneously, a multispectral band image sensor can be used to record the set of phase-shifted interferograms.

9.7.2 Achromatic Phase Shifting

Another way to solve these problems is by using an achromatic phase shifter operating on the Pancharatnam phase (see Appendix E.2 and Section 6.7.2). A typical experimental arrangement used to demonstrate the application of achromatic phase

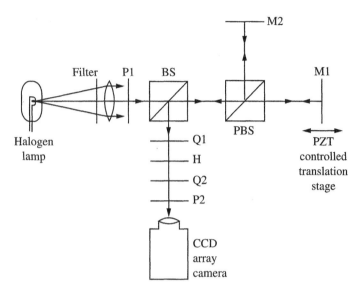

Figure 9.10. Schematic of an experimental arrangement used to demonstrate the application of achromatic phase shifting to white-light interferometry [Hariharan and Roy, 1994].

shifting to white-light interferometry is shown schematically in Fig. 9.10 [Hariharan and Roy, 1994]. It consists essentially of a Michelson interferometer illuminated with a collimated beam of white light. In this interferometer, a linearly polarized beam is divided at the polarizing beamsplitter PBS into two orthogonally polarized beams which are incident on the mirrors M1 and M2, respectively. After reflection at M1 and M2, these beams return along their original paths to a second beamsplitter BS, which sends them through a second polarizer P2 to the CCD array camera.

The phase difference between the beams is varied by a phase shifter consisting of a rotating $\lambda/2$ plate (H) mounted between two fixed $\lambda/4$ plates (Q1, Q2). If the $\lambda/2$ plate H is rotated through an angle θ, an additional phase difference 4θ, which is very nearly independent of the wavelength, is introduced between the beams.

With the phase shifter in the optical path, the total phase difference between the beams is

$$\Delta\phi_{\text{total}} = \Delta\phi_D + \Delta\phi_G, \qquad (9.8)$$

where $\Delta\phi_D$ is the dynamic phase due to the optical path difference and $\Delta\phi_G$ is the geometric (Pancharatnam) phase. However, unlike the dynamic phase, which depends on the wavelength, the geometric phase is the same for all wavelengths. The effect of a change in the geometric phase of $2m\pi$ is therefore to move the fringes formed at all wavelengths by an integral number of fringe spacings, so that the interference pattern formed with white light returns to its original configuration. As a result, the intensity at any point in the interference pattern varies between its maximum and minimum values, but the position of the fringe envelope in the field of view remains unchanged [Hariharan, Larkin, and Roy, 1994].

At each step, four measurements are made of the intensity in the fringe pattern corresponding to additional phase differences of 0°, 90°, 180°, and 270°. Since the additional phase differences introduced are the same for all wavelengths, the visibility of the interference fringes at any given point in the field can be calculated from these data using the relation

$$g(p) = \frac{2[(I_0 - I_{180})^2 + (I_{90} - I_{270})^2]^{1/2}}{I_0 + I_{90} + I_{180} + I_{270}}, \tag{9.9}$$

which, normally, is valid only for monochromatic light. The value of z corresponding to the peak visibility of the interference fringes can then be obtained directly, from a curve fit, to within a few nanometres.

In addition, it has been shown that with achromatic phase shifting, when the optical path difference is less than $\pm\bar{\lambda}$, where $\bar{\lambda}$ is the design wavelength of the $\lambda/2$ and $\lambda/4$ plates, we can calculate $\phi(\bar{\lambda})$, the phase difference at this wavelength, from the relation

$$\tan(\phi, \bar{\lambda}) = \frac{I_{90} - I_{270}}{I_0 - I_{180}} \tag{9.10}$$

with an error less than $2\pi/180$ [Hariharan and Roy, 1995].

Accordingly, if $\bar{\lambda} = 550$ nm, and the object is scanned in depth by moving it along the z-axis in steps of (say) 250 nm, the intensity values obtained for any point on the object, at the position nearest to the fringe-contrast maximum, can be used in Eq. (9.10) to find the value of the residual optical path difference at this position with an error less than ± 3 nm.

9.7.3 Spectrally Resolved Interferometry

Yet another technique that can be used with white light is spectrally resolved interferometry [Calatroni *et al.*, 1996; Sandoz, Tribillon, and Perrin, 1996]. In this technique a spectroscope is used to analyze the light from each point on the interferogram. The optical path difference between the beams at this point can then be obtained from the intensity distribution in the resulting channeled spectrum (see Section 2.11) or by phase shifting [Helen, Kothiyal, and Sirohi, 2001]. The value of the surface height obtained in this manner is completely independent of the height of neighboring points and free from 2π phase ambiguities. A resolution of 1 nm in height is possible with this technique.

An advantage of this technique is that measurements need be made at only one position of the object. There is no need to translate the specimen along the height axis, and the method is not sensitive to mechanical vibration. However, each interferogram only yields a profile along a single line.

9.7.4 Sources of Errors

With white light, it is necessary to equalize the geometrical path lengths in dispersive (glass) elements in both arms to avoid ghost steps in the measured profile. A technique for constructing a beamsplitter cube that has closely balanced paths has been described by Farr and George [1992]. An algorithm that minimizes the effects of residual deviations from this condition has also been described by Pförtner and Schwider [2001].

Another problem is skewing of the coherence envelope due to diffraction effects at steps whose height is less than the coherence length of the radiation [Harasaki and Wyant, 2000]. This problem can be overcome by using values of the phase obtained by phase-stepping [Harasaki, Schmit, and Wyant, 2000]. In addition, the presence of surface films, as well as the variation of the phase change on reflection with the wavelength, can result in a shift of the position of the peak of the coherence envelope along the axis [Hariharan and Roy, 1996b; Helen, Kothiyal, and Sirohi, 1999] which can cause errors when adjacent areas of the test object are made of different materials [Harasaki, Schmit, and Wyant, 2001].

Chapter 10

Interference Spectroscopy

The instrumental function of a spectroscope is defined by its response $a(v - v')$ to a perfectly monochromatic spectral line $\delta(v - v')$. When the spectroscope is illuminated by a spectral line with a spectral energy distribution $b(v)$, the spectral energy distribution recorded is

$$b'(v) = a(v) * b(v). \tag{10.1}$$

Two monochromatic lines are resolved when their separation $v_2 - v_1$ is greater than a certain value Δv which is determined by the width of $a(v)$. The resolving power of the spectroscope is then

$$\mathcal{R} = v/\Delta v,$$
$$= \lambda/\Delta\lambda. \tag{10.2}$$

The resolving power available with prisms and gratings is limited to about 10^6. Higher resolving powers can be achieved only by interference spectroscopy.

10.1 Etendue of an Interferometer

An important characteristic of an interferometer, when it is used for the study of faint spectra, is its light-gathering power or "throughput," which is defined by an invariant known as its *etendue*. In the arrangement shown in Fig. 10.1, the effective areas A_S and A_D of the source and the detector are images of each other. These images subtend at the lenses L_S and L_D, respectively, the same solid angle

$$\Omega = A_S/f_S^2 = A_D/f_D^2. \tag{10.3}$$

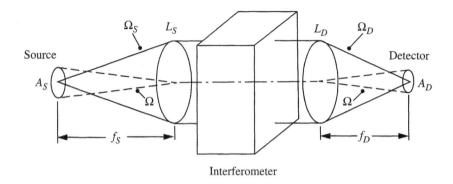

Interferometer

Figure 10.1. Etendue of an interferometer.

The amount of radiation accepted by the lens L_S is proportional to the product of the area of the source and the solid angle Ω_S subtended by L_S at a point on the source. Similarly, the amount of radiation reaching the detector is proportional to the product of the area of the detector and the solid angle Ω_D subtended at a point on it by the lens L_D. Since the two lenses have the same effective area A,

$$\Omega_S = A/f_S^2 \tag{10.4}$$

and

$$\Omega_D = A/f_D^2. \tag{10.5}$$

Accordingly, we have

$$A_S \Omega_S = A_D \Omega_D = A\Omega = E, \tag{10.6}$$

where E is known as the etendue of the interferometer.

10.2   The Fabry–Perot Interferometer

Among the instruments developed for direct observation of a spectrum at high resolution, Michelson's echelon and the Lummer–Gehrcke plate were widely used earlier [see Candler, 1951], but they have now been replaced by the Fabry–Perot interferometer (FPI) [see Hernandez, 1986; Vaughan, 1989]. This instrument, which has been briefly discussed earlier in Section 4.1, typically consists of two fused-silica plates which may have either a fixed or a variable separation. The inner faces of the plates are worked flat and are usually coated with highly reflecting multilayer dielectric stacks ($R > 0.95$). However, metal coatings are also used in applications, such as wavelength comparisons, where it is necessary to minimize the dispersion of the phase change on reflection. The plates themselves are slightly wedged, so that the beams reflected from the outer surfaces can be eliminated by a suitably positioned aperture.

If the surfaces are separated by a distance d, the instrumental function of the FPI is the Airy function,

$$a(v) = T^2/[(1 - R)^2 + 4R \sin^2(\psi/2)], \tag{10.7}$$

where $\psi = (4\pi vnd/c)\cos\theta + \Delta\phi$, θ being the angle of incidence of a ray and $\Delta\phi$ the additional phase shift on reflection at the plates. This function, which is shown in Fig. 4.2, exhibits sharp peaks separated by a frequency difference $c/2nd$, which corresponds to the free spectral range that can be handled without successive orders overlapping. The half-width of the peaks Δv is obtained by dividing the free spectral range by the finesse F, so that

$$\Delta v = (1/F)(c/2nd). \tag{10.8}$$

For an ideal FPI, the finesse is given by the relation (see Eq. 4.14)

$$F = \pi R^{1/2}/(1 - R). \tag{10.9}$$

However, in any practical case, the finesse is always lower because of deviations of the surfaces from flatness and the finite aperture of the plates; it can be estimated from more precise formulas given by Ramsay [1969].

In one mode of operation, the spacing of the plates is fixed (the Fabry–Perot etalon). Each wavelength in the source then gives rise to a system of rings centered on the normal to the plates. Because of the limited free spectral range of the interferometer, it is necessary, with a multiwavelength source, to separate the fringes formed by different spectral lines. This is done by imaging the rings on the slit of a spectrograph, as shown in Fig. 10.2, so that each line in the spectrum contains a narrow strip of the circular fringe pattern corresponding to that line. A single photograph of the spectrum can then give information on the structure of a number of spectral lines.

10.3 The Scanning FPI

Because of its circular symmetry, the FPI has a much larger etendue than the echelon and the Lummer–Gehrcke plate. However, in the configuration shown in Fig. 10.2, its etendue is limited by the narrow slit of the spectrograph. To obtain its full etendue, it is necessary to make use of all the rays having the same angle of incidence. For this

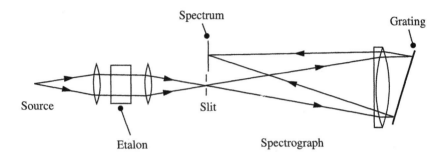

Figure 10.2. Separation of the Fabry–Perot fringes for different spectral lines by a spectrograph.

reason, the FPI is now commonly used as a scanning spectrometer. In this mode of operation, a small aperture is placed in the focal plane of a lens behind the FPI, and the transmitted intensity is recorded with a photomultiplier as the effective spacing of the plates is varied [Jacquinot and Dufour, 1948]. The size of this aperture is determined by $\mathcal{R}$, the resolving power of the instrument, the maximum solid angle which the aperture can subtend at the source without significantly degrading the resolution being given by the relation

$$\Omega_m = 2\pi/\mathcal{R}. \tag{10.10}$$

The FPI can be made to scan a selected region of the spectrum by changing either the spacing of the plates or the refractive index of the medium separating them. The latter system has the advantage of simplicity. Typically, the FPI is placed in an airtight box which is evacuated. Air is then allowed to enter at a controlled rate so that the pressure increases linearly with time. However, the spectral range which can be covered by pressure scanning is limited ($\Delta\lambda/\lambda = 3 \times 10^{-4}$ per atmosphere for air) and can be increased only by using higher pressures and denser gases [Jacquinot, 1960]. This limitation can be overcome with mechanical scanning. Although a number of techniques have been described for this purpose, the most common method, with spacings less than a few centimetres, is to use piezoelectric spacers, parallelism of the plates being maintained by means of a servo system [Ramsay, 1966; Hicks, Reay, and Scaddan, 1974; Sandercock, 1976; Atherton *et al.*, 1981]. Herbst and Beckwith [1989] have described an active stabilization system using a He-Ne laser and two position-sensitive detectors to monitor and stabilize the cavity spacing of a Fabry–Perot interferometer. An alternative, which has been used for studies of solar oscillations, is a thin, solid etalon made of lithium niobate [Burton, Leistner, and Rust, 1987]. An electric field imposed across the etalon varies its refractive index through the electrooptic (Pockels) effect, making it possible to use it as a tunable filter.

Systems have also been described which utilize a computer to ensure a strictly linear scan, as well as for the acquisition and storage of intensity data in a digital form (see Wood [1978]; Yamada, Ikeshima, and Takahashi [1980]).

10.4   The Spherical-Mirror FPI

It is apparent from Eqs. (10.6) and (10.10) that, for plates of a given diameter, the etendue of a FPI with plane mirrors (plane FPI) is inversely proportional to its resolving power. This limitation is overcome in the spherical-mirror Fabry–Perot interferometer (spherical FPI) described by Connes [1956, 1958]. This consists, as shown in Fig. 10.3, of two spherical mirrors whose separation is equal to their radius of curvature r, so that their paraxial foci coincide, giving an afocal system. Any incident ray, after traversing the interferometer four times, falls back on itself, thus giving rise to an infinite number of outgoing rays which are coincident (and not just parallel, as with the plane FPI), with an increment in optical path $\Delta p = 4r$ between successive rays.

The etendue of the spherical FPI is limited only by the spherical aberration of the mirrors, which restricts their useful aperture. This restriction is less serious with high-resolution instruments in which the radius of curvature of the mirrors is much greater than their diameter, and the superiority of the spherical FPI over the plane FPI is most marked under these conditions. This is apparent from a comparison of a plane FPI of diameter D and thickness d with a spherical FPI having mirrors of the same aberration-limited diameter and a separation $r = d/2$. Both these interferometers have the same free spectral range and, if the reflectance of the mirrors are properly chosen, the same finesse and, hence the same resolving power. However, the spherical FPI exhibits a gain in etendue over the plane FPI

$$M = 2(d/D)^2. \tag{10.11}$$

This gain is quite high when the separation of the mirrors is large compared to their diameter.

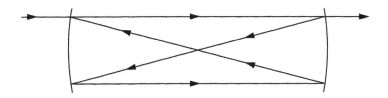

Figure 10.3. Spherical-mirror Fabry–Perot interferometer.

A further advantage of the spherical FPI is that, once the separation of the mirrors has been set, any angular misalignment merely redefines the optical axis of the system. With the plane FPI, a similar misalignment is equivalent to an imperfection in the surfaces of the mirrors.

A spherical FPI is usually operated as a scanning instrument [Herriot, 1963; Fork, Herriott, and Kogelnik, 1964]. While it has the disadvantage of a fixed free-spectral range, this is more than offset by its increased etendue and ease of adjustment. The optimization of the design of a spherical FPI has been discussed in detail by Hercher [1968]. Scanning spherical FPIs are now used commonly to examine the pattern of longitudinal and transverse modes in the output of a cw laser. With pulsed lasers, where the FPI cannot be operated in a scanning mode, the mirror separation can be made slightly less than the common radius of curvature. Such a defocused spherical FPI produces a pattern of concentric rings, with quasi-linear spectral dispersion, which can be photographed conveniently [Bradley and Mitchell, 1968].

10.5   The Multiple FPI

To study a complex spectrum with a scanning FPI, it is necessary to use a monochromator which can supply the FPI with a narrow enough spectral band to avoid overlapping of successive orders. If the etendue of the combination is not to be limited by the monochromator, it must have an etendue equal to or greater than that of the FPI. This condition cannot be met with a conventional grating or prism monochromator, but can be satisfied by using two or more FPIs in series [Chabbal, 1958]. If multiple reflections between the FPIs are eliminated, so that the FPIs are effectively decoupled (either by tilting them slightly so that they are not parallel, or by introducing a neutral density filter between them), the overall instrument function is simply the product of the instrumental functions for the individual FPIs.

In such an arrangement, the second FPI serves to eliminate the unwanted transmission peaks of the first FPI over a relatively wide range. The free spectral range can be increased by a factor m (typically $m \leq 10$) simply by making the separation of the plates in the second FPI $(1/m)$ times that in the first. However, an alternative which gives a narrower half-width and higher efficiency is to make the spacing of the second FPI $(m + 1)/m$ times that of the first [Mack *et al.*, 1963].

To explore a spectrum, the pass bands of all the FPIs must be shifted simultaneously. A simple way of doing this is by pressure scanning, maintaining a constant pressure difference between the individual FPIs [Mack *et al.*, 1963; Ramsay, Kobler, and Mugridge, 1970]. A more complex system with a computer-controlled scan has also been described [Winter, 1984]. A tunable Fabry–Perot spectrometer consisting of two lithium niobate etalons with thicknesses in a vernier ratio, operated in a double-pass configuration to eliminate the effects of birefringence, has been described by Netterfield *et al.* [1997].

10.6   The Multiple-Pass FPI

The contrast factor of a FPI, defined by the ratio of the intensities at the maxima and the minima in the instrumental function, is

$$C = [(1 + R)/(1 - R)]^2, \qquad (10.12)$$

where R is the reflectance of the mirrors. With typical coatings ($R \approx 0.95$) this figure is not adequate to ensure that a weak satellite is not masked by the background due to a neighboring strong spectral line. A much higher contrast factor, close to the square of that given by Eq. (10.12), can be obtained by passing the light twice through the same interferometer [Dufour, 1951; Hariharan and Sen, 1961d]. This technique was extended by Sandercock [1970], who used corner-cube prisms to route the beam three to five times through different areas of the mirrors. Multiple passing has been applied very successfully to Brillouin spectroscopy, and resolving powers of 10^8 with contrast factors greater than 10^{10} have been obtained in scanning FPIs [Lindsay and Shepherd, 1977; Vacher, Sussner, and Schickfus, 1980]. The use of a tandem multipass system to obtain a significantly larger free spectral range as well as higher contrast has been described by Dil *et al.* [1981].

10.7   Birefringent Filters

An interferometer consisting of a plate of a birefringent material such as quartz or calcite, with a thickness d, cut with its faces parallel to the optic axis and set between parallel polarizers with its optic axis at 45° to the axis of the polarizers, has a transmission function

$$a(\nu) = (1/2) \cos^2 \pi \nu T, \qquad (10.13)$$

where $T = (n_o - n_e)d/c$ is the delay introduced in the interferometer. A single transmission peak can be obtained by using a number of such interferometers in series, each one with a delay twice that of the preceding one. This is the principle of the birefringent filter developed by Lyot [1944] and, independently, by Öhman [1938] for studies of the solar surface. The filter can be made tunable over one order and a wider field obtained by splitting each element into two plates of equal thickness with their axes crossed, and interposing between them a rotatable half-wave plate between two fixed quarter-wave plates. A detailed description of such a filter with a pass band only 0.0125 nm wide has been given by Steel, Smartt, and Giovanelli [1961].

An alternative type of birefringent filter described by Šolc [1965] consists of m plates with equal retardations set between two polarizers. In one (fan) form, the polarizers are set parallel and the plates have their optic axes at angles $\pi/4m, 3\pi/4m, \ldots, (2m-1)\pi/4m$ to that of the polarizers, while in another (folded) form the polarizers are crossed, and alternate plates are set at angles of $+\pi/4m$ and $-\pi/4m$.

Detailed theories for both these types of birefringent filters have been given by Harris, Ammann, and Chang [1964] and Ammann and Chang [1965]. These theories can be used to design filters with any desired periodic transmittance.

A major application of such birefringent filters is as wavelength-selection elements in the resonators of tunable dye lasers [Yarborough and Hobart, 1973; Bloom, 1974]. Prisms and gratings have the disadvantage that a large beam diameter is required to attain high resolution, while etalons have a very limited tuning range. These problems are avoided with a birefringent filter.

The simplest arrangement for such a wavelength selector consists, as shown in Fig. 10.4, of a birefringent plate cut with its faces parallel to the optic axis and mounted in the laser cavity so that the beam is incident on it at the Brewster angle θ_B. If the optic axis does not lie in the plane of polarization of the laser beam (the yz plane) the transmittance of the plate is unity only for wavelengths satisfying the condition

$$[(n_0 - n_e)d \sin^2 \eta]/\sin \theta_B = m\lambda, \tag{10.14}$$

where n_0 and n_e are the ordinary and extraordinary refractive indices of the plate, d is its thickness, η is the angle between the ray in the crystal and the optic axis, and m is an integer. At other wavelengths, the polarization of the beam is modified,

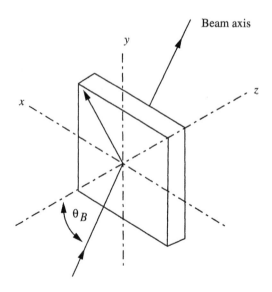

Figure 10.4. Birefringent tuning element for a dye laser.

as a result of which it suffers increased losses due to reflection at the surface of the plate. The filter can be tuned by rotating the plate in its own plane. This changes the angle between the optic axis and the axis of the laser beam and, hence, the angle η in Eq. (10.14).

The bandwidth of such a filter consisting of a stack of plates (usually three) whose thickness are in integer ratios is narrower than that of a Lyot filter with the same free spectral range. Even though the transmittance does not go to zero at the minima, the drop in transmittance is enough to ensure that the laser oscillates at only one frequency. Increased discrimination against secondary peaks can be obtained by adding glass plates at the Brewster angle [Holtom and Teschke, 1974].

A systematic design procedure which yields a filter with the maximum tuning range and the minimum number of plates has been described by Hodgkinson and Vukusic [1978]. Typically, a bandwidth of 0.025 nm and a tuning range of 100 nm can be obtained with a three-element filter.

10.8 Wavelength Meters

The increasing use of tunable dye lasers in spectroscopy has led to the need for an instrument which can be used to measure their output wavelength, in real time, with an accuracy commensurate with their line width (<1 part in 10^8).

Interference wavelength meters can be divided into two broad categories: dynamic wavelength meters, in which the measurement involves the movement of some element or group of elements, and static wavelength meters, which have no moving parts (other than those needed for initial alignment of the instrument). Dynamic wavelength meters offer the advantage of greater accuracy, but can only be used with cw lasers, while static wavelength meters can also be used with pulsed lasers.

10.8.1 Dynamic Wavelength Meters

The most common type of dynamic wavelength meter uses a two-beam interferometer in which the number of fringes is counted as the optical path is changed by a known amount. One form [Kowalski, Hawkins, and Schawlow, 1976] is shown in Fig. 10.5. In this instrument, two beams, one from the dye laser whose wavelength λ_1 is to be determined and the other from a reference laser (a frequency-stabilized He-Ne laser) whose wavelength λ_2 is known, traverse the same two paths in opposite directions. Another form [Hall and Lee, 1976] uses a folded Michelson interferometer in which the two end reflectors are mounted back-to-back on a single carriage, so that, when the carriage is moved, one optical path increases while the other decreases by the same amount. In both instruments, the beams from the dye laser and the reference laser emerge separately, and the fringe systems formed by the two wavelengths are imaged on separate detectors.

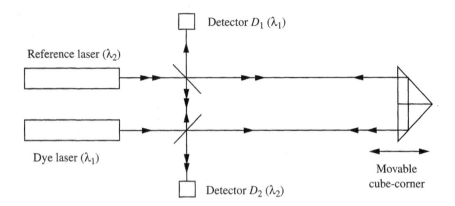

Figure 10.5. Dynamic wavelength meter [Kowalski *et al.*, 1976].

If, in the interferometer shown in Fig. 10.5, the end reflector is moved through a distance d, the number of fringes seen by the detector D_1 is

$$N_1 = 2n_1d/\lambda_1, \tag{10.15}$$

where n_1 is the refractive index of air at the wavelength λ_1. Similarly, the number of fringes seen by the detector D_2 is

$$N_2 = 2n_2d/\lambda_2, \tag{10.16}$$

where n_2 is the refractive index of air at the wavelength λ_2. Accordingly,

$$\lambda_1 = (N_2n_1/N_1n_2)\lambda_2. \tag{10.17}$$

If the interferometer is operated in a vacuum, the ratio (n_1/n_2) is equal to unity, and the wavelength of the dye laser can be determined directly from Eq. (10.17) by counting fringes simultaneously at both wavelengths. If the interferometer is operated at normal atmospheric pressure, a correction must be made for the ratio (n_1/n_2) which can be calculated from established formulas (see Section 7.5).

A convenient feature of these instruments is that the result is independent of the velocity of the end reflectors. In its simplest form, in which only integer changes in the order number are counted, the precision is approximately $1/N_2$, where N_2 is the total change in the integer order for the reference laser. Typically, for a change in the optical path difference of 1 metre, $N_2 \approx 1.6 \times 10^6$, giving a precision of about 7 parts in 10^7.

Higher precision can be obtained by phase-locking an oscillator to an exact multiple m of the frequency of the ac signal from the reference channel and measuring the frequency of this oscillator. This technique allows the fractional order number for the reference laser to be determined to $\pm 1/m$ [Hall and Lee, 1976; Kowalski *et al.*, 1978]. Higher resolution can also be obtained by digital averaging of the two modulation frequencies. This allows interpolation of the fractional order number to $N_2^{-1/2}$. A precision of the order of 1 part in 10^9, with a measuring time of a few seconds, is possible, though the accuracy may be less because of systematic errors [Monchalin *et al.*, 1981]. A third alternative is a vernier method in which the counting cycle starts and stops when the phases of the two signals coincide [Kahane *et al.*, 1983]. Improvements to these systems have been described that correct for interference fringe dropouts [Kowalski *et al.*, 1985] and ensure a constant velocity of the reflectors [Braun, Maier, and Liening, 1987].

A more compact design for a dynamic wavelength meter is based on a scanning spherical FPI [Salimbeni and Pole, 1980]. If the separation of the mirrors in such an interferometer is changed while it is illuminated by a single wavelength, the transmitted intensity consists of a series of pulses, each pulse corresponding to a change in the interference order of unity. With two lasers, coincidences will occur between the transmission peaks corresponding to the two wavelengths at intervals which satisfy the condition

$$m_1 \lambda_1 = m_2 \lambda_2 = p, \tag{10.18}$$

where m_1 and m_2 are the changes in the integer orders and p is the change in the optical path difference. If the output pulses due to the two wavelengths are detected separately and counted over this interval, the wavelength of the dye laser can be calculated from Eq. (10.18). In this arrangement, the precision is enhanced by a factor equal to the finesse of the FPI over that obtained by fringe counting over the same range in a Michelson interferometer. However, defocusing (see Section 10.4) limits the range of movement of the mirrors. A precision of 1 part in 10^7 is possible with a range of movement of 25 mm, in a time of 0.8 s.

10.8.2 Static Wavelength Meters

A number of static wavelength meters have been developed based on the Michelson, Fabry–Perot, and Fizeau interferometers.

The sigma-meter [Juncar and Pinard, 1975, 1982] uses four Michelson interferometers with optical path differences of 0.5 mm, 5 mm, 50 mm, and 500 mm sharing a common beamsplitter and reference mirror. A prism in the reference arm acts as an achromatic $\lambda/4$ plate to produce two orthogonally polarized reference waves with a phase difference of 90°. These two reference waves give rise to two separate fringe systems in each interferometer, in which the intensity distributions are, respectively,

$$I_1 = I_0(1 + \cos 2\pi \nu \tau), \tag{10.19}$$

and

$$I_2 = I_0(1 + \sin 2\pi\nu\tau), \qquad\qquad (10.20)$$

so that the fractional fringe order can be determined from these intensities. If the wavelength of the dye laser is known approximately from measurements with a monochromator, the integer order of the first interferometer can be calculated. The wavelength determined with this interferometer is then used to calculate the integer order for the next interferometer. This process is repeated successively to obtain the interference order with the last interferometer and, hence, the actual wavelength. An accuracy of 1 part in 10^8 is possible with this instrument.

Another type of static wavelength meter used three FPIs with free spectral ranges of 3 GHz, 30 GHz, and 300 GHz, respectively, along with a monochromator as a prefilter [Byer, Paul, and Duncan, 1977; Fischer, Kullmer, and Demtroder, 1981]. In this instrument, the intensity distributions in the fringes are read simultaneously by three photodiode arrays, and the fractional order for each interferometer is calculated with an accuracy of 0.01. As with the sigma meter, the approximate wavelength is given by the setting of the monochromator, and the final wavelength is found by a sequence of approximations, using each of the interferometers in turn. This instrument has a potential precision of about 5 parts in 10^9.

One of the simplest static wavelength meters is that developed by Snyder [1977], which uses a single Fizeau interferometer. This consists of two fused-silica optical flats, about 1 mm apart, making a small angle ϵ ($\approx$3 arcminutes) with each other. Light from the dye laser is collimated by an off-axis parabolic mirror and used to illuminate the interferometer. Fringes of equal thickness are then formed, which run parallel to the apex of the wedge. The intensity distribution in these fringes, along a line perpendicular to the apex of the wedge, is given by the relation

$$I(x) = I_0[1 + \cos(2\pi x/\lambda\epsilon)]. \qquad\qquad (10.21)$$

This intensity distribution is recorded directly by a 1024-element, linear photodiode array which receives the light reflected from the interferometer. The spatial period of the fringe pattern can then be used to evaluate the integer part of the interference order, while the fractional part can be calculated from the positions of the minima and the maxima with respect to a reference point on the wedge.

Errors due to imperfect collimation as well as small changes in the angle of incidence can be minimized in this arrangement by positioning the detector array so that the shear between the beams reflected from the two surfaces of the wedge is zero at the surface of the detector [Snyder, 1981; Gardner, 1983]. A precision of the order of 1 part in 10^7 with an update rate of 15 Hz is possible with this instrument.

10.9 Heterodyne Techniques

More precise measurements of the frequency of a laser and, hence, of its wavelength can be made by heterodyne techniques.

In one experiment [Bay, Luther, and White, 1972], electro-optic modulation of a laser beam (frequency ν_L) at a microwave frequency $\nu_M (\approx 10^{10}$ Hz) produced two side-band frequencies $\nu_L \pm \nu_M$. The length L of a Fabry–Perot interferometer, as well as the modulation frequency ν_M, were then adjusted so that both side bands were transmitted by the interferometer with maximum intensity. Under these conditions, it can be shown that

$$\nu_L = (N/\Delta N)2\nu_M, \tag{10.22}$$

where N is the order of interference for the frequency ν_L, and ΔN is the difference of the interference orders for the two side bands. This technique gave the frequency of the laser directly to a few parts in 10^8.

Another technique for the comparison of the frequencies of two lasers [Baird, 1983b] makes use of two slave lasers which can be tuned around the frequencies of the two lasers. These slave lasers are locked to two transmission peaks of a very stable Fabry–Perot interferometer so that their wavelengths are integral submultiples of the optical path difference in the interferometer. The frequency offsets of the lasers whose wavelengths are to be compared are then determined by measuring the beat frequency produced by each laser with the corresponding slave laser.

10.10 Measurement of Laser Line Widths

The widths of laser lines are too small to be measured by conventional methods. A method which has been developed for such measurements involves mixing light from the laser at a photodetector with a reference beam derived from the same laser which has undergone a frequency shift and a delay. Since the two signals are statistically identical, the self-coherence function of the original signal can be obtained from measurements in the radio-frequency region on the beat [Halmos and Shamir, 1982; Abitbol *et al.*, 1984].

In the experimental arrangement shown in Fig. 10.6, the frequency of the reference beam is shifted by diffraction at an acousto-optic modulator, and a suitable delay is introduced by a length of single-mode fiber. If the field due to the test beam is

$$E_1(t) = E_0 \cos[\omega t + \phi(t)], \tag{10.23}$$

the field due to the reference beam, which has undergone a frequency shift Ω and a delay τ, is

$$E_2(t) = E_0 \cos[(\omega + \Omega)t + \phi(t + \tau)]. \tag{10.24}$$

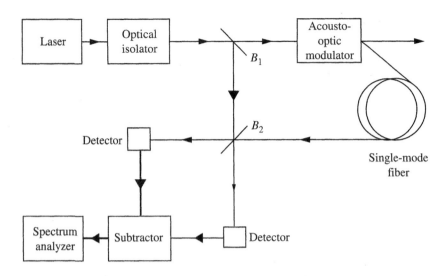

Figure 10.6. Experimental arrangement used for measurements of laser line width by heterodyne interferometry [Abitbol *et al.*, 1984].

The beat frequency signal is, therefore, proportional to

$$\cos[\Omega t + \phi(t - \tau) - \phi(t)]. \tag{10.25}$$

If $\tau \gg (1/\Delta v)$, where Δv is the width of the laser line, $\phi(t - \tau)$ and $\phi(t)$ are statistically independent. The bandwidth of the beat signal, which can be measured with a radio-frequency spectral analyzer, is then twice the width of the laser line.

10.11   Laser Frequency Standards

It is now possible to observe extremely narrow optical resonances in cold atoms or single trapped ions, so that a laser locked to such a narrow optical resonance can serve as an optical frequency standard. Several transitions with intrinsic offsets estimated at the 10^{-16} level have been studied, including the $^3P_1 - {}^1S_0$ transition at $\lambda = 657$ nm of laser-cooled neutral ^{40}Ca atoms [Schnatz *et al.*, 1996] and the $5s^2 S_{1/2} - 4d^2 D_{5/2}$ transition at 674 nm of a single, trapped and laser-cooled ^{88}Sr$^+$ ion [Bernard *et al.*, 1999], both of which have been recommended by the CIPM for the realization of the metre.

A problem in implementing such an optical frequency standard has been the need for a reliable and simple system that can measure optical frequencies. Frequency chains (see also Section 7.8) starting with a cesium atomic clock and generating higher harmonics in nonlinear diode mixers and crystals [Schnatz *et al.*, 1996; Bernard *et al.*, 1999] are quite complex and require substantial resources to build and operate.

A completely new approach to optical frequency measurements is the use of the comb of modes of a mode-locked laser to measure, or synthesize, optical frequencies with extreme precision. Since the mode spacing is equal to the pulse repetition rate with an experimental uncertainty of a few parts in 10^{16}, such a comb can be used to measure large optical frequency differences by counting modes, if the spacing between them is known [Eckstein, Ferguson, and Hänsch, 1978]. In addition, since the spectral width of the comb scales inversely with the pulse duration, the advent of femtosecond lasers has made it possible to provide a direct link between optical and microwave frequencies [Udem *et al.*, 1999a; Diddams *et al.*, 2000].

We consider a a laser cavity (length L) with a pulse circulating in it. The output consists of a carrier frequency f_c modulated by an envelope function $A(t)$. This envelope function defines the pulse repetition time T and the pulse repetition frequency $f_r = T^{-1}$. Fourier transformation of $A(t)$ shows that the resulting spectrum consists of a comb of laser modes, separated by the pulse repetition frequency f_r and centered at the carrier frequency f_c [Reichert *et al.*, 1999]. The spectral width of the frequency comb can be broadened, while maintaining the mode spacing, by self phase modulation in a nonlinear optical material.

Since f_c is not necessarily an integer multiple of f_r, the modes are shifted from frequencies that are exact harmonics of f_r by an offset f_o, so that the frequency of the nth mode is

$$f_n = nf_r + f_o, \tag{10.26}$$

where n is a large integer. This equation maps two radio frequencies, f_r and f_o, onto the optical frequencies f_n.

The repetition frequency f_r is readily measurable. In addition, with a frequency comb extending over more than one octave, it is possible, as shown in Fig. 10.7, to

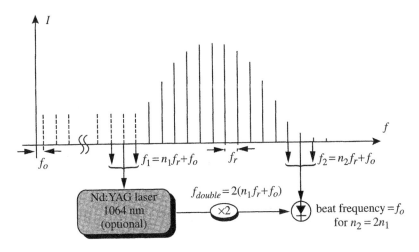

Figure 10.7. Measurement of the beat frequency between a frequency-doubled "red" component of a frequency comb and a "blue" component yields the frequency offset f_o [Holzwarth *et al.*, 2000].

obtain the frequency offset f_o from measurements of the beat frequency between a frequency-doubled "red" component and a properly chosen "blue" component. It is then possible to relate all optical frequencies f_n contained within the comb to the radio frequencies f_r and f_o, which may be locked to an atomic clock. This allows the direct comparison of radio and optical frequencies, with an estimated precision of 5×10^{-16}, without the need of additional optical frequency dividers or nonlinear optics [Udem *et al.*, 1999b; Holzwarth *et al.*, 2000].

Chapter 11

Fourier-Transform Spectroscopy

As outlined in Sections 3.6 and 3.12, the spectral energy distribution of a source can be obtained from measurements of the complex degree of coherence in a two-beam interferometer, as the path difference between the interfering wavefronts is varied.

Michelson made use of this fact when he inferred the structure of several spectral lines from a study of their visibility curves. However, measurements of the visibility only give the modulus of the Fourier transform of the spectral energy distribution, and a unique spectral distribution can be obtained from the visibility curve only when the spectral distribution is symmetrical with respect to a central frequency.

In Fourier-transform spectroscopy (FTS), the intensity at a point in the interference pattern is recorded as a function of the delay in the interferometer (the interference function). The variable part of the interference function (called the interferogram), on Fourier transformation, then gives the spectrum without any such ambiguity. One of the earliest spectra obtained in this manner is shown in Fig. 11.1, along with the corresponding interferogram [Fellgett, 1958]. This technique has been described in detail in a number of reviews and books [see Vanasse and Sakai, 1967; Bell, 1972; Chamberlain, 1979; see also Traub, Winkel, and Goldman, 1996] and is now used very widely because of the etendue and multiplex advantages.

11.1   The Etendue and Multiplex Advantages

An interferometer in which the interference pattern exhibits radial symmetry, such as the FPI, has the advantage over a prism or grating spectrograph of a much greater etendue (the etendue, or Jacquinot, advantage; see Section 10.3). However, when such an interferometer is operated in the scanning mode, to make full use of the available etendue, the total scanning time T is divided between, say, m elements of the spectrum, so that each element is observed only for a time equal to (T/m). If the

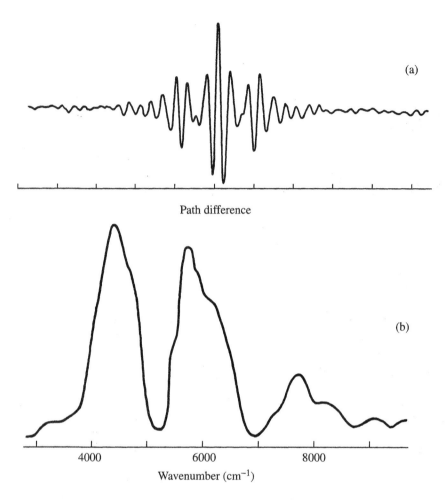

Figure 11.1. (a) Interferogram and (b) spectrum obtained with the 3.5 magnitude star α Herculis [Fellgett, 1958].

main source of noise is the detector, which is often the case in the far infrared, the noise power is independent of the signal. This results in a reduction of the signal-to-noise (S/N) ratio by a factor $m^{1/2}$, compared to the situation if each spectral element were recorded over the full time T.

The Michelson interferometer has the same etendue advantage as the FPI. In addition, when the delay is made to vary linearly with time, each element of the spectrum gives rise to an output which is modulated at a frequency inversely proportional to its wavelength. It is therefore possible to record all these signals simultaneously (or, in other words, to multiplex them) and decode them later to obtain the spectrum. Since each spectral element is now recorded over the full scan time T, an improvement in the S/N ratio by a factor of $m^{1/2}$ over a conventional scanning instrument is obtained. This is known as the multiplex (or Fellgett) advantage.

11.2 Theory

The main component of a Fourier-transform spectrometer is typically, as shown in Fig. 11.2, a Michelson interferometer illuminated with an approximately collimated beam. With monochromatic light, the output from the detector is

$$G(\tau) = g(\nu)(1 + \cos 2\pi \nu \tau)$$
$$= g(\nu) + g(\nu) \cos 2\pi \nu \tau, \tag{11.1}$$

where τ is the delay, and

$$g(\nu) = l(\nu)T(\nu)d(\nu) \tag{11.2}$$

is the product of three spectral distributions: the radiation studied $l(\nu)$, the transmittance of the spectroscope $T(\nu)$, and the detector sensitivity $d(\nu)$. When τ varies

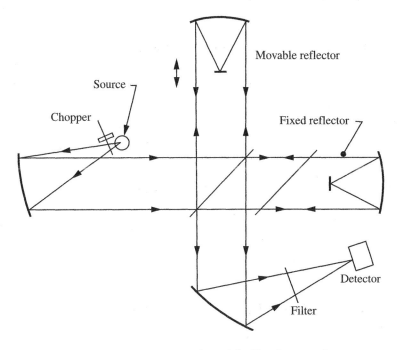

Figure 11.2. Michelson interferometer adapted for Fourier-transform spectroscopy.

linearly with time, $G(\tau)$ exhibits a cosinusoidal variation of constant amplitude extending to $\tau = \pm\infty$.

If, however, the source has a large spectral bandwidth, the output of the detector is

$$G(\tau) = \int_0^\infty g(v)dv + \int_0^\infty g(v) \cos 2\pi v \tau dv. \tag{11.3}$$

Since the first term on the right-hand side of Eq. (11.3) is a constant, the variable part of the output, which constitutes the interferogram, is

$$F(\tau) = \int_0^\infty g(v) \cos 2\pi v \tau dv. \tag{11.4}$$

The interferogram initially exhibits large fluctuations when $|\tau|$ is increased from 0, since all the spectral components are in phase when $\tau = 0$, but the amplitude of these fluctuations drops off rapidly as $|\tau|$ increases.

The actual spectral distribution $g(v)$ exists only for $v > 0$. However, for convenience, we can represent this asymmetrical function, as shown in Fig. 11.3, as the sum of an even component $g_e(v)$ defined by the relation

$$g_e(v) = g_e(-v) \tag{11.5}$$

and an odd component $g_o(v)$ defined by the relation

$$g_o(v) = -g_o(-v), \tag{11.6}$$

so that

$$g(v) = g_e(v) + g_o(v). \tag{11.7}$$

In addition, since the two components are numerically the same,

$$|g_e(v)| = |g_o(v)|. \tag{11.8}$$

Accordingly, Eq. (11.4) can be rewritten as

$$F(\tau) = \int_{-\infty}^\infty [g_e(v) + g_o(v)] \cos 2\pi v \tau dv$$

$$= 2 \int_0^\infty g_e(v) \cos 2\pi v \tau dv. \tag{11.9}$$

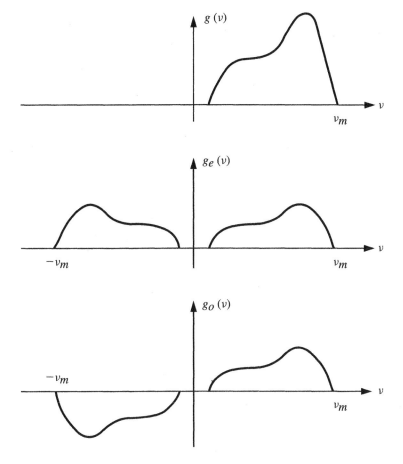

Figure 11.3. Decomposition of the physical spectrum $g(\nu)$ into an even component $g_e(\nu)$ and an odd component $g_o(\nu)$.

Fourier inversion of Eq. (11.9) then gives

$$g_e(\nu) = 2 \int_0^\infty F(\tau) \cos 2\pi \nu \tau \, d\tau. \qquad (11.10)$$

However, if we consider only positive values of ν, $g_e(\nu)$ and $g_o(\nu)$ are identical and their sum is equal to $g(\nu)$. Hence, over this range ($\nu > 0$),

$$g_e(\nu) = (1/2)g(\nu), \qquad (11.11)$$

so that

$$g(v) = 4 \int_0^\infty F(\tau) \cos 2\pi v\tau d\tau. \qquad (11.12)$$

11.3 Resolution and Apodization

The calculation of $g(v)$ from Eqs. (11.10) and (11.12) requires a knowledge of the interferogram over an infinite range of values of τ, but, in any practical case, the range of the interferogram is limited by the distance over which the end mirror of the interferometer can be moved. The effect of limited mirror travel can be taken into account by rewriting Eq. (11.10) as

$$g_e(v) = 2 \int_0^\infty A(\tau) F(\tau) \cos 2\pi v\tau d\tau, \qquad (11.13)$$

where $A(\tau)$ is a function which drops to zero at the maximum allowed value of τ. In the frequency domain, Eq. (11.13) becomes

$$g_e'(v) = a(v) * g_e(v), \qquad (11.14)$$

where (see Appendix A.1)

$$a(v) \leftrightarrow A(\tau). \qquad (11.15)$$

A comparison of Eq. (11.15) with Eq. (10.1) shows that $a(v)$ is the instrumental function of the spectrometer.

For the simplest case in which the interferogram is truncated at $\pm\tau_m$, $A(\tau)$ is the rectangular function

$$A(\tau) = \begin{cases} 1, & \text{when } |\tau| \leq \tau_m, \\ 0, & \text{when } |\tau| > \tau_m, \end{cases} \qquad (11.16)$$

and the instrumental function is

$$a(v) = \text{sinc}(2v\tau_m). \qquad (11.17)$$

If we take the resolution limit Δv to be the separation of the first zero of this function from its peak, we have

$$\Delta v = 1/2\tau_m. \qquad (11.18)$$

The sinc function is not a desirable instrumental function, since it does not drop smoothly to zero, and the side lobes due to a strong spectral line could be mistaken for other weak spectral lines. These side lobes can be eliminated by a process known as apodization [Jacquinot and Roizen-Dossier, 1964], which involves multiplying the interferogram with a weighting function that progressively reduces the contribution of greater delays and thereby eliminates the sharp cut-off at $\tau = \tau_m$. A linear taper is commonly used, in which case,

$$A(\tau) = \begin{cases} 1 - |\tau|/\tau_m, & |\tau| \leq \tau_m \\ 0, & |\tau| > \tau_m, \end{cases} \tag{11.19}$$

and

$$a(v) = \text{sinc}^2(v\tau_m). \tag{11.20}$$

The side lobes are reduced appreciably, but at the expense of a doubling of the width of the instrumental function, and a corresponding loss in resolving power. Other apodization functions and their effects on the instrumental function have been reviewed by Vanasse and Sakai [1967].

11.4 Sampling

When calculating the Fourier transform of an interferogram by means of a digital computer, it is necessary to know the minimum number of points at which the interferogram must be sampled as well as the minimum number of points for which the transform must be computed to recover an undistorted spectrum. This problem was analyzed by Strong and Vanasse [1959] and, later, by Connes [1961]. We make use of the fact that if an interferogram recorded over a range of delays from 0 to τ_m is sampled at M points, that is to say with a sampling interval τ_m/M, the recovered spectrum is repeated, as shown in Fig. 11.4, at a frequency interval M/τ_m. For a physical spectral distribution covering the range of frequencies 0 to v_m, the recovered spectrum $g_e(v)$, defined by Eq. (11.10), has a spectral bandwidth of $2v_m$. Accordingly, to ensure that there is no overlap (aliasing) of successive repetitions of the recovered spectrum, we must have

$$2v_m \leq M/\tau_m, \tag{11.21}$$

so that the minimum number of points at which the interferogram must be sampled is

$$M = 2v_m\tau_m. \tag{11.22}$$

In the spectral domain, the number of points N for which the transform must be computed is given by the bandwidth v_m divided by the resolution limit Δv. From Eq. (11.18),

$$N = v_m\Delta v$$
$$= 2v_m\tau_m. \tag{11.23}$$

Hence, from Eq. (11.22), N is equal to M.

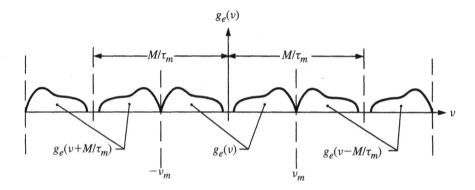

Figure 11.4. Repetition of the recovered spectrum $g_e(\nu)$ when the interferogram is sampled at an interval $\tau_m/M = 1/2\nu_m$.

11.5 Effect of Source and Detector Size

Equation (11.18), which defines the resolution limit, is valid only when the interferometer is illuminated with a perfectly collimated beam. If the source and the detector subtend a finite angle 2α at the entrance pupil and the exit pupil, respectively (see Fig. 10.1), there will be, between the axial ray and a limiting off-axis ray, an additional phase difference

$$\Delta\phi = 2\pi\nu\tau(1 - \cos\alpha)$$
$$\approx \pi\nu\tau\alpha^2$$
$$\approx \nu\tau\Omega, \tag{11.24}$$

since $\Omega = \pi\alpha^2$. As a result, the calculated frequency ν' is no longer the true frequency ν, but is given by the relation

$$\nu' = \nu(1 - \Omega/4\pi). \tag{11.25}$$

In addition, there is a progressive decrease in the visibility $\mathcal{V}$ of the fringes, with increasing delay. It can be shown [Terrien, 1958] that

$$\mathcal{V}(\tau) = \text{sinc}(\Omega\nu\tau/2\pi), \tag{11.26}$$

so that the fringes disappear when

$$\Omega = 2\pi/\nu\tau. \tag{11.27}$$

The maximum solid angle which the source and the detector can subtend at the collimating optics is then, from Eqs. (11.27) and (11.18),

$$\Omega_m = 2\pi/\mathcal{R}, \qquad (11.28)$$

where $\mathcal{R} = \nu/\Delta\nu$ is the resolving power.

11.6 Field Widening

It is possible to increase Ω_m beyond the limit given by Eq. (11.28) by using a system in which, as in the spherical FPI, emerging rays derived from the same incident ray are coincident and not merely parallel. This permits the delay to be increased, while keeping the shift (see Section 3.10) zero. Systems for this purpose have been described by Connes [1956] and by Bouchareine and Connes [1963]. Such systems can give an increase of etendue by a factor as large as 100, but have other drawbacks which limit their use.

11.7 Phase Correction

If the interferometer is not perfectly compensated (or, in other words, has residual dispersion in one path), Eq. (11.4) is no longer valid and the interferogram is represented by the relation

$$F_a(\tau) = \int_0^{\nu_m} g(\nu) \cos[2\pi\nu\tau + \phi(\nu)] d\nu. \qquad (11.29)$$

In such a case, the modulus $g_e(\nu)$ can be recovered by taking the absolute value of the Fourier transform of $F_a(\tau)$. A better method is to generate the required symmetric interferogram from the measured asymmetric interferogram by convolution with a phase correction function. Since the phase error is a smooth, slowly varying function, it can be computed separately from a short, two-sided interferogram [Forman, Steel, and Vanasse, 1966].

11.8 Noise

Noise can be classified in two categories: noise arising in the detector or interferometer, which is added to the signal (additive noise), and noise associated with the incident radiation, whose power varies with the signal level (multiplicative noise). The latter can again be subdivided into photon noise and signal noise. Photon noise arises from the random fluctuations in the number of photons arriving in a given time interval and is therefore proportional to the square root of the signal; signal noise, on the other hand, arises from fluctuations in the transmittance of the medium between the source and the detector, or from fluctuations in the output of the source itself, and is, therefore, proportional to the signal.

As shown in Section 11.1, the multiplex advantage applies for additive noise, which is independent of the signal power. With photon noise, it can be shown that multiplexing results in no advantage, while with signal noise, multiplexing is at a disadvantage when compared to sequential measurement. Accordingly, the multiplex advantage disappears in the visible and ultraviolet regions of the spectrum, where the noise is largely either photon noise or signal noise. Fourier-transform spectroscopy is therefore most useful in the far infrared, where the energy of individual photons is low, and detector noise is usually the limiting factor.

11.9   Prefiltering

With a source emitting over a wide spectral band, the dynamic range required to record an interferogram is very large. This is because all the frequencies contribute to a grand maximum at zero path difference, while relatively small variations of power at large path differences correspond to weak absorption lines. To ensure that information on such weak features is not lost, the dynamic range of the recording system must be greater than the S/N ratio in the interferogram. If this condition cannot be met, it is necessary to use optical filters to limit the spectral bandwidth studied and thereby reduce the dynamic range required.

Dynamic range is not so much of a problem now, since digital recording systems with a discrimination better than 1 part in 10^5 are available. However, even with such systems, a problem which can be encountered is aliased noise.

If the detector is sensitive up to a frequency ν_D which is higher than the maximum frequency ν_{max} up to which the spectrum is to be computed, the interferogram will contain a significant amount of photon noise and signal noise contributed by higher frequencies which are not of interest. When the interferogram is sampled and Fourier transformed, noise belonging to the adjacent repetitions of the spectrum, or aliased noise (see Fig. 11.4), will then overlap the recovered spectrum and degrade it.

Aliased noise can be eliminated, in principle, by the use of a filter which has unit transmittance between frequencies of ν_{min} and ν_{max} and zero transmission elsewhere. The Fourier transform of such a filter is a sinc function. Accordingly, if the original interferogram is convolved with this sinc function before it is transformed, all noise at frequencies outside the range of interest will be eliminated and cannot be aliased back. This procedure is called mathematical filtering [Connes and Nozal, 1961].

11.10   Interferometers for FTS

The most widely used interferometer for FTS is the Michelson interferometer. In the near infrared, thin films of Ge or Si on BaF or CaF_2 plates are used as beamsplitters.

Because of the difficulty of finding suitable beamsplitters which could be used in the far infrared, Strong and Vanasse [1958, 1960] developed an alternative in the lamellar-grating interferometer. This consists essentially of two mirrors broken up into strips so that one mirror can pass through the other. However, this interferometer must be used with a slit source narrow enough to ensure that adjacent strips are coherently illuminated; in addition, at longer wavelengths, the delay is determined by the velocity of the radiation in the slots in the mirrors. Subsequently, it was shown that a thin film of Mylar could be used as a beamsplitter in the far infrared; wire-mesh beamsplitters have also been used. A theoretical study of some new designs for multilayer beamsplitters based on coated pellicles as well as solid substrates has been made by Dobrowolski and Traub [1996].

In either case, the slide carrying the moving mirror must be of very high quality to avoid tilting of the mirror; this problem can be minimized in the Michelson interferometer by replacing the mirrors by cube corners or by "cat's-eye" reflectors [Connes and Connes, 1966].

Two approaches to the movement of the mirror (scanning) have been followed. In one, known as periodic generation, the mirror is moved repeatedly over the desired scanning range at a sufficiently rapid rate that the fluctuations of the output due to the passage of the fringes occur at an audio frequency. This technique has the advantage that ac amplification of the signal is possible without any additional modulation. A number of values of the output obtained for each value of the delay are then averaged digitally to give the final interferogram. In the other approach, known as aperiodic generation, the mirror is moved only once, relatively slowly, over the scanning range and the detector output is recorded at regular intervals. In this mode, the signal level is essentially steady during each observation.

Aperiodic generation is possible either with a continuous movement of the mirror or with a stepped movement. In the former case, the average value of the detector output is recorded during a brief time interval centered on each sampling point. This has the advantage that the drive system can be very simple. In the latter case, the mirror remains stationary during the sampling time and then moves rapidly to the next position, thus making maximum use of the observing time.

Most systems have used motorized drives combined with voice coils or piezo-electrics. An electromechanical linear actuator using a piezoelectric device which overcomes the disadvantages of such drives has been described by Sandford, Luck, and Rohrbach [1996].

With either system of aperiodic generation, it is necessary to use some form of flux modulation so that ac amplification and synchronous detection can be used. The earliest spectrometers used amplitude modulation, usually by means of a chopper, but a better method is phase modulation [Chamberlain, 1971; Chamberlain and Gebbie, 1971]. In this technique, the mirror, which usually is fixed, is made to oscillate to and fro at a frequency f so that, at any position of the stepped mirror, the delay varies with time according to the relation

$$\tau(t) = \tau + \Delta\tau \cos 2\pi f t. \tag{11.30}$$

For any input frequency ν, the output from the detector is then

$$I(t) = I_0\{1 + \cos 2\pi\nu[\tau + \Delta\tau \cos 2\pi f t]\}. \tag{11.31}$$

The right-hand side of Eq. (11.31) can be expressed as the sum of a number of harmonics; with a filter that passes only the fundamental, the output signal is

$$I'(\tau) = 2I_0 J_1(2\pi\nu\Delta\tau) \sin 2\pi\nu\tau. \tag{11.32}$$

Phase modulation has the advantage over amplitude modulation that, since the beam is not interrupted by the chopper, the reduction in output is minimal.

An adaptive digital filter permitting flexible fringe subdivision (to increase the free spectral range) as well as compensation for variations in the drive velocity has been described by Brault [1996].

Analog-to-digital (A-D) conversion with a high dynamic range and low quantization error is essential for measurements of weak emission and absorption signals. Usually, these requirements are satisfied through the use of high bit-number amplitude sampling A-D converters. An alternative, which permits the use of a 1-bit A-D converter, involves locating the intersections $\{z_i\}$ of the modulation term $s(z)$ of the interferogram and a reference sinusoid $r(z) = A \cos(2\pi f_r z)$, where z is the optical path difference [Daria and Saloma, 2000].

11.11   Computation of the Spectrum

The total number of operations involved in computing a Fourier transform by conventional routines is approximately $2M^2$, where M is the number of points at which the interferogram is sampled. When M is large, this leads to quite long computing times. Because of this, there were several early experiments exploring the use of analog techniques. However, digital computing has now become so much faster and cheaper that these early attempts are only of historic interest. In addition, the fast Fourier-transform (FFT) algorithm [Cooley and Tukey, 1965; Forman, 1966] has reduced the number of operations to $3M \log_2 M$. In an extreme case, when $M \approx 10^6$, the computing time is reduced by a factor of 25,000, so that the computation of even very complex spectra becomes feasible. Procedures for computation have been discussed in detail by Bell [1972].

11.12   Applications

Rapid scanning FTS has found many applications in infrared spectroscopy, including emission spectra and chemiluminescence, absorption spectra of aqueous solutions and transient species, and studies of the kinetics of chemical reactions [Bates, 1976]. In addition, because of their high etendue, Fourier-transform spectrometers can be used to record spectra from very faint sources. For this reason they have been used extensively in studies of planetary atmospheres [Connes and Connes, 1966; Hanel and Kunde, 1975] and night sky emission [Baker, Steed, and Stair, 1973]. Because of their high resolving power, they have also been used in studies of the molecular spectra of gases at low pressures [Guelachvili, 1978].

Other applications include contactless temperature measurements, measurements of spectral emissivity, and measurements of the concentrations of pollutants. Calibration of an FTS for such applications using three blackbody sources has been described by Lindermeir et al. [1992].

More recently, FTS has been used for measurements from space of the infrared solar spectrum [Abrams et al., 1996a] and remote sensing of 30 constituents of the earth's atmosphere [Abrams et al., 1996b]. A possibility which is being explored is the exploitation of the multiplex advantages of FTS to detect the acoustic oscillations of giant planets and stars [Maillard, 1996]. Fast time-resolved FTS has been used to record emission spectra from the products of laser-initiated transient chemical reactions [Carere, Neil, and Sloan, 1996].

Even though the multiplex advantage does not hold in the visible region, Fourier-transform spectroscopy retains the advantages over a prism or grating of high resolving power and high accuracy of measurement of wavelengths, and has been found very useful for the study of complex molecular spectra [Luc and Gerstenkorn, 1978]. Figure 11.5, which shows a small section of the absorption spectrum of iodine obtained with a Fourier transform spectrometer having an effective resolving power of 5×10^5, gives an idea of the reproducibility and low noise possible with FTS.

Excellent wavelength accuracy is available with FTS, since wavelength uncertainties arise primarily from the finite size of the aperture and detector and imperfect alignment of the sample and reference beams. Some practical aspects of calibration with wavelength standards to eliminate errors due to these causes are discussed by Salit, Travis, and Winchester [1996].

The combined advantages of high throughput, broad spectral coverage, high spectral resolution, and accurate wavelength calibration make ultraviolet and visible FTS an attractive technique for measurements of minor tropospheric constituents [Vandaele and Carleer, 1999] including the OH radical, which is a key intermediate in the catalytic ozone destruction cycle [Cageao et al., 2001].

Another interesting application has been in absolute radiometry. Mather et al. [1990] used a Martin–Puplitt polarizing Michelson interferometer as a far infrared absolute spectrophotometer (FIRAS), on the Cosmic Background Explorer (COBE), to record the spectrum of the background radiation between 1 and 20 cm^{-1} (1 cm to 0.5 mm wavelength), with a resolution of 1 cm^{-1}, from regions near the north galactic pole. The spectrum is well fitted by a blackbody with a temperature of 2.735 ± 0.06 K, with a deviation less than 1% of the peak intensity.

The FIRAS instrument also carried out the first all-sky spectral line survey in the far infrared region, as well as mapping spectra of the galactic dust distribution at $\lambda > 100$ μm with a 7° beam. The spectrum observed by FIRAS contained two major

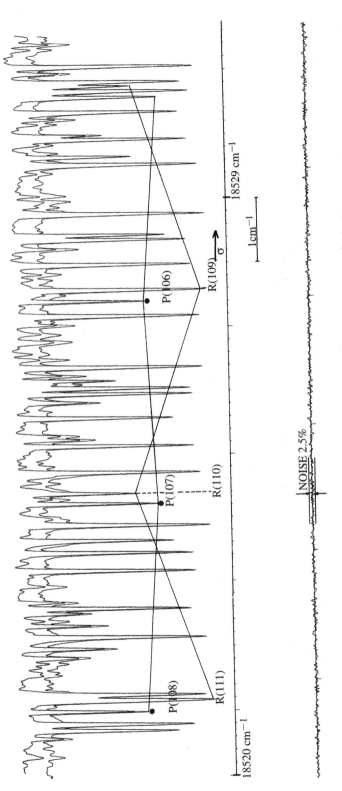

Figure 11.5. A small section of the absorption spectrum of iodine in the visible ($\lambda \approx 540$ nm) obtained with a Fourier transform spectrometer (effective resolving power about 5×10^5). The two independent traces demonstrate the reproducibility and low noise of the method [Luc and Gerstenkorn, 1978].

components: a continuous spectrum due to interstellar dust heated by starlight, and a line spectrum dominated by the strong 158-μm line from singly ionized carbon, with a spatial distribution similar to the dust distribution and a luminosity of 0.3% of the dust luminosity [Boggess *et al.*, 1992].

FTS has also been used to obtain accurately calibrated absolute brightness temperature spectra of planets [Serabyn and Weisstein, 1996].

Chapter 12

Interferometric Sensors

Advances in the technology of lasers and optical fibers have opened up the use of interferometers as sensors for a number of physical quantities.

12.1 Rotation Sensors

The use of an interferometer to detect rotation in an inertial frame dates back to experiments by Sagnac and by Michelson, using an interferometer with two beams traveling around the same closed circuit in opposite directions (see the reviews by Post [1967]; Hariharan [1975c]; Anderson, Bilger, and Stedman [1994]). In this case, both classical theory and the theory of relativity predict a fringe shift with rotation.

Sagnac was the first to demonstrate, in 1913, the feasibility of such an experiment with an interferometer of the form shown in Fig. 2.13b. When the whole interferometer, including the light source, was set rotating with an angular velocity ω about an axis making an angle θ with the normal to the plane of the interferometer, a fringe shift was observed corresponding to the introduction of a delay between the two beams

$$\tau = 4\omega A \cos\theta / c^2, \tag{12.1}$$

where A was the area enclosed by the light path. The fringe shift given by Eq. (12.1) could be doubled by comparing the fringe positions for rotation at the same speed in opposite directions. Sagnac also established that the effect did not depend on the shape of the loop or the position of the center of rotation.

Subsequently, in 1925, Michelson and Gale succeeded in demonstrating the rotation of the Earth by means of a similar experiment. To obtain the necessary sensitivity, they used a rectangular, 60 m × 15 m path, enclosed in evacuated tubes. Since the rate of rotation could not be changed, the fringe shift was measured by

189

comparing the position of the fringes in the main circuit with the fringes formed in a comparison circuit enclosing a smaller area.

12.1.1 Ring Lasers

Much higher sensitivity can be obtained by measurements of the optical beat frequency in a ring laser [Rosenthal, 1962; Macek and Davis, 1963], such as that shown schematically in Fig. 12.1. Rotation of a ring laser shifts the frequencies of the clockwise-propagating (CW) and counterclockwise-propagating (CCW) modes by equal amounts in opposite senses. A very stable beat can be obtained, because the two modes use the same optical cavity and are affected equally by any temperature changes and mechanical disturbances.

Because of the presence of the active medium within the ring and back scattering at the mirrors, the two modes in a ring laser are prone to locking at very low rotation rates [Aronowitz, 1971]. Some solutions to this problem have been discussed by Roland and Agrawal [1981]. A technique commonly used to eliminate mode locking is to dither the beat frequency by an angular oscillation of the ring laser with a constant small amplitude [Killpatrick, 1967]. An alternative is the four-frequency ring laser in which a magnetic field forces the cavity to support two pairs of counterpropagating waves with opposite circular polarizations. The difference between the beat

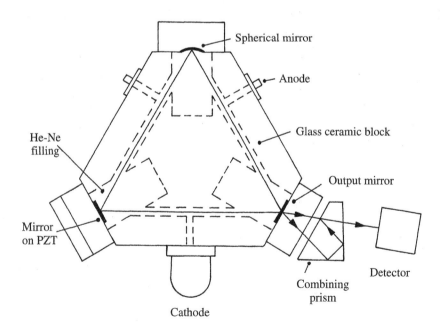

Figure 12.1. Ring laser used for rotation sensing [Roland and Agrawal, 1981].

frequencies then corresponds to the rotation signal [Chow *et al.*, 1980; Roland and Agrawal, 1981; Statz *et al.*, 1985].

The theoretical limit of sensitivity for the ring laser is given by the relation [Dorschner *et al.*, 1980]

$$\Delta\Omega = \lambda_0 P \Gamma_c / 4A(n_{ph}\tau)^{1/2}, \tag{12.2}$$

where λ_0 is the vacuum wavelength, P is the optical perimeter and Γ_c is the line width of the cavity, n_{ph} is the photon flux in the laser beam, and τ is the averaging time. With care, performance close to the sensitivity limit set by spontaneous emission in the gain medium can be obtained.

Ring-laser rotation sensors have the advantages over mechanical gyroscopes of fast warm-up, rapid response, large dynamic range, insensitivity to linear motion, and freedom from cross-coupling errors in multiaxis sensing. For these reasons they have been used widely in missile guidance and inertial navigation systems [see Chow *et al.*, 1985].

12.1.2 Ring Interferometers

The problem of mode locking can be avoided by using a passive ring interferometer with an external laser and measuring the differential shift in the resonant frequency for the two directions of propagation [Ezekiel and Balsamo, 1977]. One scheme uses only one laser frequency and a Faraday cell within the cavity to cancel out any difference in the optical path lengths for the counterpropagating beams. Another uses a single laser and two acousto-optic modulators to generate two independently controlled optical frequencies which are locked to the CW and CCW resonant frequencies of the ring [Sanders, Prentiss, and Ezekiel, 1981].

The relative merits of these and other techniques have been reviewed in a number of papers [see Ezekiel and Knausenberger, 1978; Ezekiel and Arditty, 1982].

12.2 Fiber Interferometers

The use of single-mode optical fibers to build analogs of conventional two-beam interferometers became practical with the development of lasers. Since such a fiber does not allow higher-order modes to propagate, a smooth wavefront is obtained at the output end. Fiber interferometers can give very high sensitivity for some types of measurements, because they make possible very long paths with very low noise, in a small space, permitting the use of sophisticated detection techniques.

12.2.1 Rotation Sensors

The first application of fiber interferometers was in rotation sensing, where a closed multiturn fiber-optic loop was used, instead of a conventional cavity with mirrors,

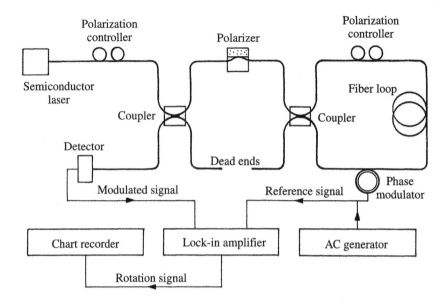

Figure 12.2. All-fiber interferometric rotation sensor [Bergh, Lefevre, and Shaw, 1981a].

to increase the effective area of the ring [Vali and Shorthill, 1976; Leeb, Schiffner, and Scheiterer, 1979].

Very small phase shifts can be measured and the sense of rotation determined by introducing a nonreciprocal phase modulation and using a phase-sensitive detector [Ulrich, 1980]. A wider measurement range can be obtained with a closed loop system, in which the phase difference caused by rotation is compensated by a nonreciprocal phase shift generated within the ring by suitably positioned acousto-optic frequency shifters [Cahill and Udd, 1979; Davis and Ezekiel, 1981].

A typical system based entirely on optical fibers [Bergh, Lefevre, and Shaw, 1981a, 1981b] is shown in Fig. 12.2. In this system, normal beamsplitters are replaced by optical couplers, and a few turns of the fiber wound around a piezoelectric cylinder are used as a phase modulator.

Fiber-optic rotation sensors have the advantages of small overall size and low cost and are an attractive alternative to ring lasers for many applications. An analysis of the limits to their sensitivity has been made by Lin and Giallorenzi [1979]. One of the factors ultimately limiting their sensitivity is noise due to back scattering, which can be controlled by a phase modulator placed in one of the beams before it enters the resonator [Sanders, Prentiss, and Ezekiel, 1981] or by using a broadband source such as a superluminescent diode [Bergh, Lefevre, and Shaw, 1984]. Another is nonreciprocal effects due to fiber birefringence and the earth's magnetic field, which can be minimized by using a single-polarization fiber [Burns et al., 1984]. Very small phase shifts can be measured by using heterodyne techniques, in conjunction with a piezoelectric phase modulator placed near one end of the fiber coil [Kim and Shaw, 1984b; Kersey, Lewin, and Jackson, 1984]. If precautions are taken to minimize the effects of vibration, local temperature variations, and external magnetic fields, performance close to the limit set by photon noise can be attained [Bergh, Lefevre, and Shaw, 1984].

Another method of nulling the Sagnac phase shift is by applying a sawtooth modulation to the phase modulator and adjusting its amplitude or frequency [Kim and Shaw, 1984a]. This method can be implemented with an integrated-optic phase modulator [Ebberg and Schiffner, 1985; Chien, Pan, and Chang, 1991]. The phase shift can also be measured by mixing a train of square pulses with the photodetector current and adjusting the pulse spacing, using an electronic closed loop [Toyama *et al.*, 1991].

12.2.2 Sensors for Physical Quantities

Other applications of fiber interferometers use the fact that the optical path length changes when a fiber is stretched and is also affected by the ambient pressure and temperature, so that an optical fiber can be used as a sensor for a number of physical quantities [Culshaw, 1984; Culshaw and Dakin, 1989] (see also Giallorenzi *et al.* [1982]; Kyuma, Tai, and Nunoshita [1982]).

While early fiber interferometers used gas lasers, diode lasers are now employed widely because, apart from their small size and high output, they can operate at a wavelength ($\lambda \approx 0.9$ μm) at which the losses in silica fibers are much lower than in the visible region. Optical arrangements similar to the Michelson and Fabry–Perot interferometers have been used [Imai, Ohashi, and Ohtsuka, 1981; Yoshino *et al.*, 1982a], but the optical layout shown in Fig. 12.3, which is analogous to the Mach–Zehnder interferometer, has the advantage that it avoids light being reflected back to the laser, and therefore eliminates optical feedback [Giallorenzi *et al.*, 1982].

In this arrangement, normal beamsplitters have been replaced by twisted-fiber optical couplers [Sheem and Giallorenzi, 1979], permitting an all-fiber arrangement with a considerable reduction in noise. Measurements are usually carried out by a phase-tracking system or by a heterodyne system [Jackson, Dandridge, and Sheem, 1980; Dandridge and Tveten, 1981]. Phase modulation is introduced by an integrated-optic phase shifter or by a fiber stretcher [Jackson *et al.*, 1980] in the reference path and the resultant output signal is picked up by a photodetector followed by a demodulator. Slow drifts can be eliminated by a phase-tracking system using an electronic feedback loop, or by tuning the frequency of the diode laser source [Dandridge and Tveten, 1982]. Optical phase shifts as small as 10^{-6} radian can be detected with such a system [Dandridge and Tveten, 1981].

Other techniques which have been used include a modulated laser source [Gilles *et al.*, 1983] and laser-frequency switching [Kersey, Jackson, and Corke, 1983a]. A variety of modulation and signal processing schemes based on measurements of

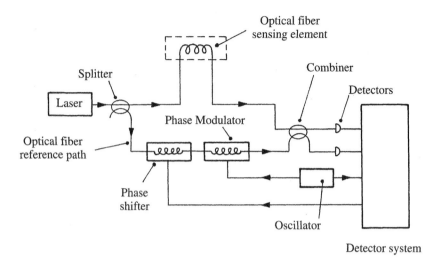

Figure 12.3. Schematic of an interferometer using a single-mode fiber as a sensing element [Giallorenzi *et al.*, 1982]. © 1982, IEEE; reproduced with permission.

the amplitude, phase or frequency of the output signal have been described [see Chien, Chang, and Chang, 1997].

Miniature Fabry–Perot fiber optic sensors have also found many applications [Yoshino *et al.*, 1982b]. They have the advantages of high sensitivity and lead insensitivity, since light is transmitted to and from the sensor through the same fiber [Kersey, Jackson, and Corke, 1983b; Jackson and Jones, 1987]. With low-reflectance mirrors, the output exhibits a cosine variation with cavity length, simplifying data processing. Extrinsic fiber-optic sensors use an air gap between the fiber end and a reflector [Murphy *et al.*, 1991]. Intrinsic fiber-optic sensors use internal mirrors formed in a continuous length of fiber by fusion-splicing a second fiber onto a mirror-coated fiber end [Inci *et al.*, 1992]. A simple signal processing technique that can be used with miniature low-finesse Fabry–Perot sensors is based on obtaining quadrature outputs by using two different wavelengths [Wright, 1991; Ezbiri and Tatam, 1995; Potter, Ezbiri, and Tatam, 1997].

A technique which has received considerable attention is fiber-optic low-coherence interferometry [Rao and Jackson, 1996]. In the most commonly used configuration, shown schematically in Fig. 12.4, light from a broadband source is launched into one arm of a bidirectional fiber coupler. The output optical signal from the sensing interferometer is coupled via a fiber link into a reference interferometer. The optical path differences (OPDs) in the two interferometers are larger than the coherence length of the light but, since they are almost equal, an interference signal is obtained which is a function of the difference of the OPDs. Any change in the OPD in the sensing interferometer results in a change in the fringe visibility as well as

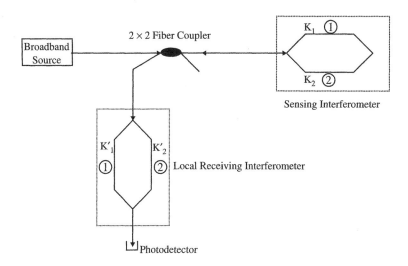

Figure 12.4. Schematic of a fiber-optic low-coherence interferometer [Rao and Jackson, 1996].

a change in the phase of the signal, and can be evaluated by tracking the reference interferometer.

Sensors using fiber Bragg gratings have the advantage that the output is encoded directly as a wavelength shift. As a result, the output is not affected by changes in the intensity. The wavelength shift induced in the sensing element can be measured by placing the sensing element and a reference fiber-optic grating in the output legs of an unbalanced Mach–Zehnder interferometer. With a broadband source, each grating reflects a narrow spectral band, and the wavelength shift can be evaluated from measurements of the changes in the phases of the two interference signals [Kersey, Berkoff, and Morey, 1993].

The earliest applications of fiber interferometer sensors were in the measurement of strains and temperature differences [Bucaro, Dardy, and Carome, 1977; Butter and Hocker, 1978; Hocker, 1979; Jackson, Dandridge, and Sheem, 1980; Lacroix *et al.*, 1984].

A device which has been found very suitable for temperature measurements is the reflective fiber Fabry–Perot interferometer shown in Fig. 12.5 [Leilabady *et al.*, 1986]. This instrument uses a section of a highly birefringent single-mode fiber as the sensing element, so that two separate interferometers are formed with the same fiber. The outputs from the two detectors correspond to the phase retardations for the fast and slow eigenmodes of the fiber. Since the phase change in each interferometer induced by a temperature change in the fiber, as well as the differential phase change, can be recovered separately, it is possible to obtain high resolution along with a greatly increased ambiguity-free measurement range. With a 70-mm sensing element, a phase resolution of 1 mrad corresponds to a change in temperature of less than 0.1 mK; at the

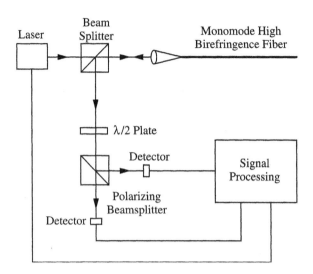

Figure 12.5. Interferometric sensor using a single highly birefringent single-mode fiber [Leilabady *et al.*, 1986].

same time, the ambiguity-free measurement range, corresponding to a differential phase shift of 2π, is about 9 K.

Another application is in acoustic pressure sensing. The sensitivity of a fiber sensing element for this application can be increased by a factor of almost 100, at low frequencies, by using a compliant coating [Hocker, 1979]. Acoustic sensitivities of -30 dB (relative to 1 μPa) have been achieved with a 10-m sensing element.

A compact ultrasound sensor with high sensitivity can be obtained by using a Fabry–Perot cavity mounted on the fiber tip [Beard and Mills, 1996]; a more versatile system uses a metal-coated fiber tip, which forms the mirror in the measurement arm of a Michelson interferometer [Koch, 1999].

All-fiber interferometers in which a single-mode fiber is the sensing element have been used as accelerometers [Tveten *et al.*, 1980; Kersey, Jackson, and Corke, 1982; Bucholtz, Kersey, and Dandridge, 1986]. A miniature hemispherical Fabry–Perot interferometer addressed through a single-mode optical fiber has also been used for this purpose [Gerges *et al.*, 1989]. The sensing element in this arrangement is a spherical mirror attached to the center of a loaded elastic diaphragm, and the end of the fiber constitutes the other mirror of the interferometer. A resolution of 5 μGal was obtained with this system.

Fiber interferometers can be used to measure magnetic fields by placing a magnetostrictive jacket on the fiber sensor, or by bonding it to a magnetostrictive element [Yariv and Winsor, 1980; Dandridge *et al.*, 1980; Rashleigh, 1981; Willson and Jones, 1983]. Nickel was used initially as the magnetostrictive material, but several metallic glasses have higher magnetostrictive coefficients [Koo and Sigel, 1984]. A sensitivity of 4×10^{-7} A m^{-1} can be obtained with a 1-m sensing element. Similarly, electric fields can be measured by using a fiber bonded to a piezoelectric film, or with a jacket made of a piezoelectric polymer [Koo and Sigel, 1982; De Souza and Mermelstein, 1982]. Currents in high-voltage transmission lines can be measured by using a current transformer terminated with a small resistive load. The resulting potential difference drives a piezoelectric ring which, in turn, applies a strain, proportional to the current, to an optical fiber [Ning and Jackson, 1992]. Higher stability and simpler signal processing is possible with a system using a fiber ring interferometer that detects the derivative of the current [Swart and Spammer, 1996].

12.2.3 Multiplexed Fiber-Optic Sensors

Where measurements have to be made of various quantities at a particular location, or a particular quantity at different locations, it is advantageous to multiplex several optical-fiber sensors in a single system. Such a multiplexed system avoids duplication of light sources, fiber transmission lines, and photodetectors. The techniques developed for this purpose include frequency-division multiplexing, time-division multiplexing, and coherence multiplexing.

In frequency-division multiplexing the frequency of the source is ramped linearly, and the system is set up so that the difference in the lengths of the two optical paths is different for each sensor. When the outputs arrive at the detector, they produce a unique heterodyne frequency for each sensor [Gilles *et al.*, 1983; Sakai, 1986; Sakai, Parry, and Youngquist, 1986].

We consider the case when the angular frequency $\omega(t)$ of the laser source is a saw-tooth function of time given by the relation

$$\omega(t) = \omega_0 + a(t - nT_s), \tag{12.3}$$

where ω_0 is the initial angular frequency, a is the sweep rate, n is an integer, and T_s is the period of the sweep. If Δt is the differential delay between the two paths in the interferometer, a pseudo-heterodyne output is obtained from the detector at a frequency $a\,\Delta t$ with a phase $\omega_0 \Delta t$. A small change in the optical path difference in the interferometer results in a negligible change in the frequency of this beat signal, but a measurable change in its phase [Jackson *et al.*, 1982]. To avoid a discontinuity at the end of each sweep, it is convenient to make the beat frequency an integral multiple of the sweep frequency. This condition can be met for all the interferometers if the optical path differences are proportional to a set of integers. The outputs from the individual interferometers can then be recovered by bandpass filtering of the multiplexed output at the appropriate frequencies.

In time-division multiplexing, the optical input is pulsed and each sensor produces an output pulse which, as a consequence of the system geometry, is separated in time from the other sensor signals. The pulses travel back on the return fiber bus to the central processor, where they are separated and processed [Brooks *et al.*, 1987].

A typical arrangement [Farahi *et al.*, 1988a] uses a diode laser source which is driven by a gated saw-tooth current to produce pulses during which the laser frequency

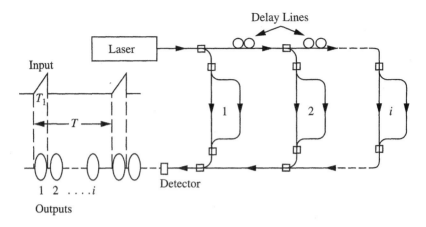

Figure 12.6. Time-division multiplexed system using modulation of the laser injection current by a gated ramp [Farahi *et al.*, 1988a].

is ramped linearly. As shown in Fig. 12.6, delay lines are introduced between the interferometers to separate the pulses by intervals greater than their duration. If the time between successive pulses is chosen to be greater than the sum of all the delays, the outputs from the individual interferometers can be separated in the time domain. Information on the phase can then be recovered from each of the individual heterodyne signals. Frequency-division multiplexing can also be combined with time-division multiplexing to increase the number of interferometric sensors in a multiplexed system [Farahi, Jones, and Jackson, 1988].

In coherence multiplexing, a continuous source of light with a short coherence length is used with a set of sensing interferometers with different optical path differences, all much greater than the coherence length of the light. As a result, interference cannot take place between the two waves from each sensing interferometer. Phase information is recovered by a set of interferometers in the receiver, whose path differences are matched, on a one-to-one basis, with those of the sensing interferometers [Brooks *et al.*, 1985]. Coherence multiplexing can be used with a variety of signal processing techniques including homodyne and pseudo-heterodyne techniques [Farahi *et al.*, 1988b].

A number of multiplexing techniques that can be used for fiber-optic sensors in which the sensing element is a ring interferometer have been described by Farahi *et al.* [1989].

12.3   Laser-Feedback Interferometers

Laser-feedback interferometers [Ashby and Jephcott, 1963; Clunie and Rock, 1964] make use of the fact that if, as shown in Fig. 12.7, a mirror M_3 is used to reflect the

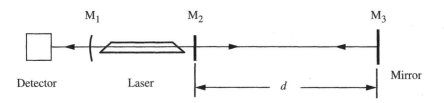

Figure 12.7. Laser feedback interferometer.

output beam back into the laser cavity, the laser output varies cyclically with d, the distance of M_3 from M_2, the laser mirror nearest to it.

This effect, commonly termed self-mixing interference (SMI), can be analyzed [Clunie and Rock, 1964; Peek, Bolwijn, and Alkemade, 1967] by considering M_2 and M_3 as a Fabry–Perot interferometer, which effectively replaces M_2 as the output mirror of the laser. The reflectance of this Fabry–Perot interferometer is

$$R = \frac{r_2^2 + r_3^2(1 - A_2)^2 - 2r_2r_3(1 - A_2)\cos\phi}{1 + (r_2r_3)^2 - 2r_2r_3\cos\phi}, \tag{12.4}$$

where r_2 and r_3 are, respectively, the reflectances for amplitude of M_2 and M_3, A is the loss due to scattering and absorption in M_2, and $\phi = 4\pi nd/\lambda$ is the phase difference between successive interfering beams. It follows from Eq. (12.4) that the value of R is a maximum when $nd = m\lambda/2$, where m is an integer, and drops to a minimum when $nd = (2m + 1)\lambda/4$. Typically, with a laser mirror having a transmittance of 0.008, the output can be made to vary by a factor of four by using an external mirror with a reflectance of 0.1.

The response of such a system decreases at high frequencies because of the finite time required for the amplitude of the laser oscillation to build up within the cavity. Measurements with a He–Ne laser and a spinning reflector have shown that the depth of the modulation decreases by 50% at a modulation frequency of 100 kHz. A detailed analysis of the factors contributing to the finite response time has been made by Hooper and Bekefi [1966].

If the laser is oscillating on a single transition, the resultant change in the gain merely enhances or diminishes the laser power on this transition. If, however, the laser is oscillating simultaneously on two transitions which share a common upper or lower level, and if M_3 reflects only one of these wavelengths, the output power at the other wavelength will vary in antiphase to the power at the first wavelength. This effect can be observed with a He–Ne laser and has been used to measure plasma densities at 3.39 μm with a detector sensitive only to radiation at 633 nm [Ashby et al., 1965].

An n-fold increase in sensitivity can be obtained by passing the beam back and forth n times within the external cavity by means of a spherical external mirror [Gerardo and Verdeyen, 1963]. A system that is easier to align and allows the isolation of a selected higher-order beam consists of a focusing lens used in conjunction with a plane laser-output mirror and a plane return mirror [Heckenberg and Smith, 1971]. Increased sensitivity can be obtained for measurements on plasmas by using the 10.6-μm line from a CO_2 laser, to take advantage of the larger dispersion of the electrons at this longer wavelength [Herold and Jahoda, 1969]. Another application has been in measurements of low-frequency changes in the optical path length [Bearden *et al.*, 1993] where high sensitivity can be obtained by using a set of phase-shifted intensity measurements [Ovryn and Andrews, 1999].

The main advantages of laser-feedback interferometers are very easy alignment, inherent mechanical stability, and high sensitivity, since a small change in the gain results in a large change in the output. A problem is that when the phase change exceeds π, it is difficult to determine its sign. One solution is to use a laser oscillating on two longitudinal modes that are orthogonally polarized. If the ratio of the lengths of the external cavity and the laser cavity are properly chosen, these signals can be separated and used to obtain phase data in quadrature over a limited displacement range [Timmermans, Schellekens, and Schram, 1978].

12.3.1   Diode-Laser Interferometers

A very compact laser-feedback interferometer can be set up with a single-mode GaAlAs laser and an external mirror whose position is to be monitored [Dandridge, Miles, and Giallorenzi, 1980]. In one mode of operation, the laser current is held constant, and the output power is monitored by a photodiode; alternatively, the drive current to the laser to maintain a constant output power is measured. Additional flexibility in applications is possible by using a single-mode fiber to couple the external mirror to the laser.

Since the response of such a system is not linear, its useful range is limited. Greater dynamic range, as well as higher sensitivity, can be obtained by mounting the mirror on a PZT and using an active feedback loop to hold the optical path from the laser to the mirror constant, at a suitable fixed point on the linear section of the power vs distance curve [Yoshino *et al.*, 1987].

Another way to determine the absolute distance to a target is by varying the frequency of the laser, as a linear function of time, by modulating the drive current, and measuring the frequency of the beat signal produced by self-mixing. de Groot, Gallatin, and Macomber [1988] have shown that this beat signal cannot be explained simply in terms of the coherent mixing of two light waves with different frequencies and have presented an analysis of the actual processes involved. de Groot *et al.* [1989] have also described a very simple interferometer, using self-mixing in a diode laser, for measuring the aberrations of optical components and optical systems.

A phase-locked diode laser interferometer using self-mixing interference has been described by Suzuki *et al.* [1999a]. In this system, absolute distance was measured by using a triangular modulation of the injection current and measuring the frequency of the SMI signal under feedback control. Small displacements were measured by changing the laser wavelength, using the feedback control on the injection current, so as to cancel the phase change.

The self-mixing effect can also be used for measurements of distances with a three-electrode semiconductor laser, from observations of the modulation of the optical power as the output wavelength of the laser is continuously tuned [Mourat, Servagent, and Bosch, 2000].

12.4 Doppler Interferometry

12.4.1 Laser-Doppler Velocimetry

Laser-Doppler velocimetry makes use of the fact that light scattered from tracer particles introduced in a moving fluid has its frequency shifted. This frequency shift can be measured using the beats produced by interference between the scattered light and a reference beam [Yeh and Cummins, 1964], or between the scattered light from two illuminating beams incident at different angles [Durst and Whitelaw, 1971]. With the optical system shown in Fig. 12.8, the frequency of the beat signal is given by the relation

$$f = \frac{2|v| \sin \theta}{\lambda}, \tag{12.5}$$

where v is the component of the velocity of the particle, in the plane of the beams, at right angles to the direction of observation. If the frequency of one of the beams is offset initially by a known amount by means of an acousto-optic modulator, it is possible to distinguish between positive and negative flow directions [Eberhardt and Andrews, 1970]. It is also possible to measure the velocity components along two orthogonal directions simultaneously by using two pairs of illuminating beams in orthogonal planes. In this case, orthogonal polarizations, or different laser wavelengths, are used for the two pairs of beams to avoid any interaction between them.

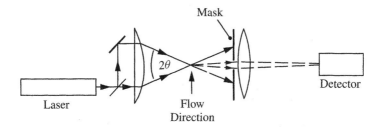

Figure 12.8. Optical arrangement used for laser-Doppler velocimetry.

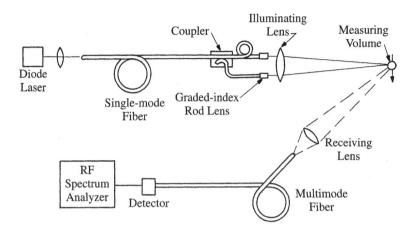

Figure 12.9. All-fiber system for laser-Doppler velocimetry.

Laser-Doppler velocimetry can be used for measurements of the instantaneous flow velocity at any point without inserting a probe, which could disturb the flow. In addition, it permits measurements in a hostile environment, such as, for example, in studies of combustion. For these reasons, laser-Doppler velocimetry is used widely for studies of fluid flows in industry as well as in research [Durst, Melling, and Whitelaw, 1976; Adrian, 1993]. Techniques for estimating velocity spectra from laser-Doppler data have been reviewed by Benedict, Nobach, and Tropea [2000].

Very compact systems for laser-Doppler velocimetry can be built using optical fibers [Sasaki *et al.*, 1980]. A single-mode fiber is used, as shown in Fig. 12.9, to connect the laser source to the probe, and beamsplitters are replaced by directional couplers. A frequency offset can be introduced between the beams by using a piezo-electric phase shifter, driven by a saw-tooth waveform, in one path [Chan, Jones, and Jackson, 1985]. Alternatively, a diode laser can be used as the source, and its injection current can be swept linearly [Jones *et al.*, 1982].

The self-mixing effect has also been used for laser-Doppler velocimetry with a He–Ne laser [Rudd, 1968] and with a CO_2 laser [Churnside, 1984]. Measurements of higher velocities can be made with a very simple arrangement using the self-mixing effect in a diode laser [Shinohara *et al.*, 1986; Jentink *et al.*, 1988].

12.4.2   Surface Velocities

Laser-Doppler velocimetry has also been used for noncontact measurements of surface velocities by making the polished surface of the moving specimen one of the end mirrors in a Michelson interferometer [Barker and Hollenbach, 1965]. The output is then modulated at a frequency corresponding to the Doppler shift. However, such a simple setup is limited to low velocities.

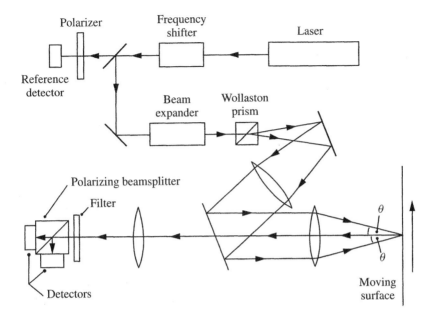

Figure 12.10. Optical system of a laser-Doppler interferometer used to measure the velocity of moving surfaces [Truax, Demarest, and Sommargren, 1984].

The modulation frequency can be reduced by using an arrangement in which the laser beam is reflected off the specimen before entering the interferometer, where it is split into two beams, one of which is subjected to a delay before it is recombined with the other [Barker, 1971]. The total fringe count at any time is then proportional to the average velocity over the delay interval, while the fringe frequency is proportional to the acceleration [Kamegai, 1974]. High velocities, as well as rapid variations in velocity, can be measured by using interference between two beams incident on the surface at different angles [Maron, 1977, 1978].

Figure 12.10 shows a system used to measure the velocity of moving materials, such as hot-rolled steel, in order to cut it to given lengths [Truax, Demarest, and Sommargren, 1984]. In this system, two orthogonally polarized beams produced by a Wollaston prism are focused on the surface of the moving material, by a lens, at angles of incidence of $\pm\theta$. The scattered light from the surface goes to a polarizing beamsplitter oriented at 45° which sends half of each polarization to two detectors. The frequency of the beat signal at the two detectors is then

$$\nu_s = (2/\lambda)|\upsilon_p| \sin \theta, \tag{12.6}$$

where υ_p is the component of the velocity of the material parallel to its surface; however, their phase differs by 180°. Accordingly, if the difference of the two signals

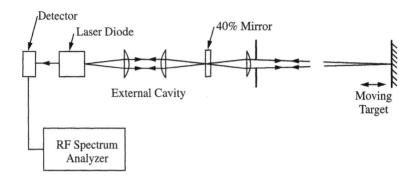

Figure 12.11. Self-mixing configuration using a diode laser for Doppler velocimetry [de Groot and Gallatin, 1989].

is taken, the effects of variations of surface reflectivity are eliminated, while the Doppler signal is doubled.

A system using a Fourier-transform lens and two detectors that permits simultaneous measurement of the velocity and the displacement of a moving rough surface has been described by Matsubara *et al.* [1997].

It is also possible to use a self-mixing configuration for velocimetry, in which the light reflected from the moving object is returned to the laser and mixed with the original oscillating wave inside the laser cavity. The beam leaving the rear end of the laser is incident on a photodetector, whose output contains the beat signal. This technique was first studied with the He–Ne laser and then with the CO_2 laser [Churnside, 1984]. However, a more compact system is possible with a diode laser, and measurements can be made at higher velocities [Shinohara *et al.*, 1986]. In addition, the shape of the output wave form can be used to discriminate between positive and negative Doppler shifts [Shimizu, 1987].

A greatly enhanced modulation depth leading to much greater sensitivity can be obtained if the diode laser is operated near threshold, but a problem then is a sharp reduction in the coherence length and, hence, in the operational range of the velocimeter. This problem can be overcome by using, as shown in Fig. 12.11, an external cavity to ensure single-mode operation of the diode laser [de Groot and Gallatin, 1989]. Measurements can be made with such a system at ranges up to 50 m.

12.5 Vibration Measurements

Laser-Doppler techniques can also be used for noncontacting measurements of vibration amplitudes. Typically, one of the beams in a Michelson interferometer is reflected from a point on the vibrating surface, while the other is reflected from a fixed mirror [Deferrari, Darby, and Andrews, 1967]. Measurements are facilitated by offsetting the frequency of one beam, by a known amount, by diffraction at an acousto-optic modulator. The output from a photodetector then consists of a component at the off-set frequency (the carrier) and two side bands. The amplitude of the vibration can be determined by a comparison of the amplitudes of the carrier and the side bands, while the phase of the vibration can be obtained by comparison of the carrier with a reference signal [Puschert, 1974].

Heterodyne techniques make it possible to measure vibration amplitudes down to a few thousandths of a nanometre at frequencies above 50 kHz. At lower frequencies, the accuracy is reduced by low-frequency $(1/f)$ noise. However, the effects of noise can be minimized by shifting the desired signal to a higher frequency. One way to do this is, as shown in Fig. 12.12, to use a laser beam containing two frequency-shifted components produced by an acousto-optic modulator [Ohtsuka and Sasaki, 1977; Ohtsuka and Itoh, 1979]. The output from the interferometer is then given by an expression of the form

$$I(t) = (1/2)\{I_R + I_S + 2(I_R I_S)^{1/2} \cos[\phi_R - \phi_S(t)]\}$$
$$+ (1/2)\{I_R + I_S + 2(I_R I_S)^{1/2} \cos[\phi_R - \phi_S(t)]\} \cos 2\pi f_m t, \qquad (12.7)$$

where I_R and I_S are the intensities of the reference beam and the signal beam, ϕ_R is the constant part and $\phi_S(t)$ the time-varying part of the phase difference between

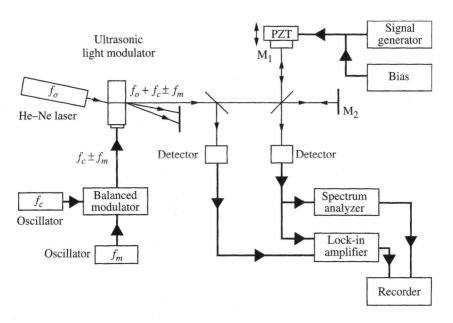

Figure 12.12. Laser interferometer for measurements of low-frequency vibrations [Ohtsuka and Itoh, 1979].

the two beams, and f_m is the frequency difference between them. The second term of Eq. (12.7) corresponds to a signal at the frequency f_m, which can be separated from the low-frequency noise. Information on the low-frequency displacements of the object is contained in the variations of the amplitude of this signal. Vibration amplitudes of 0.3 nm can be measured at a frequency of 0.2 Hz [Ohtsuka and Tsubokawa, 1984].

Optical fibers can be used to set up very compact systems for noncontacting measurements of vibration amplitudes. However, conventional frequency-shifting devices cannot be used conveniently with optical fibers. One method of avoiding this problem is to use an optical arrangement similar to a Mach–Zehnder interferometer in which the phase of one beam is modulated sinusoidally by wrapping the fiber around a piezoelectric cylinder [Lewin, Kersey, and Jackson, 1985]. If the amplitude of the phase modulation is properly chosen, it is possible to recover a signal of exactly the same form as a conventional heterodyne signal from the output of the detector. With this setup, the highest vibration velocity that can be measured is limited by the modulation frequency applied to the piezoelectric cylinder. Operation at a much higher modulation frequency is possible by using a diode laser as the light source and sinusoidally modulating the drive current. This technique also makes it possible to use a very simple optical system, analogous to a Fabry–Perot interferometer, formed by the free end of the fiber and the target. A feedback loop controls the amplitude of the modulation applied to the drive current to produce the required frequency excursion of the laser for the actual working distance of the probe from the target [Laming *et al.*, 1985].

12.6   Interferometric Magnetometers

As described in Section 12.3, fiber interferometers can be adapted to measure magnetic fields by using a magnetostrictive jacket on the fiber sensor or by bonding it to a magnetostrictive element. However, interferometric sensing of displacements, combined with microelectromechanical-systems (MEMS) technology has now made possible compact interferometric magnetometers with extremely high sensitivity.

The active element in these magnetometers is a silicon bar (length $l_b = 0.5$ mm) supported at the nodes of its fundamental vibration mode. An alternating current I whose frequency matches the fundamental vibration frequency of the bar passes through the bar. An external magnetic field B then generates a Lorentz force

$$F = l_b\, I\, B \sin\theta, \tag{12.8}$$

where θ is the angle between I and B. This magnetic force, which is along an axis at right angles to the surface of the bar, causes the bar to vibrate at the frequency of the driving current with an amplitude that is linearly proportional to the drive current, the magnetic field, and the mechanical quality factor of the bar [Givens *et al.*, 1999].

The vibration amplitude is measured by a system similar to a Michelson interferometer. Two orthogonal silicon bars, driven at different frequencies, are used as the end mirrors, providing vectorial information on the magnetic field [Oslander *et al.*, 2001]. Light from a diode laser ($\lambda = 675$ nm) passes through a beamsplitter that transmits 50% of the incident beam to each of the silicon bars. The reflected beams from the bars are recombined at the same beamsplitter and directed to a photodetector. The optical path to one bar is modulated with an amplitude of 5 nm at a fixed frequency (f_{ref}) of 1 kHz by mounting the bar on a piezoelectric transducer.

The output of the photodetector is monitored by three independent lock-in amplifiers. The signal at the second harmonic ($2f_{ref}$) of the piezoelectric modulation frequency is used for active stabilization of the mean value of the optical path difference, while the signals at the resonant frequencies of the two bars are processed to obtain information on the two orthogonal components of the magnetic field. Very small field variations (a few nT), which are superimposed on the much larger background (10 μT) of the Earth's magnetic field, can be sensed with this system.

12.7 Adaptive Optics

Adaptive optics make it possible to sense changes in wavefront aberrations and to correct them in real time. A typical application is in correcting the distortions introduced by turbulence in the earth's atmosphere ("seeing") in the images of astronomical objects formed by a ground-based telescope.

The earliest adaptive optical systems used a lenslet array and a set of position-sensitive detectors (a Hartmann–Shack sensor) to measure distortions of the incoming wavefront, and a deformable mirror driven by an array of actuators to which correction signals could be applied [see Tyson, 1991].

An alternative approach was the interference phase loop proposed by Fisher and Warde [1979, 1983], in which the distorted wavefront was combined with a reference wavefront in an interferometer incorporating a phase modulator driven by a signal derived from the intensity at the output of the interferometer. The system operated with negative feedback so that the phase introduced by the modulator was the conjugate of that due to the distortions in the incoming wavefront.

A significant improvement was the use of a radial shearing interferometer to measure the wavefront distortions, without the requirement of a separate, aberration-free reference wavefront. With a multipixel liquid-crystal spatial light modulator (LCSLM) as the phase-modulating array, it then became possible to correct the wavefront aberrations directly, in real time, without the need for extensive external computation [Barnes, Eiju, and Matsuda, 1996].

This system had to be used with polarized light, since only the extraordinary refractive index of a nematic liquid crystal could be varied by the application of an electric field. This limitation was overcome by using the LCSLM in the reflection mode [Love, 1993; Bold *et al.*, 1998]. A quarter-wave plate (QWP), set at 45° between the LCLSM and a mirror, rotated the planes of polarization of both components by 90°, so that the component that experienced the ordinary index on the outward pass experienced the extraordinary index on the return pass, and vice versa. Compensation over a wide wavelength range, resulting in improved performance with white light, can be obtained by replacing the QWP with an achromatic combination of a QWP and a half-wave retarder [Ciddor and Hariharan, 2001].

Chapter 13

Nonlinear Interferometers

The availability of lasers capable of producing short pulses of light of extremely high intensity has opened up completely new areas of optics based on the use of nonlinear optical elements [see Butcher and Cotter, 1990; Boyd, 1992]. Two such areas are second-harmonic interferometry and phase-conjugate interferometry.

13.1   Second-Harmonic Interferometry

Second-harmonic interferometers produce a fringe pattern corresponding to the phase difference between two second-harmonic waves generated at different points in the optical path from the original input wave at the fundamental frequency. Several optical systems, which are analogs of the Fizeau, Twyman–Green, and Mach–Zehnder interferometers, have been described that can be used for this purpose [Hopf, Tomita, and Al-Jumaily, 1980; Hopf and Cervantes, 1982].

A second-harmonic interferometer, which may be called a nonlinear Twyman–Green interferometer, is shown schematically in Fig. 13.1a. This interferometer can be used to test an optical element made of a material, such as silicon, which transmits only in the near infrared and obtain an interferogram with visible light. In this instrument, the beam from a Q-switched Nd:YAG laser ($\lambda_1 = 1.06\,\mu$m) is incident on a frequency-doubling crystal. The green ($\lambda_2 = 0.53\,\mu$m) and infrared beams emerging from this crystal are then expanded by a pair of lenses and separated by a dichroic beamsplitter (BS_2), so that each beam traverses one arm of the interferometer. The test piece is placed in the infrared beam while the green beam serves as the reference. When the two beams return to the crystal, the green beam passes through it unchanged, while the infrared beam undergoes frequency doubling to generate a second green beam. Another dichroic beamsplitter (BS_1) then reflects the two green beams to the observation plane.

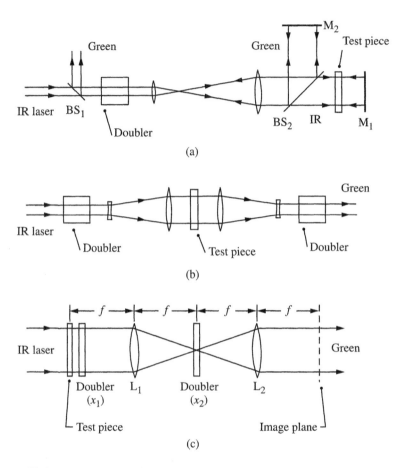

Figure 13.1. Second-harmonic interferometers: (a) nonlinear Twyman–Green interferometer, (b) nonlinear Mach–Zehnder interferometer, and (c) nonlinear point-diffraction interferometer [Hopf, Tomita, and Al-Jumaily, 1980; Liepmann and Hopf, 1985].

In this interferometer, the additional optical path difference between the beams, at any point $P(x, y)$ in the field, due to the test piece is

$$p(x, y) = 2(n_1 - 1)d(x, y), \qquad (13.1)$$

where $d(x, y)$ is the thickness of the test piece at $P(x, y)$, and n_1 is its refractive index at 1.06 μm. The resulting change in the interference order is, therefore,

$$N(x, y) = p(x, y)/\lambda_2$$
$$= 2(n_1 - 1)d(x, y)/\lambda_2. \qquad (13.2)$$

Since the same small fraction of the power in the infrared beam is doubled at each pass, the two green beams have the same intensity, and fringes with good visibility are obtained.

Another interferometer of this type, which may be called a nonlinear Fizeau interferometer, is obtained if the beamsplitter BS_2 and the mirror M_2 in Fig. 13.1a are removed. In this case, since both the infrared and the green beams go through the object, there is, at any point $P(x, y)$ in the field, a difference in their optical paths

$$p(x, y) = 2(n_2 - n_1)d(x, y), \tag{13.3}$$

where n_2 is the refractive index of the test piece at 0.53 μm. As before, when the two beams return to the crystal, the infrared beam undergoes frequency doubling to produce a second green beam which interferes with the first one.

The interference order at a point (x, y) is then

$$N(x, y) = 2(n_2 - n_1)d(x, y)/\lambda_2. \tag{13.4}$$

However, with a conventional Fizeau interferometer using green light, the interference order would have been

$$N'(x, y) = 2(n_2 - 1)d(x, y)/\lambda_2. \tag{13.5}$$

It follows that a desensitized interferogram is obtained, corresponding to an equivalent wavelength

$$\lambda_{eq} = \lambda_2(n_2 - 1)/|n_2 - n_1|. \tag{13.6}$$

An alternative form of this interferometer using two frequency-doubling crystals, shown in Fig. 13.1b, can be regarded as an analog of the Mach–Zehnder interferometer. In this interferometer, the beam from a Q-switched Nd:YAG laser ($\lambda_1 =$ 1.06 μm) is incident on a frequency-doubling crystal, and the green ($\lambda_2 = 0.53$ μm) and infrared beams emerging from this crystal traverse the test piece. When the two beams arrive at the second crystal, the infrared beam undergoes frequency doubling to produce a second green beam which interferes with the green beam produced at the first crystal. The interference order at any point $P(x, y)$ in the field is then

$$N(x, y) = (n_2 - n_1)d(x, y)/\lambda_2, \tag{13.7}$$

where $d(x, y)$ is the thickness of the test specimen and n_1 and n_2 are its refractive indices for infrared and green light, respectively.

An interferometer of this type has advantages in applications such as the study of plasmas, where the dispersion is proportional to the electron density. Alum,

Koval'chuk, and Ostrovskaya [1984] have analyzed some optical configurations and discussed the advantages of such interferometers for this purpose.

Another type of second-harmonic interferometer [Liepmann and Hopf, 1985] has some similarities to the point-diffraction interferometer. In this interferometer, as shown in Fig. 13.1c, the test object is placed just in front of the first frequency-doubling crystal X_1, in the front focal plane of the lens L_1, while the second frequency-doubling crystal X_2 is placed in its back focal plane. The second-harmonic wave generated by the crystal X_2 is very nearly a plane wave, while the second harmonic wave that is generated by the frequency-doubling crystal X_1 is a faithful reproduction of the transmitted wave. However, since its wavelength is half that of the fundamental wave, the phase shifts are doubled. The interferogram obtained is a direct representation of the shape of the object wavefront, with twice the sensitivity of a conventional interferometer.

13.1.1   Critical Phase-Matching

Another group of second-harmonic interferometers uses the phase mismatch resulting from a tilt of the object wavefront in a critically phase-matched frequency-doubling crystal to produce an interference pattern [Hopf et al., 1981].

The complex amplitude of the second harmonic produced by a frequency-doubling crystal is given by the relation

$$a_2 = \frac{8\pi^2 L}{\lambda_1 n_1} q_{\text{eff}}\, a_1^2 \left[\frac{\sin(\Delta k L/2)}{\Delta k L/2} \right] \exp(i\Delta k L/2), \qquad (13.8)$$

where a_1 is the complex amplitude of the incident fundamental wave and λ_1 its wavelength, L is the length of the crystal and n_1 its refractive index at the fundamental wavelength, q_{eff} is the nonlinear coefficient, and the factor Δk is given for a type-I doubler by the expression

$$\Delta k = 2\pi (n_2 - n_1)/2, \qquad (13.9)$$

where n_2 is the refractive index for the second harmonic. For critical phase matching, the angle θ between the normal to the wavefront and the optic axis is set so that $\Delta k = 0$. Any change in the angle θ will then result in a phase mismatch.

The phase mismatch arising from a tilt of the object wavefront can be used to produce a fringe pattern in two ways. One method makes use of the fact that the value of the term $[\sin(\Delta k L/2)/(\Delta k L/2)]$ in Eq. (13.8), and hence, the amplitude of the second harmonic, drops to zero in the directions for which $\Delta k L/2 = m\pi$, where $m\ (\neq 0)$ is an integer, giving rise to dark fringes.

Another method uses two frequency-doubling crystals. If the first crystal is non-critically phase matched, it generates a green wavefront that is a faithful replica of the original infrared wavefront. However, the green wavefront generated by the second, critically phase matched crystal has a phase distortion superimposed on it because of the factor $\exp(i\Delta kL/2)$ on the right-hand side of Eq. (13.8). The interference fringes produced by this phase distortion are a measure of the local changes in slope of the wavefront.

Because of its sensitivity to phase gradients, this type of interferometer can be used to study localized distortions of a wavefront, in cases where the actual phase deviations are too small to be measured by conventional interferometric techniques.

13.2 Phase-Conjugate Interferometers

The conjugate of a light wave may be defined as a wave exhibiting phase aberrations of the opposite sign and traveling in the opposite direction. For example, the conjugate of a wave diverging from a point source would be a wave converging back to the same point. In phase-conjugate interferometers, the wavefront under study is made to interfere with its complex conjugate. As a result, no reference wavefront is necessary, and the sensitivity is doubled.

Two types of interferometers based on the generation of a conjugate wave in real time were proposed by Hopf [1980]. In the first, shown schematically in Fig. 13.2a, a nonlinear crystal located in the signal beam in the upper arm of the interferometer is also illuminated by a coherent pump beam from the same laser. The conjugate wave generated by four-wave mixing in the crystal (see Yariv [1978]) travels down the lower arm of the interferometer to the second beamsplitter, where it is combined with the signal beam to produce the interference pattern.

The other type of interferometer, shown in Fig. 13.2b, uses degenerate three-wave mixing. In this system, the signal wave is imaged by the lens L_1 on a noncritically phase-matched crystal, where it is mixed with an orthogonally polarized pump beam generated from the same laser in a frequency doubler. The interference pattern formed in the output from the crystal is imaged by the lens L_2 into the observation plane.

13.3 Phase-Conjugating Mirrors

Phase-conjugate interferometers similar to Michelson and Mach–Zehnder interferometers, in which one of the mirrors was replaced by a phase-conjugating mirror, were subsequently described by Bar-Joseph *et al.* [1981], Feinberg [1983], and Howes [1986a, 1986b]. A compact design for a phase-conjugate interferometer that is an analog of the Fizeau interferometer is shown in Fig. 13.3 [Gauthier *et al.*, 1989]. As can be seen, the interferometer consists of a conventional partially reflecting mirror placed in front of a phase-conjugate mirror. A single crystal of barium titanate is used as an internally self-pumped phase conjugator.

An early application of a phase-conjugate Twyman–Green interferometer was for testing the collimation of a laser beam [Shukla *et al.*, 1991]. In this case,

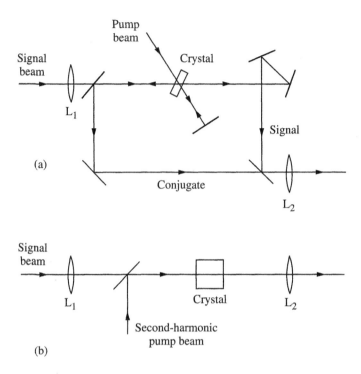

Figure 13.2. Phase-conjugate interferometers using (a) four-wave mixing, and (b) degenerate three-wave mixing [Hopf, 1980].

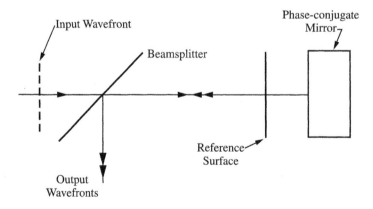

Figure 13.3. Phase-conjugate Fizeau interferometer [Gauthier *et al.*, 1989].

with a diverging wavefront, the reflected wavefront from the plane mirror would also be diverging, while the reflected wavefront from the phase-conjugate mirror would be converging. Interference takes place between two wavefronts of the same radius of curvature but of opposite sign. An improvement is the use of a double mirror to give a split field and increased sensitivity [Darlin, Kothiyal, and Sirohi, 1998].

The nature of the phase shifts of conjugate waves from a self-pumped phase-conjugate mirror (SPPCM) has been investigated by Tomita, Yahalom, and Yariv [1989], who have developed a phenomenological model to explain the experimental fact that spatially uniform phase shifts due to delays are not reversed in sign by the SPPCM, while the relative phases of the partial (plane) waves making up the distorted input wave are reversed. They also showed that the phase shifts of conjugate waves are nearly linear in response to a uniform phase change of one of the inputs, while the rate of change depends on the relative power of the two inputs, indicating cross talk between the input amplitudes and phases and the output phases.

A sinusoidal phase-modulating interferometer using a SPPCM has been described by Wang et al. [1994]. In this arrangement, the image of the object field is formed on the surface of a glass plate, where it interferes with its phase conjugate, and a sinusoidal phase modulation is introduced by vibrating the glass plate. Advantages of this interferometer are that it is self referencing and unaffected by vibration of the object. In addition, its sensitivity is twice that of a conventional interferometer.

The application of a Michelson interferometer with an SPPCM to measure small vibration amplitudes of a rough surface has also been described by Wang et al. [2000]. The distorted wavefront that is reflected from the rough surface is reversed in sign by the SPPCM to cancel the effects of the irregularities of the surface and yield interference with a high signal-to-noise ratio.

Interferometers in which both mirrors have been replaced by phase-conjugating mirrors have some unique properties. The interference pattern obtained in such an interferometer is unaffected by misalignment of the mirrors, and disturbances due to air currents are canceled out. However, the delay in the response of a phase conjugator, such as barium titanate, can be used to display dynamic changes in the optical path difference. An interferometer using a single phase conjugator for this purpose is shown in Fig. 13.4 [Anderson, Lininger, and Feinberg, 1987]. In the steady state, this interferometer is insensitive to spatial variations in the optical path difference, and the field of view is completely dark. However, any sudden local change in the optical path difference results in the appearance of a bright spot in the field. This spot fades away as the phase conjugator responds to the change.

The ability of photorefractive crystals such as BSO to generate a phase-conjugate image in real time has been exploited in the study of vibrating objects [Marrakchi, Huignard, and Herriau, 1980]. In this case, time-average fringes are formed because the crystal integrates over the characteristic time τ_H required to build up a hologram, which is long compared to the period of the vibration [see Hariharan, 1996d].

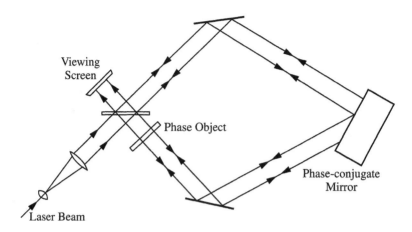

Figure 13.4. Phase-conjugate interferometer used to display dynamic changes in the optical path difference [Anderson, Lininger, and Feinberg, 1987].

13.4   Interferometers with Active Elements

An analysis based on the theory of Lie groups shows that conventional interferometers using only passive elements can be characterized by an SU(2) group symmetry. An alternative class of interferometers using active elements characterized by an SU(1,1) group symmetry [Yurke, McCall, and Klauder, 1986] exploits the fact that the output of an active element, such as a four-wave mixer or degenerate parametric amplifier, depends on the relative phases of the pump and the incoming signal. It should therefore be possible to design interferometers in which beamsplitters are replaced by such active elements, and no light is fed into the input ports.

In one such arrangement, the two beamsplitters of a conventional Mach–Zehnder interferometer are replaced by four-wave mixers, and a phase shifter in one path is adjusted so that the transformation performed by the second four-wave mixer is the inverse of that performed by the first four-wave mixer. When the phase difference between the two beams in the interferometer is zero, no light is delivered to the photodetectors, since the pairs of pump photons converted into pairs of four-wave output photons at the first four-wave mixer are absorbed by the second four-wave mixer and converted back into pump photons. An alternative arrangement uses two degenerate parametric amplifiers. The output of this device is sensitive to the difference between the phases accumulated by the signal and pump beams.

In principle, such interferometers offer the possibility of making measurements of phase changes that are smaller than the standard quantum limit (see Section 15.7).

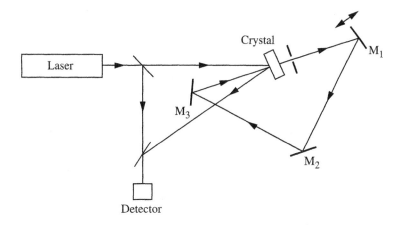

Figure 13.5. Experimental arrangement used to measure frequency detuning of a photorefractively pumped, unidirectional ring oscillator [Kwong *et al.*, 1985].

13.5 Photorefractive Oscillators

The phenomenon of two-beam coupling in a photorefractive crystal has been used to obtain oscillation in a ring cavity [White *et al.*, 1982]. The frequency of this oscillation is, in general, slightly different from that of the pump beam [Feinberg and Bacher, 1984; Rajbenbach and Huignard, 1985].

This effect has been exploited in the interferometer shown in Fig. 13.5 [Kwong *et al.*, 1985], in which the frequency of oscillation varies linearly with the displacement of one of the mirrors over each period, equal to one wavelength of the pump laser, of the change in the optical path.

13.6 Measurement of Third-Order Susceptibility

Interferometric methods have been used for studies of nonlinear susceptibilities, including measurements of the sign and magnitude of the third-order susceptibility $\chi^{(3)}$. Initially, measurements were made with the material under investigation placed in a Mach–Zehnder interferometer [Bliss, Speck, and Simmons, 1974; Moran, She, and Carman, 1975]. More recently, measurements have been made with modified Twyman–Green interferometers [Olbright and Peyghambarian, 1986]. A modified Twyman–Green interferometer which can be used to measure the relative phase shift between two phase conjugators and the ratio of their susceptibilities is shown in Fig. 13.6 [Saltiel, Van Wonterghem, and Rentzepis, 1989].

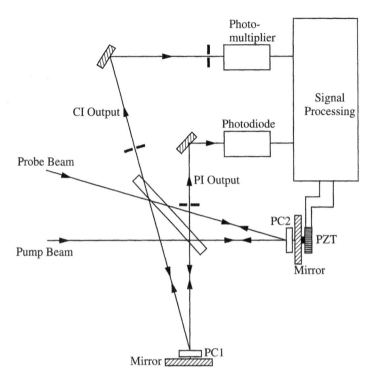

Figure 13.6. Modified Twyman–Green interferometer used for measurements of third-order susceptibility [Saltiel, Van Wonterghem, and Rentzepis, 1989].

In this interferometer, the pump and probe beams are divided at a beamsplitter and interact at the two phase-conjugate mirrors. The signal produced by the recombination of the reflected pump beams at the beamsplitter corresponds to that from a conventional Twyman–Green interferometer (the pump interferometer, PI). In addition, the two phase-conjugate beams generated at the phase-conjugate mirrors interfere to produce a signal corresponding to that from a phase-conjugate interferometer (CI). Alignment of the PI to produce a uniform field automatically aligns the CI.

If a phase difference α_{PI} is introduced between the two arms of the PI, it can be shown that the phase difference between the two beams in the CI is

$$\alpha_{CI} = \alpha_{PI} + \phi_{0,2} - \phi_{0,1}, \tag{13.10}$$

where $\phi_{0,1}$ and $\phi_{0,2}$ are the phase shifts at the two phase-conjugate mirrors. Accordingly, if the absolute value of the phase shift in one of the materials used as a phase conjugator is known, the phase shift in the other can be calculated. In addition, the ratio of the values of $\chi^{(3)}$ for the two materials used as phase conjugators can be determined from the ratio of the maximum and minimum intensities in the interference pattern. High sensitivity is possible with weak signals, since the output corresponds to the interference of the two phase-conjugate waves. If the ratio of the intensities of the two phase-conjugate signals $I_{c,2}/I_{c,1} = q \ll 1$, the variation in the intensity at the CI output is

$$\frac{(I_{\text{Max}} - I_{\text{Min}})}{I_{\text{Av}}} \approx 4\sqrt{q}. \tag{13.11}$$

In a typical case where the ratio of the intensities of the phase-conjugate signals $I_{c,2}/I_{c,1} = 2 \times 10^{-4}$, a modulation of the CI output of 6% is obtained, equivalent to a signal amplification of 300.

13.6.1 Optical Fibers

The change in the refractive index of an optical fiber as a function of the incident optical intensity can be measured using the self-phase modulation effect between the counterpropagating beams in a Sagnac loop [Garvey *et al.*, 1996]. A simpler method which can be implemented with a cw light source such as a laser diode, using the cross-phase modulation effect in a Sagnac fiber loop, has been described by Li *et al.* [1997].

13.7 Optical Switches

Nonlinear optical effects have been exploited to develop high-speed interferometric optical switches.

Silica fiber is the most widely studied material for optical switching, since it has a very short relaxation time, permitting very high switching rates, but its nonlinearity is weak, requiring long interaction lengths. On the other hand, semiconductors exhibit nonlinearities that are typically four orders of magnitude greater, permitting interaction lengths of the order of millimetres. However, free-carrier recombination times normally limit operating rates.

One of the first demonstrations of all-optical switching was by Lattes *et al.* [1983] using an integrated Mach–Zehnder interferometer. Other demonstrations have used a Sagnac interferometer as a nonlinear optical loop-mirror [Doran and Wood, 1988; Jinno and Matsumoto, 1991].

In these interferometric switches, one of the optical inputs is a low-power clock stream (the signal stream) that is split into the two arms of the interferometer and recombined at the output. The other is a high-power stream that is used to induce transmission and phase changes in one arm. For the AND configuration, the interferometer is biased OFF (minimum signal pulse transmission) in the absence of the control stream. The signal pulse is turned ON by the control pulse. In the inverting NOT configuration, the interferometer is biased ON in the absence of the control stream, and switched OFF by the control pulse.

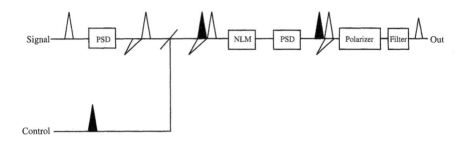

Figure 13.7. Block diagram of a single-arm interferometer. PSDs, polarization-sensitive delays; NLM, nonlinear medium; Filter, used to eliminate the control stream [Patel, Hall, and Rauschenbach, 1998].

Such two-arm interferometers have two problems. The first is that gain differentials between the two light paths can degrade the output contrast. The other is that components of the nonlinear response that relax on longer time scales can make the output dependent on the pattern of the control bits, resulting in bit errors.

The effects of long-lived refractive-index nonlinearities can be circumvented with a single-arm interferometer (SAI) [LaGasse *et al.*, 1989]. As shown schematically in Fig. 13.7, an input-signal pulse enters the device and is split, by a polarization-sensitive delay element, into orthogonal polarizations separated by a pulse width. These two signal components traverse a nonlinear medium into which a control pulse, temporally coincident with the lagging signal pulse, is coupled by a beamsplitter. The two signal components are retimed to overlap by a second polarization-sensitive delay, and then made to interfere by a polarizer set at 45° to the orthogonal signal polarizations. The control pulse is eliminated by a filter at the output.

Since both signal components traverse the same nonlinear medium in the SAI, both of them experience any phase changes that occur over time scales longer than twice the pulse width, whereas only the lagging signal pulse experiences the fast components of the refractive-index nonlinearity induced by the control pulse. As a result, the effects of long-lived refractive-index nonlinearities are eliminated.

In an improved version of the SAI, the control pulse is coupled into the nonlinear medium in a counterpropagating configuration [Patel, Hall, and Rauschenbach, 1996]. This arrangement obviates the need for a filter at the output, allows cascadability, and offers the possibility of higher switching rates [Patel, Hall, and Rauschenbach, 1998].

Chapter 14

Stellar Interferometry

A problem which has attracted the attention of astronomers for centuries is the measurement of stellar diameters. The resolution of conventional telescopes is inadequate for this purpose, since even the largest stars have angular diameters of the order of 10^{-2} arcsecond. On the other hand, interferometric methods are very suitable for such measurements, since a star can be modeled as a planar uncorrelated source. Some of these methods are described in this chapter.

14.1 Michelson's Stellar Interferometer

Since the dimensions of a star are very small compared to its distance from the earth, it follows from the van Cittert–Zernike theorem (see Section 3.5) that the complex degree of coherence between the light vibrations from the star reaching two points on the earth's surface is given by the normalized Fourier transform of the intensity distribution over the stellar disc. The angular diameter of the star can, therefore, be obtained from a series of measurements of the complex degree of coherence at points on the earth separated by different distances. This can be done, as discussed in Section 3.4, by observations of the visibility of the fringes in an interferometer which samples the wave field from the star at these points.

In Michelson's stellar interferometer [Michelson and Pease, 1921], as shown schematically in Fig. 14.1, four mirrors M_1, M_2, M_3, M_4 were mounted on a 6-m long support on the 2.5-m (100 inch) telescope at Mt. Wilson. The light from the star received by the two mirrors M_1, M_2, whose spacing could be varied, was reflected by the fixed mirrors M_3, M_4, to the main telescope mirror, which brought the two beams to a focus at O.

In this arrangement, when the two images of the star are superimposed, and the two optical paths are equalized, straight, parallel interference fringes are seen, whose

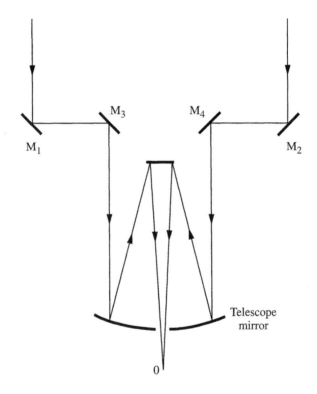

Figure 14.1. Michelson's stellar interferometer.

spacing is

$$\Delta x = \lambda f/d, \tag{14.1}$$

where f is the focal length of the telescope and d is the separation of M_3 and M_4, while their visibility is

$$\mathcal{V} = |\mu_{12}|, \tag{14.2}$$

where μ_{12} is the complex degree of coherence of the wave fields at M_1 and M_2. When M_1, M_2 are close together, $\mathcal{V}$ is nearly equal to unity, but as D, the separation of M_1 and M_2, is increased, $\mathcal{V}$ decreases until, eventually, the fringes vanish.

If we assume the stellar disc to be a uniform circular source with an angular diameter 2α, the visibility of the fringes is, from Eq. (3.60),

$$\mathcal{V} = 2J_1(u)/u, \tag{14.3}$$

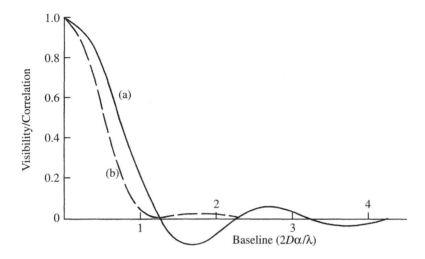

Figure 14.2. Variation with the length of the baseline, for a uniform circular source, of (a) the visibility of the fringes in Michelson's stellar interferometer, and (b) the measured correlation in an intensity interferometer.

where $u = 2\pi\alpha D/\lambda$. The fringe visibility then varies with the separation of the mirrors as shown by the solid line in Fig. 14.2, dropping to zero when

$$D = 1.22\lambda/2\alpha. \tag{14.4}$$

Measurements with Michelson's stellar interferometer present serious difficulties because of two very stringent requirements that must be met to be sure that the observed disappearance of the fringes is actually due to the finite diameter of the star. One is that the optical path difference between the two beams must be small compared to the coherence length of the light. The other is that the optical path difference must be stable to a fraction of a wavelength. The latter condition is very difficult to satisfy, even with a rigid structure, since atmospheric turbulence produces rapid random changes in the two optical paths.

Because of these problems, an instrument with a baseline of 15 m did not give consistent results and, finally, had to be abandoned.

14.2 The Intensity Interferometer

The problems of maintaining equality of the two optical paths to less than a few wavelengths and minimizing the effects of atmospheric turbulence were eliminated in the intensity interferometer [Brown and Twiss, 1956, 1957a, 1957b, 1958a, 1958b].

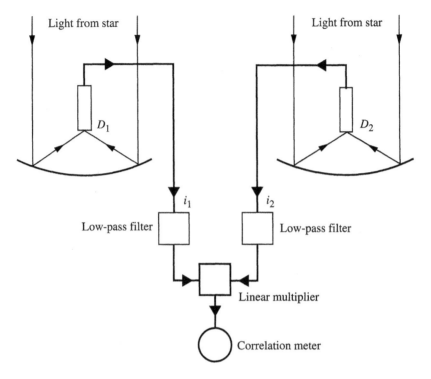

Figure 14.3. Simplified scheme of an intensity interferometer [Brown, 1974].

In this instrument, as shown schematically in Fig. 14.3, light from a star was focused on two photoelectric detectors, which were separated by a distance which could be varied, and the correlation between the fluctuations in the output currents from the two detectors was measured.

The fluctuations in the output currents from the two detectors consist of two components. One is the shot noise associated with the current, while the other, which is smaller and may be called the wave noise, is due to fluctuations in the intensity of the incident light. The shot noise in the two detectors is not correlated, but the wave noise exhibits a correlation which depends on the degree of coherence of the light at the two detectors. To evaluate this correlation, we consider, as shown in Fig. 3.2, two points, P_1 and P_2, illuminated by an extended uncorrelated source. The size of the source and the separation of P_1 and P_2 are assumed to be very small compared to the distance of the source from P_1 and P_2. The intensities at P_1 and P_2 at time t can then be written as

$$I_1(t) = V_1(t)V_1^*(t) \tag{14.5}$$

and

$$I_2(t) = V_2(t)V_2^*(t), \tag{14.6}$$

where $V_1(t)$ and $V_2(t)$ are the analytic signals corresponding to the wave fields at P_1 and P_2. If the interferometer introduces a delay τ between these signals, the correlation (see Appendix A.4) of these signals is

$$\langle I_1(t)I_2(t+\tau)\rangle = \langle V_1(t)V_1^*(t)V_2(t+\tau)V_2^*(t+\tau)\rangle. \tag{14.7}$$

However, since $V_1(t)$ and $V_2(t)$ are complex Gaussian processes, it follows from a theorem of Reed [1962] that

$$\begin{aligned}
\langle V_1(t)&V_1^*(t)V_2(t+\tau)V_2^*(t+\tau)\rangle \\
&= \langle V_1(t)V_1^*(t)\rangle\langle V_2(t+\tau)V_1^*(t+\tau)\rangle \\
&\quad + \langle V_1(t)V_2^*(t+\tau)\rangle\langle V_1^*(t)V_2(t+\tau)\rangle.
\end{aligned} \tag{14.8}$$

Accordingly, from Eqs. (3.7) and (3.19),

$$\langle I_1(t)I_2(t+\tau)\rangle = I_1 I_2 + |\Gamma_{12}(\tau)|^2, \tag{14.9}$$

where I_1 and I_2 are the time-averaged intensities at P_1 and P_2, and $\Gamma_{12}(\tau)$ is the mutual coherence function of the wave fields at P_1 and P_2.

The second term on the right-hand side of Eq. (14.9) corresponds to $R_{12}(\tau)$, the cross-correlation of $\Delta I_1(t)$ and $\Delta I_2(t)$, the fluctuations of intensity about the mean values I_1 and I_2, so that, from Eqs. (14.9) and (3.20), we have

$$\begin{aligned}
R_{12}(\tau) &= \langle \Delta I_1(t)\Delta I_2(t+\tau)\rangle \\
&= |\Gamma_{12}(\tau)|^2 \\
&= I_1 I_2 |\gamma_{12}(\tau)|^2, \tag{14.10}
\end{aligned}$$

where $\gamma_{12}(\tau)$ is the complex degree of coherence of the wave fields at P_1 and P_2. Equation (14.10) shows that the correlation observed at any given separation of the two detectors is proportional to the square of the modulus of the complex degree of coherence of the wave fields at these points. When $\tau = 0$, the normalized value of the correlation reduces to $|\mu_{12}|^2$; its variation with the length of the baseline for a uniform circular source is then shown by the broken line in Fig. 14.2.

It should be noted that the analysis leading to Eq. (14.10) applies only to linearly polarized light. When dealing with unpolarized light, it is necessary to take into account the fact that the orthogonal components of the field are uncorrelated. Equation (14.10) then becomes [Mandel, 1963]

$$R_{12}(\tau) = (1/2)I_1 I_2 |\gamma_{12}(\tau)|^2. \tag{14.11}$$

As we have seen in Section 14.1, the principal disadvantage of Michelson's stellar interferometer is that small, random changes in the optical paths traversed by the two beams, due to atmospheric turbulence, result in movements of the fringes which make observations difficult. These movements correspond to changes in the phase of the complex degree of coherence (see Section 3.4).

However, the intensity interferometer measures the square of the modulus of the complex degree of coherence, which is completely unaffected by changes in its phase. As a result, atmospheric turbulence has a negligible effect on the measured correlation in the intensity interferometer.

In addition, it can be shown that it is only necessary to maintain equality of the two paths to a few centimetres to obtain satisfactory results. If we define the Fourier transforms of the fluctuations of the electrical signals as

$$g_1(f) \leftrightarrow \Delta_{i_1}(t) \tag{14.12}$$

and

$$g_2(f) \leftrightarrow \Delta_{i_2}(t), \tag{14.13}$$

it follows from the Wiener–Khinchin theorem (see Appendix A.2) that

$$\langle \Delta_{i_1}(t)\Delta_{i_2}(t+\tau) \rangle = \int_0^\infty g_1(f)g_2(f)\exp(-\mathrm{i}2\pi f\tau)\mathrm{d}f$$

$$= \int_0^\infty S_{12}(f)\exp(-\mathrm{i}2\pi f\tau)\mathrm{d}f, \tag{14.14}$$

where $S_{12}(f)$ is the cross-spectral density of the fluctuations. Accordingly,

$$\frac{R_{12}(\tau)}{R_{12}(0)} = \frac{\int_0^\infty S_{12}(f)\exp(-\mathrm{i}2\pi f\tau)\mathrm{d}f}{\int_0^\infty S_{12}(f)\mathrm{d}f}, \tag{14.15}$$

so that the variation of the correlation with the delay is given by the normalized Fourier transform of the cross-spectral density of the fluctuations.

In this instrument, the bandwidth of the electrical signals $\Delta i_1(t)$ and $\Delta i_2(t)$ at the correlator is much smaller than the optical bandwidth and is determined by the low-pass filters in the two channels. If, therefore, we consider the simplest case, where the filters are identical and have a rectangular pass band extending from $f = 0$ to $f = f_m$, Eq. (14.15) reduces to

$$R_{12}(\tau)/R_{12}(0) = \mathrm{sinc}(f_m\tau). \tag{14.16}$$

Equation (14.16) shows that with a bandwidth of 100 MHz, the correlation drops to 90% of the maximum value for a delay corresponding to an optical path difference of 300 mm. It follows that the intensity interferometer is about 10^{-5} times less sensitive to the effects of optical path differences, or time delays in the two paths, than Michelson's stellar interferometer. As a result, the light collectors need not be finished to normal optical tolerances, and it is possible to work with much longer baselines.

In a large instrument of this type [Brown, 1964, 1974] the light collectors had a diameter of about 6.5 m and were made up of 252 hexagonal glass mirrors. These collectors were mounted on trucks running on a circular railway track with a diameter of 188 m. To follow a star, the two trucks moved around the track, while maintaining the desired separation. The beam from each collector, after passing through an interference filter with an 8-nm-wide pass band centered on a wavelength of 433 nm, was focused on a photomultiplier.

The main drawback of this instrument was the very small signal obtained, because of the narrow bandwidth of the low-pass filters. This limited its use to stars brighter than +2.5 magnitude. Even with these stars, the correlation signal had to be integrated over a period of several hours to obtain a satisfactory S/N ratio. However, measurements were made successfully on 32 stars with angular diameters down to 0.42×10^{-3} arcsecond.

14.3   Heterodyne Stellar Interferometry

The use of a laser as a local oscillator for heterodyne detection of light from a star was first studied by Nieuwenhuizen [1970]. In trials with a 2-m telescope, light from a star was mixed with light from a He-Ne laser at a photodetector, and radio-frequency signals arising from interference between the light from the laser and components of the light from the star at very nearly the same frequency were observed. Heterodyne detection also has the advantage that the spectral resolution is defined by the bandwidth of the detector; however, much higher sensitivity is obtained over this limited bandwidth than with direct detection, because the output is proportional to the product of the intensities of the laser and the star. Since the relative sensitivity of optical heterodyne detectors improves directly with increasing wavelength [Siegman, 1966], their use at a wavelength of 10.6 μm is attractive because of the 8–14 μm atmospheric transmission window and the availability of a powerful source of coherent radiation in the CO_2 laser.

Gay and Journet [1973] were the first to use a CO_2 laser and two HgCdTe photodiodes to make measurements of the angular diameter of the sun at a wavelength of 10.6 μm. Correlation of the amplified currents from the detectors gave a sinusoidal output, the equivalent of fringes in an interferometer, because of the diurnal motion of the sun. The fringe amplitude was found to be a function of the separation of the detectors, and the angular diameter of the sun, calculated from the separation of the detectors corresponding to the first zero of the fringe amplitude, was 33 arcminutes.

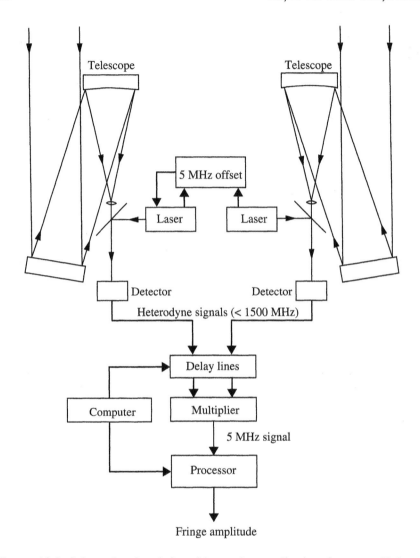

Figure 14.4. Schematic of an infrared heterodyne stellar interferometer [Johnson, Betz, and Townes, 1974].

Subsequently, Johnson, Betz, and Townes [1974] constructed a heterodyne spatial interferometer with a baseline of 5.5 m, operating at 10.6 μm, and used it to make measurements on a number of astronomical sources. As shown in Fig. 14.4, this instrument used two independent telescopes with an effective aperture of about 0.8 m, each consisting of a steerable coelostat followed by a fixed off-axis focusing mirror that directed the beams to a fixed focus. The two coelostats were situated on an east–west baseline with a center-to-center separation of 5.5 m and gave, at an operating wavelength of 10.6 μm, an angular resolution of about 0.5 arcsecond.

A high-speed Ge-Cu photoconducting detector, located at the focus of each of the telescopes, mixed the light from the star with the beam from a stabilized, 1-W, CO_2 laser, which acted as the local oscillator. The amplified signals from the two detectors went to a correlator where they were multiplied together. The amplitude of the sinusoidal output signal from the correlator was proportional to the degree of coherence between the fields at the two telescopes, while its frequency was determined by the motion of the star across the field of view. For a horizontal east–west baseline of length D, this frequency was

$$f_n = [\Omega D \cos \delta \cos H]/\lambda, \qquad (14.17)$$

where Ω is the rate of rotation of the earth, δ the declination, and H the hour angle of the source. To produce a signal at a convenient frequency for further processing, and also to avoid interaction between the lasers, the two lasers were phase locked with a frequency difference of 5 MHz. This 5-MHz carrier frequency was finally removed in a single-sideband demodulator to obtain the natural fringe signal.

To observe interference of two beams over a frequency bandwidth $\Delta \nu$, the difference in the length of the optical paths must be small compared with $c/\Delta\nu$. The amplified signals from the photodetectors were therefore passed through adjustable radio-frequency delay lines, which compensated for the changes in the two optical paths as the star was tracked across the sky. In this case, the 1500-MHz bandwidth of the radio-frequency signals required the path lengths to be equalized only to within a few centimetres.

Observation have been made with this interferometer on a number of infrared sources, including M-type supergiants and Mira variables, to obtain information on the temperature and spatial distribution of circumstellar dust shells [Sutton *et al.*, 1977; Sutton, 1979].

14.3.1 Large Heterodyne Interferometer

Since the sensitivity of such an interferometer is proportional to the available collecting area, it is advantageous to use larger telescopes. This is possible in the infrared, since, for a given degree of atmospheric turbulence, the diameter of a diffraction-limited telescope increases as $\lambda^{6/5}$. It follows that, at a wavelength of 10.6 μm, with reasonably good seeing conditions, the useful diameter of each telescope could be as much as 3.8 m. Another desirable feature would be the possibility of extending the baseline up to, say, 100 m and also changing its orientation. This would give, in addition to better angular resolution, information on any departure from circular symmetry of the source. Finally, measurements of the phase of the signal would permit making accurate positional measurements as well as mapping regions that are not symmetrical with respect to inversion.

A large heterodyne stellar interferometer designed to meet these requirements has been described by Townes [1984]. This instrument is basically made up of two telescope units, each of which consists of a 2-m flat mirror rotating about two axes and a 1.65-m paraboloid with its axis horizontal, yielding a compact system which can be mounted on a trailer and moved, when required, to a new site to change the baseline. Once in position, each mirror rests on kinematic mounts on a concrete pad set into the ground.

During observations, He-Ne laser interferometers are used to monitor the positions of the telescopes with respect to invar posts set in bedrock, as well as the optical path lengths within the telescope. In addition, the large flat mirrors are pointed using interferometric measurements at four positions around the edge of each mirror to determine its angular position relative to the horizontal optic axis of the telescope. The two CO_2 lasers are phase locked by means of a fast feedback circuit, and the effects of any changes in the optical path between the two telescopes are eliminated by a variable element in the path of one laser beam. Very fast HgCdTe detectors with a bandwidth of 4 GHz are used, and the path lengths are equalized to better than 1 cm with a variable-length, radio-frequency cable.

This interferometer was initially operated with baselines up to 13 m [Bester, Danchi, and Townes, 1990] and provided angular resolution of 0.001 arcsecond, as well as astrometric precision of 0.010 arcsecond. More recently, it has been operated with baselines up to 56 m, and the addition of a third telescope should make it possible to use it as an imaging interferometer (see Section 14.10). A filter bank allows measurements on a number of spectral lines between 9 and 12 μm, opening up new areas of study, including the spatial distribution of dust shells, the location of molecular formation, and the distribution of ions around high-temperature stars [Hale *et al.*, 2000].

14.4   Long-Baseline Stellar Interferometers

With the development of modern detection, control and data-handling techniques, it became possible to overcome the difficulties encountered with Michelson's stellar interferometer [Tango and Twiss, 1980], and improved instruments of this type, designed to make measurements over baselines up to 640 m, have been constructed [Liewer, 1979; Davis, 1979, 1984].

The optical system of one of these instruments [Davis, 1984; Davis *et al.*, 1999] is shown schematically in Fig. 14.5. Two siderostats C, C, on concrete plinths at the ends of a north–south baseline, direct light from the star *via* two beam-reducing telescopes (BRT) and an optical path-length compensator (OPLC) to the beamsplitter B. Error signals from two quadrant detectors (Q) in each channel, viewing the image of the star, are used to control the siderostats and the two piezoelectric-actuated tilting mirrors

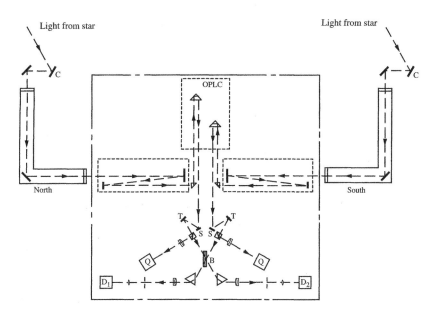

Figure 14.5. Schematic diagram of a long-baseline stellar interferometer. (Courtesy J. Davis, University of Sydney.)

(T), respectively, to keep these images exactly on the axis. Interference therefore takes place between two pairs of nominally parallel wavefronts leaving the beamsplitter B.

In the original instrument, two photon-counting detectors D_1, D_2 measured the total flux in a narrow spectral band in the two interference patterns within a sampling time of 1–10 ms. The signals from these two detectors were then proportional to $(1 + |\mu_{12}| \cos \phi)$ and $(1 - |\mu_{12}| \cos \phi)$, where ϕ was a phase angle that varied randomly with time because of changes in the optical path through the atmosphere. However, the mirrors S, which were mounted on piezoelectric translators, enabled an additional phase difference of 90° to be introduced between the two interfering beams in alternate sampling periods. As a result, the signals from the two detectors during the next period were proportional to $(1 + |\mu_{12}| \sin \phi)$ and $(1 - |\mu_{12}| \sin \phi)$. The average value of the square of the difference between the two signals then gave the value of $2|\mu_{12}|^2 \langle \cos^2 \phi + \sin^2 \phi \rangle$ or, in other words, the square of the visibility.

To ensure that the optical path difference was small compared to the coherence length of the radiation, photon-counting array detectors (PCADs) were used to observe the channeled spectrum produced by the combined beams. The spacing of the fringes then gave a direct indication of the optical path difference. Data from the PCADs were accumulated in a switchable memory, which allowed one data set to be processed to obtain a control signal while the next was being stored [Tango *et al.*, 1994].

Some results obtained with this instrument have been presented by Davis and Tango [1986] and Davis *et al.* [1994]. Measurements made, with baselines up to 30 m, of the angular diameters of a number of stars previously measured with the intensity interferometer, were in good agreement with the earlier results.

Subsequently, the use of a switchable phase modulator operating on the geometric (Pancharatnam) phase [Hariharan and Ciddor, 1995] was studied, since it provided nearly achromatic performance over a wide bandwidth as well as two outputs, one of which could be used for measurements of fringe visibility, while the other could be used for a high-speed fringe tracker [Tango and Davis, 1996].

Recently, a major upgrade has seen the sensitivity of this instrument increased by a factor of 50. The most important changes are a new adaptive optical system, a new beam-combining optical system, using optical fibers to feed two cooled avalanche photodiode detectors, and a new fringe-detector/tracker. In the new fringe-detection system, a mirror in one arm of the interferometer is mounted on a PZT, and the optical path difference (OPD) is swept linearly over a range of 100 μm. The signal is processed by taking its Fourier transform and averaging the power spectrum for a number of such sweeps. The "fringes" then appear as a peak in the power spectrum [Baldwin *et al.*, 1994]. The position of the fringe envelope within the sweep range gives a direct indication of the OPD, which can be fed back to the OPD control system.

This instrument currently operates with baselines up to 160 m.

14.5   Stellar Speckle Interferometry

In theory, the resolution of a telescope should increase with its diameter. However, because of turbulence in the earth's atmosphere, there is little improvement in resolution when the aperture of the telescope exceeds a value of approximately 100 mm. Thus, even though the diffraction-limited resolution of a 5-m telescope should be about 0.02 arcsecond, the star images typically have an angular diameter of about 1 arcsecond. Stellar speckle interferometry [Labeyrie, 1970, 1976] is a technique which can give, within certain limitations, diffraction-limited imaging of stellar objects, in spite of image degradation by the atmosphere and telescope aberrations.

Stellar speckle interferometry makes use of the fact that the image of a star in a large telescope, when observed under high magnification, exhibits a speckle structure [Texereau, 1963]. Individual speckles in such a pattern are actually equivalent to diffraction-limited images of the star (see Appendix G.1). However, since the speckles are in continuous motion, short exposures (≈20 ms) are required to record these patterns. In addition, it is necessary to use a filter with a bandwidth of about 20 nm and compensate for the dispersion of the atmosphere to obtain sharp images of the speckles.

The technique used by Labeyrie [1970] to extract a high-resolution image from a number of speckled images is shown schematically in Fig. 14.6. If $h(u, v)$ is the amplitude transmittance function of the perturbed atmosphere at a given instant, the distribution of the complex amplitude in the image of a point source is

$$H(x, y) \leftrightarrow h(u, v). \tag{14.18}$$

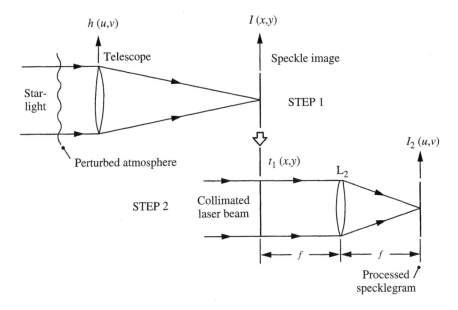

Figure 14.6. Steps involved in processing a series of speckle images to obtain a diffraction-limited image of a star.

If, then, $O(x, y)$ is the intensity distribution across a star, the intensity distribution $I(x, y)$ in a single, short-exposure speckle image can be written as

$$I(x, y) = O(x, y) * |H(x, y)|^2. \tag{14.19}$$

This image is recorded on a film which is processed so that the resultant amplitude transmittance $\mathbf{t}_1(x, y)$ is proportional to $I(x, y)$.

In the second stage, the transparency $\mathbf{t}_1(x, y)$ is placed in the front focal plane of a lens L_2 and illuminated by a collimated beam of monochromatic light. The complex amplitude in the back focal plane of L_2 is then

$$
\begin{aligned}
a_2(u, v) &= \mathcal{F}\{\mathbf{t}_1(x, y)\} \\
&= \mathcal{F}\{O(x, y) * |H(x, y)|^2\} \\
&= \mathcal{F}\{O(x, y)R_{hh}(u, v), \tag{14.20}
\end{aligned}
$$

where $R_{hh}(u, v)$ is the autocorrelation function of $h(u, v)$. The intensity in the back focal plane of L_2 is, therefore,

$$I_2(u, v) = |\mathcal{F}\{O(x, y)\}|^2 |R_{hh}(u, v)|^2. \tag{14.21}$$

To improve the S/N ratio of the final result, many of these transformed speck-legrams are averaged, for example, by making multiple exposures on the same photographic film. The resultant amplitude transmittance of this film is then

$$t_2(u, v) = \langle I_2(u, v) \rangle$$
$$= |\mathcal{F}\{O(x, y)\}|^2 \langle |R_{hh}(u, v)|^2 \rangle. \qquad (14.22)$$

The term within the pointed brackets on the right-hand side of Eq. (14.22) is the time average of the transfer function of the system (atmosphere + telescope) for incoherent light, which can be obtained from measurements carried out, under the same conditions, on a neighboring unresolved star. Since this transfer function has a finite positive value up to the diffraction limit of the telescope, the left-hand side of Eq. (14.22) can be divided by it to obtain $|\mathcal{F}\{O(x, y)\}|^2$. Fourier transformation of this result then gives the autocorrelation of the object,

$$|\mathcal{F}\{O(x, y)\}|^2 \leftrightarrow O(x, y) \star O(x, y). \qquad (14.23)$$

It should be noted that the angular resolution that can be obtained with speckle interferometry is limited by the aperture of the telescope and is, therefore, much less than what is possible with the long-baseline interferometers described in Sections 14.1 to 14.4. However, speckle interferometry has been applied successfully to a number of problems, including measurements of limb darkening and oblateness of supergiant stars, with the 5-m (200-inch) telescope at Mt. Palomar [Gezari, Labeyrie, and Stachnik, 1972; Bonneau and Labeyrie, 1973]. It is most useful in the study of close double stars [Labeyrie *et al.*, 1974]. In this case, the speckled image corresponds to the superimposition of two identical speckle patterns with a relative shift equal to the separation of the two stars. As a result, the transformed specklegram consists of a system of Young's fringes (see Appendix G.2), from which the separation and the relative positions of the two stars can be obtained [Labeyrie, 1974]. The scope and speed of the method have been progressively extended by the use of photon counters to record very faint images, and digital techniques of image processing which permit averaging a very large number of transformed specklegrams [Beddoes *et al.*, 1976; Blazit *et al.*, 1977; Arnold, Boksenberg, and Sargent, 1979]. A large number of double stars have been directly resolved, and some new components have been identified. Digital processing of four sets of observations has also made it possible to determine unambiguously the quadrant of the secondary star at these epochs [see McAlister *et al.*, 1988, 1989; McAlister, Hartkopf, and Franz, 1990].

14.6   Speckle Holography

A disadvantage of all the techniques of stellar interferometry discussed so far is that they do not give an image of the object, but only the autocorrelation of the image.

This results in a 180° ambiguity in the position angles of the components in a group of stars. One method developed for avoiding this ambiguity and reconstructing a diffraction-limited image from a number of speckled images is speckle holography [Liu and Lohmann, 1973; Bates, Gough, and Napier, 1973; Gough and Bates, 1974; Weigelt, 1978].

Speckle holography requires an unresolved star $\delta(x - x_R, y - y_R)$ fairly close to $O(x, y)$, the astronomical object under study. Speckle images are recorded of the total object

$$O_T(x, y) = O(x, y) + \delta(x - x_R, y - y_r), \tag{14.24}$$

and these images are processed, using the Labeyrie technique, to give its auto-correlation

$$\begin{aligned} O_T(x, y) \star O_T(x, y) = {} & \delta(x, y) + O(x, y) \star O(x, y) \\ & + O(-x + x_R, -y + y_R) \\ & + O(x - x_R, y - y_R). \end{aligned} \tag{14.25}$$

The distribution defined by Eq. (14.25) resembles a holographic image of the original object (see Appendix F.1) and consists of a central spot $\delta(x, y)$ on which is superposed the autocorrelation of the object, $O(x, y) \star O(x, y)$, and, away from the axis, on opposite sides of it, the desired high-resolution image $O(x - x_R, y - y_R)$ and a conjugate image $O(-x + x_R, -y + y_R)$.

14.7 The Hyper-Telescope

The possibility of directly obtaining a usable image at the combined focus of a diluted interferometric array was ruled out initially [Traub, 1986], because the sharp central interference peak would be superimposed on a large halo corresponding to the diffraction pattern of a single subpupil. However, work by Pedretti *et al.* [2001] using two micro-lens arrays to reimage the discrete exit pupils as nearly contiguous wavefront segments (a densified-pupil interferometer, or hyper-telescope) has opened up the possibility of overcoming this limitation.

14.8 Astrometry

Besides measurements of stellar diameters, another application of interferometric techniques is to astrometry—the measurement of stellar positions and motions [Shao and Staelin, 1977]. Precise measurements of stellar positions would permit matching radio sources with their optical counterparts and allow direct measurements of stellar parallaxes to establish the cosmic distance scale. In addition they would help to establish the size and shape of the orbits of binary systems. The Navy Prototype Optical Interferometer (NPOI) is designed for astrometry and uses an array of four fixed elements in a Y with 20-m arms [Armstrong *et al.*, 1998]. Another instrument, the Astronomical Studies of Extrasolar Planetary Systems (ASEPS-0) interferometer, searches for reflex motions of stars, due to planets orbiting around them, by using two elements, each of which simultaneously observes two stars separated by a few arcminutes. Since the fringes for both stars are detected at the same time, atmospheric effects should largely cancel, and it should be possible to detect extremely small changes in their relative positions [Colavita *et al.*, 1994]. A recent implementation is the Palomar Testbed Interferometer (PTI), a long-baseline infrared interferometer operating at a wavelength of 2.2 μm. This instrument has three fixed 40-cm apertures, which can be combined, pairwise, to provide baselines up to 110 m and track two stars simultaneously for phase referencing and narrow-angle astrometry [Colavita *et al.*, 1999].

14.9   Nulling Interferometers

A problem when trying to detect a planet near a star is that the star is 6–9 orders of magnitude brighter than the planet. Nulling interferometry is a method for reducing the flux from the star, relative to its surroundings, by making the light from the star interfere destructively with itself [Bracewell, 1978]. This result can be achieved with a 180° rotational shear [Shao and Colavita, 1992]. Some measurements of rejection ratios obtained in laboratory experiments with a rotational shearing interferometer have been presented by Serabyn *et al.* [1999], and the design of symmetrical optical systems for a nulling interferometer has been described by Serabyn and Colavita [2001].

A proposal for achromatic nulling interferometry with white light using the geometric (Pancharatnam) phase (see Section 16.6 and Appendix E.2) has been formulated by Baba, Murakami, and Ishigaki [2001]. Experiments with a three-dimensional Sagnac interferometer using the geometric (spin-redirection) phase to obtain achromatic nulling have also been described by Tavrov *et al.* [2002].

14.10   Telescope Arrays

The angular resolution that can be obtained with a single telescope, even with techniques such as speckle interferometry, is limited by the aperture of the telescope and is, therefore, much less than what is possible with the long-baseline interferometers described in Sections 14.1 to 14.4. One way to improve the resolution is by combining coherently the speckle images from two telescopes at the ends of a long baseline, or from several telescopes [Labeyrie, 1976, 1978; see also Breckenridge, 1994; Robertson and Tango, 1994; Lawson, 1997].

Ultimately, the biggest advance would be the ability to record images of stars. Unfortunately, since atmospheric fluctuations introduce large local phase variations, it is only possible to obtain information on the fringe amplitude with a two-element interferometer. In the absence of reliable phase measurements, the reconstruction of images is not possible.

One solution was the orbiting stellar interferometer (OSI), a concept for a first-generation space interferometer. This instrument was planned as a triple Michelson stellar interferometer, with an 18-m maximum baseline and aperture diameters of 40 cm [Colavita, Shao, and Rayman, 1993], which would be inserted into a 900-km sun-synchronous orbit. This idea and the search for exoplanets have now been taken further with the Space Interferometry Mission (SIM) project [see Shao, 2002].

Another solution is a triangular array. Correlators can then be used on each of the three interferometer baselines to provide measurements of the amplitude and phase of the spatial coherence function. The true phases ϕ_{12}, ϕ_{23}, and ϕ_{31} on the three baselines are disturbed by the unknown atmospheric and instrumental phases α_1, α_2, and α_3 at the three telescopes to give the measured values $(\phi_{12} + \alpha_1 - \alpha_2)$, $(\phi_{23} + \alpha_2 - \alpha_3)$, and $(\phi_{31} + \alpha_3 - \alpha_1)$. However, the sum of the three measured phases (the closure phase) which is $(\phi_{12} + \phi_{23} + \phi_{31})$, is a property of the coherence function alone, independent of the atmospheric and instrumental effects. As the number of elements increases, the number of triangles and closure phases increases rapidly, and the image becomes better constrained.

The application of closure-phase measurements to interferometric imaging was first suggested by Rogstad [1968], and a method that involved placing, over the mirror of a large telescope, nonredundant aperture masks with holes simulating several, small, closely spaced, individual telescopes, was discussed by Rhodes and Goodman [1973]. Measurements of fringe visibility and closure phase by this technique were successfully carried out by Baldwin et al. [1986] using a 2.25-m telescope. These measurements were extended to much lower light levels, using an image photon-counting system, by Haniff et al. [1987], who were able to produce an image of a binary system and map the surface of a red giant star.

With such large apertures, adaptive optics are needed to remove aberrations from the wavefront [Loos, 1992]. Other limitations of these instruments are the small field of view (one Airy disc) and the restriction to three beams.

To increase the field of view, two different schemes are possible: mosaicing and homothetic mapping. Mosaicing an image involves scanning the object in steps of one Airy disc and putting the individual images together. Homothetic mapping relies on reimaging the interferometric array into the entrance pupil of the instrument, thus forming on the detector a regular image of the object with a superposed fringe pattern. If images with different array configurations are recorded, it should be possible to superpose the Fourier transforms of these images and reconstruct the complete image.

The next step would be an instrument that can combine more beams (6–8), allowing for more efficient observations and producing directly an image of good quality.

Toward this end, several new, large, multielement interferometers are nearing completion. One of these, the Cambridge Optical Aperture Synthesis Telescope (COAST), intended for imaging at wavelengths of 0.8 and 2.2 μm, is an array of five telescopes with an aperture of 40 cm, in the form of a Y with arms up to 60 m long [Baldwin *et al.*, 1994]. Another, the Navy Prototype Optical Interferometer (NPOI), will consist of up to six elements that can be moved along the arms of a 250-m Y. The Keck Observatory interferometer on Mauna Kea is planned to consist of six 1.5-m telescopes separated by up to 100 m and can also make use of the two 10-m main telescopes, which are separated by a distance of 85 m. The Very Large Telescope Interferometer (VLTI) in Chile proposes to use eight 1.8-m telescopes, as well as the four 8-m telescopes available on site, to form a 198-m-wide interferometric array.

Many problems are faced in the operation of such large interferometers. One is polarization effects arising from celestial field rotation and phase delays between the *s* and *p* fringe patterns. Techniques to minimize these effects have been discussed by Rousselet-Perraut, Vakili, and Mourard [1996]. Another problem is random variations in the atmospheric delay. Real-time tracking of the delay presents problems at low signal levels; an attractive alternative is an off-line delay-tracking system that uses stored data and the statistics of the delay process to derive an estimate of the probable delay at the time of observation [Meisner, 1996].

Another problem with such instruments is the high degree of mechanical stability required in interferometry. A solution that is being actively investigated is the use of optical fiber links from the focus of each telescope to the mixing station [Froehly, 1982; Restaino *et al.*, 1996, 1997; Delage, Reynaud, and Lannes, 2000]. Single-mode optical fibers have several advantages; most importantly, after the light is coupled into the fibers, problems of misalignment, surface quality, and diffraction are eliminated. In addition, the light output has nearly perfect spatial coherence, reducing the effects of atmospheric turbulence. Finally, beam recombination is simplified by using directional couplers [Shaklan and Roddier, 1987; Shaklan, 1990]. However, a problem with optical fibers is their significant dispersion, which, if not compensated, can cause the fringe visibility curve to broaden, decrease its contrast, and shift its centroid [Dyer and Christensen, 1997].

The first images from COAST were obtained in 1996 [Baldwin *et al.*, 1996]; fringes have also been observed with the VLTI and the Keck Observatory interferometer. Recent work on these and other instruments, as well some results obtained with them, have been presented in a number of papers (see Lena and Quirrenbach [2000]; Quirrenbach [2001]; Traub [2002]).

Chapter 15

Space-Time and Gravitation

The definition of the metre discussed in Section 7.9 assumes that the speed of light is a constant. This is an axiom in the theory of relativity which is now well established, but was not apparent to scientists of the 19th century who believed that light was a transverse wave, propagated through an all-pervading medium called the "luminiferous aether." A question which arose, naturally, was what effect the movement of the earth through this medium had on the speed of light. A very important experiment, using an interferometer, carried out by Michelson in 1881 and repeated in collaboration with Morley in 1887, was designed to answer this question.

15.1   The Michelson–Morley Experiment

Consider an interferometer in which, as shown in Fig. 15.1a, the arm OM_1 is parallel to the direction in which the earth is moving with a velocity v through space. According to the classical laws of physics, the time taken for a light wave to travel from O to M_1 would be

$$\tau_1' = l/(c - v),\tag{15.1}$$

where l is the length of the arm and c is the speed of light, while the time taken for the light wave to travel from M_1 to O would be

$$\tau_1'' = l/(c + v).\tag{15.2}$$

Accordingly, the total delay for the path OM_1O would be

$$\tau_1 = \tau_1' + \tau_1''$$

239

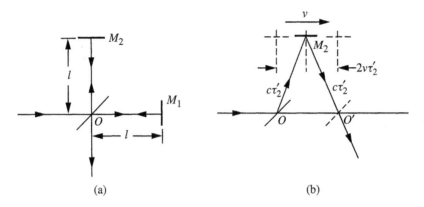

Figure 15.1. Theory of the Michelson–Morley experiment.

$$= 2l/c(1 - v^2/c^2)$$
$$\approx (2l/c)(1 + v^2/c^2), \tag{15.3}$$

since $v \ll c$.

To calculate τ_2, the delay for the path $O M_2 O'$ (see Fig. 15.1b), we have to allow for the lateral displacement of O during this time. We then have

$$c^2 \tau_2'^2 = l^2 + v^2 \tau_2'^2, \tag{15.4}$$

where τ_2' is the time taken to cover the path $O M_2$. This is also the time taken to cover the path $M_2 O'$. The total delay for the path $O M_2 O'$ is, therefore,

$$\tau_2 = 2\tau_2'$$
$$= (2l/c)(1 - v^2/c^2)^{-1/2}$$
$$= (2l/c)(1 + v^2/2c^2). \tag{15.5}$$

It follows that, even if the two paths are equal, there should be a difference in the delays

$$\Delta \tau = \tau_1 - \tau_2$$
$$= (l/c)(v/c)^2. \tag{15.6}$$

Accordingly, a change in the delay of $2\Delta \tau$ should be observed if the interferometer is rotated through $90°$, effectively shifting the additional delay from one path to the other. The absence of any such change in the delay led to the rejection of the concept of the aether [Shankland, 1973].

The Michelson–Morley experiment was repeated several times between 1905 and 1930, but always failed to exhibit an effect that was more than 10% of that expected. This null result was subsequently confirmed to a much higher degree of accuracy using laser heterodyne techniques [Jaseja *et al.*, 1964; Brillet and Hall, 1979]. In the latter experiment, the frequency of a He–Ne laser ($\lambda \approx 3.39 \ \mu$m) was locked to the resonant frequency of a very stable, thermally isolated Fabry–Perot interferometer, which was mounted along with it on a rotating horizontal granite slab. As a result, any variations in the delay within this Fabry–Perot interferometer would appear as variations of the laser frequency. Such variations could be detected with an extremely high degree of sensitivity by measuring the beat frequency generated by mixing this beam (see Section 5.3) with the beam from a stationary reference laser which was stabilized by saturated absorption in CH_4. Analysis of the results showed that the residual frequency shifts were less than 5×10^{-7} of those predicted by the aether hypothesis.

15.2 Gravitational Waves

According to the general theory of relativity, gravity is due to a curvature of space-time. A weak curvature is responsible for Newtonian gravitational fields, such as that on the earth. However, in objects such as black holes, the curvature is so strong that nonlinear effects produce a space-time singularity, which can force the curvature to evolve chaotically. The resulting rapid variations in curvature generate disturbances—gravitational waves—that propagate through the universe.

A collapsing supernova is a likely source of gravitational waves, but it is not easy to estimate their strength and frequency. A binary neutron star provides a source of gravitational waves whose intensity can be estimated theoretically. In such a system, the two stars spiral toward each other, over hundreds of millions of years, until they collide and merge. In the last few minutes before they coalesce, the frequency of the gravitational waves increases from about 10 Hz to more than 1000 Hz.

The detection of gravitational waves would lead to many new and interesting possibilities. While electromagnetic waves are easily absorbed and scattered, gravitational waves travel almost unchanged through intervening clouds of matter, making it possible to study very different aspects of the universe. The profile of a gravitational wave also contains information that can be used to map out the space-time geometry of the source. In addition, from the arrival times of electromagnetic- and gravitational-wave signals from the same stellar event, it would be possible to verify that gravitational waves are propagated at the speed of light, as predicted by the theory of relativity, and that, therefore, the rest mass of the graviton is zero.

With a network of detectors, it should be possible to determine the angular position of an event from the times of arrival of the signals at the different detectors. If then, the distance to an event can be inferred from some parameters of the signals, and the red shift of the host cluster is known, it should be possible to obtain Hubble's constant in an unambiguous way, providing a much more reliable cosmological scale. Successful detection of gravitational waves not only would help to sort out competing theories of gravity, but would open up a completely new field of observational astronomy.

15.3   Gravitational Wave Detectors

Early attempts to detect gravitational waves from cosmological sources used resonant bar detectors [Weber, 1969], but it soon became evident that a long, wide-band antenna offered advantages. Such an antenna could consist of a Michelson interferometer in which the beamsplitter and the end reflectors are attached to separate, freely suspended masses [Moss, Miller, and Forward, 1971; Weiss, 1972; Forward, 1978; Maischberger *et al.*, 1981].

A change in the difference of the lengths of the optical paths in the two arms due to a gravitational wave results in a change in the phase difference between the beams, which can be measured by the intensity change at the detector.

Because of the transverse, quadrupole nature of gravity waves, the relative motion of the test masses is the sum of the contributions from two different polarizations, $h_+(t)$ and $h_\times(t)$, each with its own wave form. As a result, an arm lying along the direction of propagation of a wave with the appropriate polarization does not experience a strain, whereas the strain produced by the gravitational wave results in a change in the length of the other arm. With a gravitational wave propagating at right angles to the plane of the interferometer, the changes in length of the two arms have opposite signs. For other directions of incidence, the fractional difference in the length of the arms caused by the wave is equal to a linear combination of the changes due to the two polarizations and can be written as

$$\frac{\Delta L(t)}{L} = F_+ h_+(t) + F_\times h_\times(t) \equiv h(t), \tag{15.7}$$

where the coefficients F_+ and $F_\times$ depend on the direction to the source and the orientation of the interferometer, and $h(t)$ is termed the gravitational strain. It follows that the relative motion of the test masses and, hence, the change in the optical path difference for a given value of the gravitational strain, is proportional to the separation of the test masses.

Theoretical estimates of the intensity of bursts of gravitational radiation, as well as an analysis of the response of an interferometric detector [Rudenko and Sazhin, 1980], suggest that a sensitivity to strains better than 10^{-21} over a bandwidth of approximately a kilohertz is needed. In a conventional Michelson's interferometer, this sensitivity would require unrealistically long arms.

One method of obtaining high sensitivity, which has been studied, involves the use of two active cavities at right angles to each other. Each pair of mirrors acts as the resonator in a laser, so that a change in their spacing results in a change in the laser frequency. Changes in the beat frequency obtained by heterodyning the two lasers are monitored [Abramovici, Vager, and Weksler, 1986]. It has been shown that, even with small cavities and low laser power, a strain sensitivity of 3×10^{-15} can be obtained, which is close to the limit set by spontaneous emission noise. Estimates suggest that an increase in sensitivity by three orders of magnitude may be obtained by reducing cavity losses and using a high-gain amplifying medium. However, the arrangements that are currently being explored involve two variants of the Michelson interferometer.

Higher sensitivity can be obtained with a Michelson interferometer by using a delay line in each arm [Billing *et al.*, 1979; Caves, 1980]. In this case, the optical path difference (expressed as a fraction of the wavelength) introduced between the beams emerging from the two arms is

$$\Delta p = B \frac{\Delta L}{\lambda} = B \frac{hL}{\lambda}, \qquad (15.8)$$

where B is the number of times the light bounces back and forth in the delay lines before exiting. The precision with which this change in the optical path difference can be monitored is ultimately limited by photon shot noise and is approximately

$$\Delta p = 1/\sqrt{N}, \qquad (15.9)$$

where N is the number of photons incident on the beamsplitter during the period over which the signal from the photodetector is integrated.

The strain due to a gravitational wave that would give a signal of the same magnitude as the measurement fluctuations would then be of the order

$$H_{\min} = \frac{\lambda}{LB\sqrt{N}}. \qquad (15.10)$$

This estimate assumes that both the integration time and the storage time in the cavities, (BL/c), are less than half the period of the gravitational wave.

Another way to obtain increased sensitivity is by using, as shown in Fig. 15.2, two identical Fabry–Perot interferometers at right angles to each other [Drever *et al.*, 1981; Weiss, 1999], with their mirrors mounted on freely suspended test masses. The frequency of a laser is locked to a transmission peak of one interferometer, while the optical path in the other is continuously adjusted by a second servo system, so that its transmission peak also coincides with the laser frequency. The corrections applied to the second interferometer then provide a signal which can be used to detect any changes in the length of one arm with respect to the other, over a wide frequency range, with extremely high sensitivity [Drever, 1983; Drever *et al.*, 1983].

A further increase in sensitivity can be obtained by recycling the available light. In a system such as that shown in Fig. 15.2, the light that returns to the beamsplitter leaves

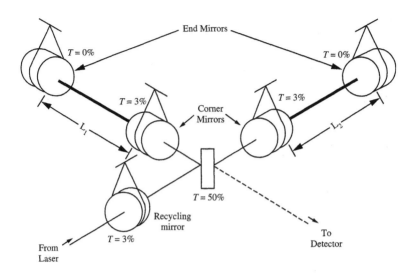

Figure 15.2. Gravitational wave detector using two Fabry–Perot interferometers [Weiss, 1999].

the interferometer in two directions. Since the interferometer is normally adjusted so that observations are made on a dark fringe, to avoid overloading the detector, most of the light is returned toward the source and is lost. If this light is reflected back into the interferometer in precisely the right phase by an extra mirror placed in the input beam, as shown in Fig. 15.2, the amount of light traversing the arms of the interferometer can be increased substantially, leading to a corresponding improvement in sensitivity over a limited frequency band [Meers, 1988; Strain and Meers, 1991]. The gain obtainable by this technique has been confirmed experimentally [Fritschel, Shoemaker, and Weiss, 1992; Sato *et al.*, 2000].

Two other techniques, which have shown promise for obtaining even higher sensitivity, are signal recycling and resonant side-band extraction. Signal recycling [Meers, 1988] uses an additional mirror placed at the antisymmetric port of the interferometer to increase the storage time and, hence, the sensitivity. Signal recycling is used most effectively with power recycling to achieve the highest sensitivity, in which case the technique is known as dual recycling. It can be used in a narrow-band mode, where gravitational wave signals are expected at a particular frequency, by adjusting the position of the signal recycling mirror. Resonant side-band extraction [Mizuno *et al.*, 1993] also employs the same optical system as signal recycling, but with arm cavities of much higher finesse, the signal extraction mirror being set to reduce the signal storage time. The sensitivity is expected to be comparable with that for dual recycling, and it also permits operation in a narrow-band mode. This technique has the advantage that the light power at the beamsplitter is much less than with power recycling.

15.4 LIGO

The Laser Interferometer Gravitational Observatory (LIGO) project is now build-
ing facilities for detecting and studying gravitational waves at two sites in the USA
[Abramovici et al., 1992, 1996; Weiss, 1999]. Similar facilities are also being
developed in other countries by the GEO 600, TAMA, and VIRGO projects.

   The LIGO project involves the construction of three laser interferometers: a single
interferometer with 4-km-long arms at one site, and two interferometers, with 4-km-
and 2-km-long arms sharing the same vacuum system, at the other site. Correlating
the outputs of the three interferometers at two widely separated sites would make it
possible to distinguish signals due to gravitational waves from bursts of instrumental
and environmental noise.

   The test masses and the optical paths in each interferometer are housed in
a vacuum ($\approx 10^{-9}$ torr) to avoid buffeting by air molecules and also to prevent
fluctuations in the number of air molecules in the beam paths from causing phase
fluctuations. The 4-km-long vacuum pipes running between the test masses have a
diameter of 1.2 m to accommodate auxiliary laser beams, as well as additional detectors. The beam from the laser entering the interferometer, after having its frequency
and amplitude stabilized, is then phase-modulated and spatially filtered to minimize
noise.

   The LIGO interferometers are expected to have a Gaussian noise spectrum $\bar{h}(f)$
shown by the upper solid curve in Fig. 15.3. The principal individual contributions
to this noise, shown as dashed curves, arise from three sources. The first, seismic
noise (man-made ground vibrations), is dominant at frequencies below approximately
70 Hz. At higher frequencies, between approximately 70 and 200 Hz, the noise is prin-
cipally due to thermally induced vibrations of the test masses and their suspensions.
Finally, above approximately 200 Hz, photon shot noise becomes the most important
factor. This curve is based on the assumption of approximately 2 W of effective laser
power, a fractional power loss of 5×10^{-5} per reflection, and a gain in effective
laser power by a factor of 30 from recycling. The lower solid curve presents the
likely values for a next-generation advanced interferometer, currently under development. Noise reduction in this interferometer would be achieved by increasing the
laser power to 60 W and reducing the mirror losses to permit recycling the light 100
times, as well as by increasing the test masses from 10 to 1000 kg, and improving the
suspension and isolation systems.

   Experiments performed with prototype interferometers with arm lengths of 10–
40 m, and other special setups, have confirmed most of the factors entering into
these calculations. Measurements with a 40-m prototype, using two Fabry–Perot
cavities, have shown that it is possible to achieve a minimum interferometer noise of
3×10^{-19} m/Hz$^{1/2}$, corresponding to an rms gravitational strain noise level ($\Delta L/L$)
of about 2×10^{-19}, around 450 Hz [Abramovici et al., 1996]. Similar results have
been obtained in a 30-m Michelson interferometer using delay lines [Robertson et al.,
1995]. In order to reach the design sensitivity (10^{-21} m/Hz$^{1/2}$), LIGO also incor-
porates an active, wave-front based, cavity-alignment system [Fritschel et al., 1998]

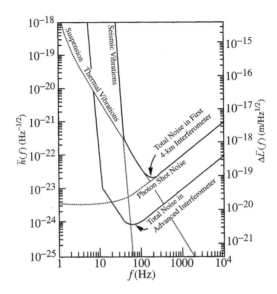

Figure 15.3. Expected noise spectrum in the LIGO 4-km interferometer (upper solid curve) and in an advanced interferometer (lower solid curve). The dashed curves show the principal contributions to the total noise in the first interferometer [Abramovici *et al.*, 1992].

as well as an electronic feedback system that keeps the cavities on resonance [Fritschel *et al.*, 2001].

A comparison of estimates of signal strengths from some possible sources with the expected noise in the first LIGO detector, and an advanced design incorporating several possible improvements, is presented in Fig. 15.4. Based on these figures, the first LIGO detectors should be able to observe gravitational waves from a few events each year.

Recent work on LIGO and other ground-based interferometric detectors as well as on the Laser Interferometer Space Antenna (LISA) project, using drag-free spacecraft for the detection of low-frequency gravitational waves, has been described in a series of papers [see Saulson and Cruise, 2002].

15.5   The Standard Quantum Limit

The limiting precision that can be attained with interferometric measurements is a critical factor in applications such as the detection of gravitational waves. After all other sources of noise are eliminated, this limit is ultimately set by the randomness in the arrival times of photons at the detector.

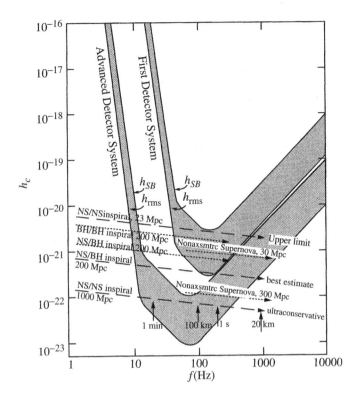

Figure 15.4. Estimates of rms noise h_{rms} in the first and advanced LIGO detectors and the characteristic amplitude h_c of gravitational-wave bursts from different sources (NS, neutron star; BH, black hole) [Abramovici *et al.*, 1992].

A value for this limit can be obtained from an argument which balances the error due to photon-counting statistics against the disturbances of the end mirrors produced by fluctuations in radiation pressure [Edelstein *et al.*, 1978; Forward, 1978]. According to this argument, since the number of photons which pass through the interferometer in the measurement time τ is

$$n = P\tau/\hbar\omega, \tag{15.11}$$

where P is the laser power, fluctuations in the laser power produce an uncertainty in n given by the relation

$$\Delta n \approx n^{-1/2}. \tag{15.12}$$

The existence of this quantum limit is now well established, but the argument leading to it has been open to question, since it relies on the assumption that the power fluctuations in the two arms are uncorrelated. A more rigorous analysis [Caves, 1980] reveals two different, but equivalent, points of view on the origin of the fluctuations. The first attributes them to the fact that each photon incident on the beamsplitter is scattered independently, thereby producing binomial distributions of photons in the two arms which are precisely anticorrelated. The second ascribes them to vacuum (zero-point) fluctuations in the field entering the interferometer from the other input port. This field acts in antiphase on the laser fields in the two arms. It follows, therefore, that the photon-counting error is an intrinsic property of the interferometer.

Heisenberg's principle states that the uncertainty in the number of quanta n in a beam of light and the uncertainty in its phase ϕ are linked through the relation

$$\Delta n \Delta \phi \geq 1/2. \tag{15.13}$$

The standard quantum limit (SQL) in interferometry is then obtained by inserting Eq. (15.11) in Eq. (15.13) to obtain the corresponding uncertainty in the measured values of the phase

$$\Delta \phi \geq 1/2\sqrt{\bar{n}}, \tag{15.14}$$

where $\bar{n}$ is the mean photon number (see Yamamoto et al. [1990]).

Two schemes for interferometric measurements with sensitivity limited only by quantum noise have been analyzed by Stevenson et al. [1993]. Direct detection is applicable to signals at modulation frequencies away from frequencies affected by classical noise. Signals obscured by classical optical noise may be recovered with a phase-modulation technique that shifts the signals to a quantum-noise-limited region of the photocurrent spectrum.

15.6   Squeezed States of Light

We can represent the electric field of a monochromatic light wave as the sum of two quadrature components in the form

$$E = E_0[X_1 \cos \omega t + X_2 \sin \omega t], \tag{15.15}$$

where X_1 and X_2 are complementary operators satisfying the commutation relation

$$[X_1, X_2] = (1/2)i, \tag{15.16}$$

and whose variances, therefore, obey the uncertainty relationship

$$\Delta X_1 \Delta X_2 \geq 1/4. \tag{15.17}$$

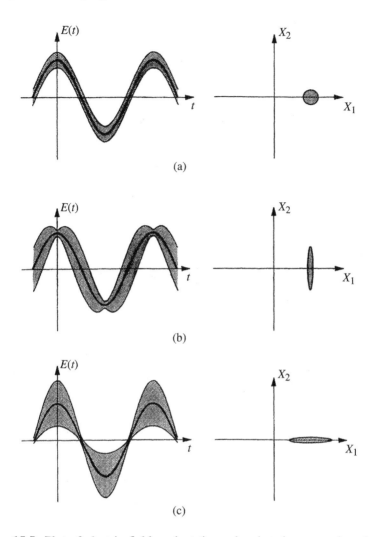

Figure 15.5. Plot of electric field against time, showing the uncertainty for (a) a coherent state, (b) a squeezed state with reduced amplitude fluctuations, and (c) a squeezed state with reduced phase fluctuations [Caves, 1981].

For normal coherent light, the variances are equal. For a squeezed state, the variances are unequal, though their product remains unchanged. Accordingly, it is possible to reduce phase fluctuations with squeezed light, as shown in Fig. 15.5, at the expense of a corresponding increase in the amplitude fluctuations [Caves, 1981; Walls, 1983].

The degree of squeezing can be measured with a balanced homodyne detector which yields a phase-sensitive measurement of the noise. The squeezed light is combined with another strong beam from the same laser, which constitutes a local

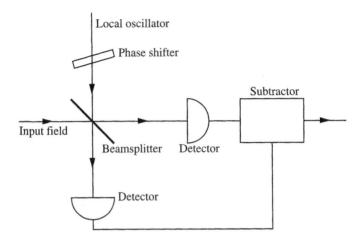

Figure 15.6. Balanced homodyne detector used for measurements of noise due to a single field-quadrature component.

oscillator, at a 50:50 beamsplitter. The beams emerging from the beamsplitter are directed, as shown in Fig. 15.6, to two photodetectors, and the difference of the two photocurrents is displayed. With this arrangement, intensity fluctuations in the local oscillator and the signal cancel out, and the output corresponds to the interference of the local oscillator with the signal. If the intensity of the local oscillator is much greater than that of the signal, the fluctuations in the output are essentially due to the signal. As the phase difference between the squeezed-light signal and the local oscillator is varied, the detector becomes sensitive first to one quadrature amplitude and then to the other, and the output noise amplitude varies accordingly.

A number of physical phenomena can be used, in principle, to generate squeezed states. The earliest and most common method has been degenerate four-wave mixing in a medium with a nonlinear susceptibility [Slusher *et al.*, 1985; Shelby *et al.*, 1986; Maeda, Kumar, and Shapiro, 1987]). In such a process, energy is transferred from two strong pump beams to two weaker beams. As a result, correlations are established between the photons in the two weaker beams. When these two beams are combined, the resulting light exhibits the characteristics of squeezed states.

Wu *et al.* [1986] have shown that a greater reduction in noise can be achieved by degenerate parametric down-conversion (see Section 16.3.1). The two photons in the pair, each with a different frequency, are strongly correlated, and this correlation is responsible for the squeezed fluctuations.

When the gain from parametric amplification in the down-conversion crystal becomes large, there is a transition from spontaneous to stimulated emission. Since the gain depends on the phase of the amplified light relative to the phase of the pump beam, vacuum fluctuations in one quadrature are squeezed [see Kimble and Walls, 1987].

A photon-number squeezed state can also be generated directly from a diode laser, at low temperatures, by operating it at a drive current several times higher than the threshold current. In one experiment, the amplitude noise of a current-driven diode laser was found to be 8.3 dB below the shot-noise limit [Richardson, Machida, and Yamamoto, 1991].

15.7 Interferometry Below the SQL

One way to overcome the SQL in an interferometer is by injecting squeezed states into one or both ports of the interferometer [Caves, 1981; Bondurant and Shapiro, 1984]. A limit

$$\Delta\phi \approx 1/\bar{n} \tag{15.18}$$

is achievable, in principle, by feeding suitably constructed squeezed states into both input ports of the interferometer. In actual experiments, an increase in the signal-to-noise ratio of 2–3 dB, relative to the shot-noise limit, has been achieved using squeezed light [Grangier *et al.*, 1987; Xiao, Wu, and Kimble, 1987]. The maximum improvement in the signal-to-noise ratio can be obtained by preparing the light delivered to the two input ports of the interferometer in a state which consists of exactly equal numbers of photons entering each of the two input ports in the interferometer [Yurke, McCall, and Klauder, 1986; Holland and Burnett, 1993; Sanders and Milburn, 1995]. States satisfying this requirement can be generated, for example, with two-mode four-wave mixers [Yurke, McCall, and Klauder, 1986]. The difference in the photocounts at the two output ports can then be processed to obtain the phase difference. However, maximum sensitivity is achieved only when the deviation of the phase difference ϕ from zero is less than $1/\bar{n}$. To meet this requirement, it would be necessary to use a feedback loop which holds ϕ at zero.

Chapter 16

Single-Photon Interference

Young's classical experiment has always been regarded as a conclusive demonstration of the wave-like nature of light. However, at low light levels, photodetectors register distinct events corresponding to the annihilation of individual photons. It follows that light cannot be either a particle or a wave, but exhibits the characteristics of both.

From a practical point of view, we can say that we are in the "single-photon" regime when, with a perfectly efficient detector, the mean time interval between the detection of successive photons is much greater than the time taken for light to travel through the system.

16.1 Interference at the "Single-Photon" Level

If interference is a phenomenon involving the interaction of at least two particles, interference effects should become weaker as the number of photons decreases and disappear completely when only one photon is in the apparatus at a time. This argument led to a series of experiments involving observations of interference patterns at extremely low light levels, all of which showed that the quality of the pattern did not depend on the intensity (see Pipkin [1978]).

However, experiments at low light levels with conventional light sources cannot be regarded as involving single-photon states. Even with a single-mode laser source, the arrival of photons at a photodetector exhibits a Poisson distribution. As a result, photodetections are more likely to occur very close together, or at the same time. This phenomenon is known as photon bunching [Mandel and Wolf, 1965]

16.2 Interference—The Quantum Picture

Interference involves mixing two fields. The result is a two-mode coherent state which, in the weak-field limit, represents the interference of a photon with itself. This self-interference of a photon can be interpreted as a sum over histories [Feynman, Leighton, and Sands, 1963].

In this picture, a photon can take either of two paths from the source to the detector. Associated with each path is a complex probability amplitude a_i, ($i = 1, 2$), whose absolute square represents the probability of the photon taking this path. The intensity at the detector is obtained by summing the probability amplitudes for the two paths and taking the square of the modulus of the sum, $|a_1 + a_2|^2$, which gives the probability of detecting a photon.

As Dirac puts it, "... each photon interferes only with itself. Interference between different photons never occurs" [Dirac, 1958].

16.3 Sources of Nonclassical Light

An approximation to a single-photon state can be prepared by generating a pair of photons. In this case, the detection of one photon acts as a signal that a second photon is present in the field. The frequency and direction of propagation of the second photon are related to those of the first photon, and can be determined by analyzing the first, or "gate," photon. The second photon field can then be regarded as being in a one-photon Fock state.

One method of generating such a pair of photons is by means of an atomic cascade. If atoms of calcium are excited to the $6^1 S_0$ level, they return to the ground state via the $4^1 P_1$ level. In this two-step process, they emit, in rapid succession, two photons with wavelengths of 551.3 nm and 422.7 nm, respectively [Aspect, Grangier, and Roger, 1981].

However, a better method of generating correlated pairs of photons is by parametric down-conversion.

16.3.1 Parametric Down-Conversion

In parametric down-conversion, a single UV photon spontaneously decays in a crystal with a $\chi^{(2)}$ nonlinearity into two photons (a *signal* photon and an *idler* photon) with polarizations orthogonal to that of the original UV photon and wavelengths close to twice the UV wavelength [Harris, Oshman, and Byer, 1967; Burnham and Weinberg, 1970]. Phase-matching is achieved by using a birefringent crystal. The two down-converted photons are emitted "simultaneously" within the resolving time of the detectors [Hong and Mandel, 1985; Friberg, Hong, and Mandel, 1985].

Since energy is conserved, we have

$$\hbar\omega_0 = \hbar\omega_1 + \hbar\omega_2, \qquad (16.1)$$

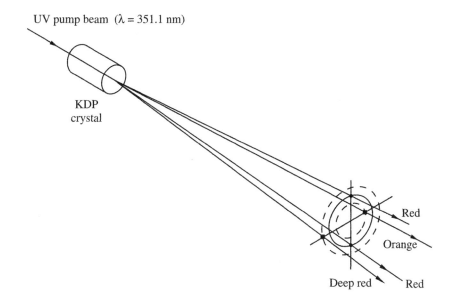

UV pump beam ($\lambda = 351.1$ nm)

KDP
crystal

Red

Orange

Deep red Red

Figure 16.1. Generation of photon pairs by parametric down-conversion of UV photons in a nonlinear crystal.

where $\hbar\omega_0$ is the energy of the UV photon, and $\hbar\omega_1$ and $\hbar\omega_2$ are the energies of the two down-converted photons. Similarly, since momentum is conserved, we have

$$\mathbf{k}_0 = \mathbf{k}_1 + \mathbf{k}_2, \qquad (16.2)$$

where $\mathbf{k}_0$ is the momentum of the UV photon, and $\mathbf{k}_1$ and $\mathbf{k}_2$ are the momenta of the down-converted photons. It follows from Eq. (16.1) that, while the frequencies of the individual down-converted photons may vary over a broad range, the sum of their frequencies is well defined. Similarly, it follows from Eq. (16.2) that the photons in each pair are emitted on opposite sides of two cones, whose axis is the UV beam, and produce, as shown in Fig. 16.1, a set of rainbow-colored rings.

Because of the additive contributions of the vacuum state, the down-converted photons carry information about the phase of the pump. They are also in an entangled state, which does not allow factorization into a product of signal and idler states [see Mandel and Wolf, 1995]. As a result, the state of the idler photon is governed by measurements made on the signal photon and vice versa.

Typically, the UV beam from an Ar$^+$ laser ($\lambda = 351.1$ nm) and a potassium dihydrogen phosphate (KDP) crystal can be used to generate pairs of photons with wavelengths around 746 and 659 nm, leaving the crystal at angles of approximately $\pm1.5°$ to the UV beam [Hong and Mandel, 1986]. Higher conversion efficiencies can be obtained with a crystal such as beta-barium borate (BBO) having a higher nonlinear coefficient. In addition, since the number of down-converted photons produced is proportional to the square of the interaction time, a further increase in down-conversion efficiency can be obtained by placing the nonlinear crystal in a resonant cavity [Hariharan and Sanders, 2000; Oberparleiter and Weinfurter, 2000].

The possibility of extending parametric down-conversion to generate entangled four-photon states from two pump photons by using two noncollinear pump beams has been discussed by Tewari and Hariharan [1997].

16.4   The Beamsplitter

With a single-photon state, quantum mechanics predicts a perfect anticorrelation between the counts at the two output ports of a beamsplitter [Clauser, 1974]. This behavior is very different from that with a thermal source, where the correlation between photons detected at the two outputs from a beamsplitter is positive, or with a coherent field from a laser, where there is no correlation between the two outputs. Such anticoincidences were observed by Grangier, Roger, and Aspect [1986], as well as by Diedrich and Walther [1987], indicating that each photon was either transmitted or reflected.

The indivisibility of the photon is due to the entanglement of the input field with the vacuum field. It follows that any analysis of the effect of a beamsplitter on a beam of light with definite photon number must take into account the vacuum field at the unused input port [Fearn and Loudon, 1987, 1989; Campos, Saleh, and Teich, 1989].

16.5   Interference with Single-Photon States

Interference effects produced by single-photon states were first studied using an atomic cascade by Grangier, Roger, and Aspect [1986]. As shown in Fig. 16.2a, the arrival of the first photon (frequency v_1) at the detector D_0 acted as a trigger for a gate generator, enabling the two detectors D_1 and D_2 on the two sides of the beamsplitter for a time $2\tau_s$. During this period, the probability for the detection of a second photon (frequency v_2) emitted by the same atom is much greater than the probability of detecting a similar photon emitted by any other atom in the source. While a classical wave would be divided between the two output ports of the beamsplitter, a single photon cannot be divided in this fashion. We can therefore expect an anticorrelation between the counts on the two sides of the beamsplitter at D_1 and D_2, measured by a parameter

$$\mathcal{A} \equiv N_{012}N_0/N_{01}N_{02},\qquad(16.3)$$

where N_{012} is the rate of triple coincidences between the detectors D_0, D_1, and D_2; N_{01} and N_{02} are, respectively, the rate of double coincidences between D_0 and D_1, and D_0 and D_2; and N_0 is the rate of counts of D_0.

For a classical wave, it follows from Schwarz's inequality that $\mathcal{A} \geq 1$. On the other hand, the indivisibility of the photon should lead to arbitrarily small values of $\mathcal{A}$. The experiment confirmed that the number of coincidences observed for the second

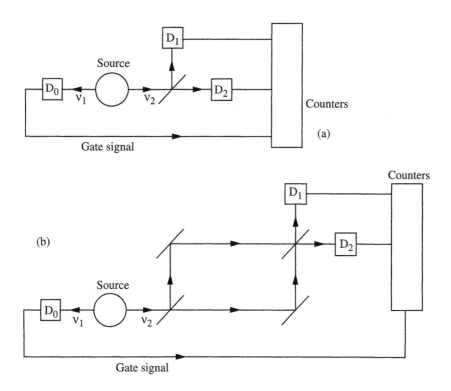

Figure 16.2. Experimental arrangement used (a) to detect single-photon states, and (b) to demonstrate interference with single-photon states [Grangier, Roger, and Aspect, 1986].

photon, with a gate time of 9 ns, was only 0.18 of that expected from classical theory, but corresponded to that predicted by quantum theory.

This source was then used in the optical arrangement shown in Fig. 16.2b, with the detectors D_1 and D_2 receiving the two outputs from a Mach–Zehnder interferometer. The interferometer was initially adjusted and checked without the gating system in operation, and interference fringes with a visibility $V > 0.98$ were obtained. In the actual experiment, with the gate on, the optical path difference was varied around zero in 256 steps, each of $\lambda/50$, with a counting time of 1 s at each step. The results of 15 such sweeps were then averaged to improve the signal-to-noise ratio. Analysis of the data showed that, even with the gate operating, values of the visibility $V > 0.98$ were obtained.

16.6   The Geometric Phase

Berry [1984] showed that the wave function of a quantum system may undergo a phase shift (a geometric phase) when the parameters of the system undergo a cyclic change.

Two manifestations of the geometric phase are observed in optics [Tiwari, 1992]:

(1) The phase change acquired due to a circuit on the sphere of directions in momentum space (the spin-redirection phase), which is equal to the solid angle subtended by the circuit. One effect of the spin-redirection phase is the rotation of the plane of polarization of a linearly polarized light beam propagating in a nonplanar path, such as an optical fiber coiled into a helix; another is the phase difference introduced between a right- and a left-handed circular polarized wave, when they travel along a nonplanar path [Tomita and Chiao, 1986; Jiao et al., 1989]. A direct demonstration uses an interferometer consisting of a single-mode fiber ring, with part of it forming a helix. The fringe shift observed when the pitch of the helix is changed is a measure of the geometric (spin-redirection) phase [Frins and Dultz, 1997].

(2) The phase acquired due to a circuit on the Poincaré sphere of polarization states (the Pancharatnam phase), which is equal to half the solid angle subtended by the circuit at the center of the Poincaré sphere (see Appendix E) [Pancharatnam, 1956; Ramaseshan and Nityananda, 1986; Berry, 1987]. Several experiments have been described to demonstrate the Pancharatnam phase [Bhandari and Samuel, 1988; Chyba et al., 1988; Hariharan and Roy, 1992; Hariharan, Mujumdar, and Ramachandran, 1999].

16.6.1   Observations at the Single-Photon Level

Observations of the geometric (Pancharatnam) phase have been made by Hariharan et al. [1993a] at the single-photon level. They used a Sagnac interferometer in which the optical paths traversed by the two beams were always equal, and a phase difference could be introduced between them only by operating on the Pancharatnam phase [Hariharan and Roy, 1992]. As shown in Fig. 16.3, light from a He-Ne laser, linearly polarized at $45°$ to the plane of the figure by a polarizer P_1, was divided at a polarizing beamsplitter into two orthogonally polarized beams that traversed the same closed triangular path in opposite directions. A second polarizer P_2, with its axis at $45°$ to the plane of the figure, brought the two beams leaving the interferometer into a condition to interfere at a photomultiplier.

The phase difference between the beams was varied by a system consisting of a rotating half-wave plate HWP located between two fixed quarter-wave plates, QWP_1 and QWP_2. Rotation of HWP by an angle θ advanced the phase of one beam by 2θ and retarded the phase of the other beam by -2θ. As a result, a phase difference $\Delta\phi = 4\theta$ was introduced between the two beams.

Measurements were made at an input power level <1 pW, corresponding to a photon flux $<3.2 \times 10^6$ photons/s. At this level, if $P(1)$ and $P(n > 1)$ are, respectively, the probabilities for the presence of one photon and more than one photon in the apparatus, we have $P(n > 1)/P(1) = 0.005$. The net counting rate, after subtracting the dark count, exhibited a sinusoidal variation, corresponding to the relation $\Delta\phi = 4\theta$. The visibility of the interference fringes was better than 0.97.

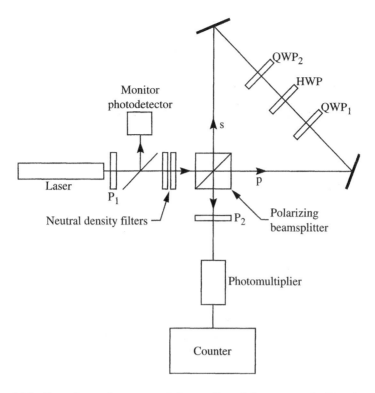

Figure 16.3. Experimental setup used for studies of the geometric (Pancharatnam) phase at the single-photon level [Hariharan *et al.*, 1993a].

16.6.2   Observations with Single-Photon States

Kwiat and Chiao [1991] performed an experiment to demonstrate the existence of a geometric phase for single-photon states (see Section 16.3) using a light source which produced pairs of photons with wavelengths centered at 702.2 nm by parametric down-conversion (see Section 16.3.1). As shown in Fig. 16.4, the idler beam was transmitted through the narrow-band filter F1 to the detector D1, while the signal beam entered a Michelson interferometer. The beam leaving the interferometer was incident on a second beamsplitter B2, from which it was transmitted to the detector D2 through the filter F2, or reflected to the detector D3 through the filter F3. The count rates for coincidences between D1 and D2 and between D1 and D3, as well as for triple coincidences between D1, D2, and D3, were recorded.

One arm of the interferometer contained a fixed quarter-wave plate Q1 with its axis at 45° as well as a quarter-wave plate Q2 which could be rotated. A rotation of Q2 through an angle θ introduced an additional phase shift $\Delta\phi = 2\theta$ in this arm. When data were recorded using filters with a bandwidth of 10 nm at F2 and F3, and an optical path difference of 220 μm, which was greater than the coherence length

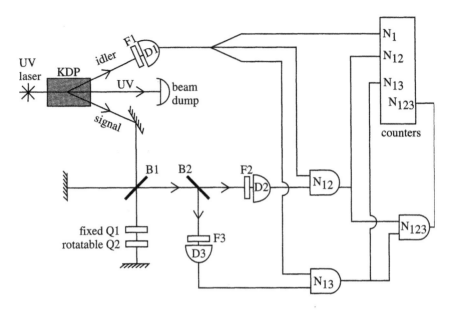

Figure 16.4. Apparatus used to demonstrate the geometric (Pancharatnam) phase for single-photon states [Kwiat and Chiao, 1991].

corresponding to this bandwidth (about 50 μm), the fringe visibility seen by the detectors D2 and D3, operating individually, was essentially zero. However, when a filter with a bandwidth of 0.86 nm was placed in front of D1, the count rate for coincidences between D1 and D3 varied sinusoidally with the angular setting θ of Q2, with a visibility of 0.60 ± 0.05. With a broad band filter at F1, the coincidence fringes disappeared.

Measurements of the anticorrelation parameter (see Section 16.5)

$$\mathcal{A} \equiv N_{123}N_1/N_{12}N_{13}, \tag{16.4}$$

where N_{123} is the rate of triple coincidences between D1, D2, and D3, N_1 is the rate of single counts by D1 alone, N_{12} is the rate of coincidences between D1 and D2, and N_{13} is the rate of coincidences between D1 and D3, confirmed that the observations essentially involved photons in $n = 1$ Fock states.

16.7   Interference with Independent Sources

Pfleegor and Mandel [1967a, 1967b; 1968] were the first to show that light beams from two independent lasers can produce interference fringes, even when the light

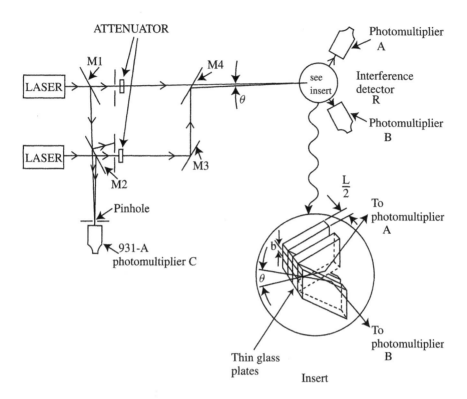

Figure 16.5. Experimental system used to demonstrate interference with two independent laser sources at low light levels [Pfleegor and Mandel, 1967b].

intensity is so low that the mean time between photons is long enough for one photon to be absorbed with high probability at the photodetector before the next photon is emitted by either of the two sources. As shown in Fig. 16.5, the beams from two He-Ne lasers were superimposed at a small angle, to produce interference fringes on the edges of a stack of glass plates whose thickness was equal to half the fringe spacing. Two photomultipliers received the light from alternate plates. To minimize effects due to movements of the fringes, an additional photodetector was used to detect beats between the beams, and observations were restricted to 20-μs intervals, corresponding to periods during which the frequency difference between the two laser beams was less than 30 kHz. The transit time was approximately 3 ns, while the photon fluxes in the two beams were around 3×10^6 photons/s and the quantum efficiency of the photomultipliers was about 0.07, so that about 10 photons were detected in each 20-μs period. The average of 400 such measurements was taken in each experiment.

In this experiment, the positions of the fringe maxima are not predictable and vary from measurement to measurement. However, there should always be a

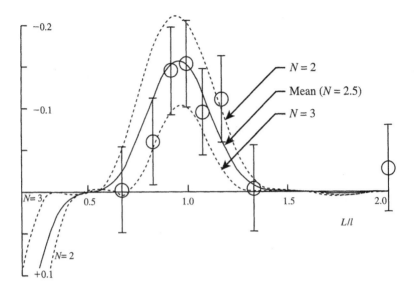

Figure 16.6. Experimental results for the normalized correlation coefficient (solid line) and theoretical curves for $N = 2$ and $N = 3$ (broken lines) [Pfleegor and Mandel, 1967b].

negative correlation between the number of photons registered in the two channels, which should be a maximum when the fringe spacing l is equal to L, the thickness of a pair of plates. Figure 16.6 shows the variation in the degree of correlation of the two counts with the ratio L/l, together with the theoretical curves for $N = 2$ and $N = 3$, where N is the number of pairs of plates in the detector array.

This experiment confirmed that interference effects were associated with the detection of each photon. However, since observations could be made only over very short time intervals, during which a very small number of photons were detected, the precision of the experiment was limited.

16.7.1 Observations in the Time Domain

More accurate observations of interference effects with two sources, at very low light levels, have been made by using beats between two laser modes [Hariharan, Brown, and Sanders, 1993]. A convenient low beat frequency was obtained by applying a transverse magnetic field to a He-Ne laser oscillating in two longitudinal modes. The laser then oscillates on a single axial cavity mode composed of two orthogonally polarized components which exhibit a small frequency difference due to the magnetically induced birefringence of the gas in the laser tube [Morris, Ferguson, and Warniak, 1975]. These two Zeeman-split components can be regarded as equivalent to beams from two independent lasers, because the coupling between them is quite

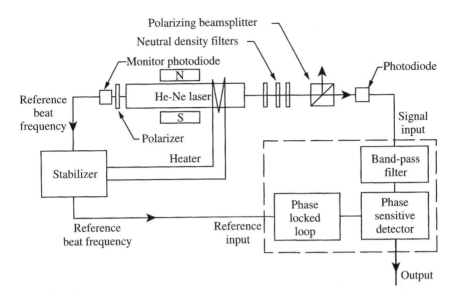

Figure 16.7. Experimental arrangement used to measure the amplitude of beats at very low light levels [Hariharan, Brown, and Sanders, 1993].

weak. In addition, with normal excitation, there is no coherence between the two upper states for the lasing transitions.

The experimental arrangement is shown schematically in Fig. 16.7. The beat frequency was stabilized by mixing the two orthogonally polarized components in the back beam of the laser at a monitor photodiode, and using the output to control the length of the cavity through a servo amplifier and a heating coil on the laser tube [Ferguson and Morris, 1978]. A beat frequency of 80 kHz, with a frequency bandwidth estimated at 1 Hz, was obtained. Neutral-density filters were used to reduce the intensity of the output beams from the laser, in accurately known steps over a range of 10^8:1. The attenuated beams were incident on a photodiode, after passing through a polarizing prism which brought them into a condition to interfere. The signal from this photodiode was taken through a band-pass filter to a homodyne detector which received a reference signal from the monitor photodiode. Because the variations in the frequency of the beat signal were small and were tracked by the reference signal from the monitor photodiode, measurements could be made with integrating times up to 100 seconds, ensuring a good signal-to-noise ratio, even at the lowest light levels.

Observations were made with the photodiode at a distance of 0.2 m from the laser as the incident power was varied from 1.0 μW down to 4.8 pW, corresponding to values of the incident flux ranging from 3.18×10^{12} photons/s to 1.53×10^7 photons/s, respectively. At the lowest power level, the probability for the presence of more than one photon in the apparatus at any time, relative to that for the presence of a single photon, was less than 0.005.

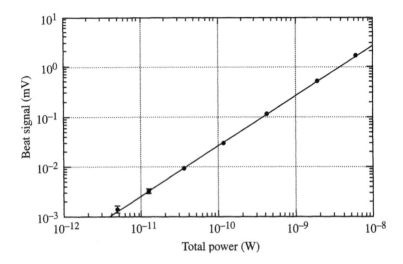

Figure 16.8. Output signal from the homodyne detector as a function of the total power in the laser beams [Hariharan, Brown, and Sanders, 1993].

Figure 16.8 shows the output from the homodyne detector plotted as a function of the power incident on the photodiode. The measurements showed no significant deviations from a straight line, confirming that the interference phenomena remained unchanged down to power levels at which there was a very high probability that one photon was absorbed before the next one was generated.

Observations of beats produced by a He-Ne laser operating in three axial modes were also made by Hariharan *et al.* [1993b, 1995]. In this case a low-frequency beat was observed, corresponding to the second difference between the frequencies of the three modes. The amplitude of this beat was found to vary linearly with the power down to a power level of 0.19 nW. At this power level, the ratio of the probability for the presence of at least one photon from all three modes to the probability for the presence of at least one photon from any mode was only 0.0009, and the mean time interval between the arrival of successive photons at the detector was greater than the period of the beat.

16.8 Superposition States

It follows that the interference phenomena observed in all these cases are associated with the detection of each photon and not with the interference of one photon with another. Earlier studies [Jordan and Ghielmetti, 1964; Mandel, 1964] suggested that for interference effects to become observable, the average number of photons in the same spin state falling on a coherence area in a coherence time would have to be appreciably greater than 1. However, it is now clear that interference phenomena are produced by a sequence of photons, each one of which is in a superposition state that originates from the modes involved [Mandel, 1976; Walls, 1977; Hariharan, Brown, and Sanders, 1993]. An explicit description of the interference effects produced by two independent laser beams, using the techniques of quantum field theory, has been presented by Agarwal and Hariharan [1993].

Chapter 17

Fourth-Order Interference

Measurements of fourth-order coherence can be made by correlating signals from two photodetectors that are separated in space or time. The first studies of fourth-order coherence with the intensity interferometer [Brown and Twiss, 1956] used spatially separated detectors to measure the angular diameters of stars (see Section 14.2). Subsequently, it was shown that with a time delay produced by an optical path difference that was much greater than the coherence length of the radiation, the effects of such correlated intensity fluctuations could still be observed as a spectral modulation [Alford and Gold, 1958; Mandel, 1962].

Fourth-order interference provides a means for distinguishing classical and nonclassical light, since an optical field can exhibit nonclassical fourth-order interference effects even when the usual second-order interference effects cannot be observed [Mandel, 1983; Ou, 1988]. Such fourth-order interference effects can be observed with correlated photons produced by parametric down-conversion (see Section 16.3.1).

17.1 Nonclassical Fourth-Order Interference

We consider the detection of the field produced by the signal and idler modes from a two-photon source at a point x_1 (see Fig. 17.1). Since the output from the down-converter is an approximation to the two-photon Fock state $|1, 1\rangle_{AB}$ for weak fields, the probability of detecting a photon between the arbitrary positions x_1 and $x_1 + \Delta x_1$ can be shown to be [Ghosh and Mandel, 1987]

$$\mathcal{P}_1(x_1)\Delta x_1 = 2K_1\Delta x_1, \tag{17.1}$$

where K_1 is a scale factor. A separate measurement made at x_2 with another photodetector yields a similar result. No second-order interference effects are observed because no definite phase relationship exists between the two down-converted fields.

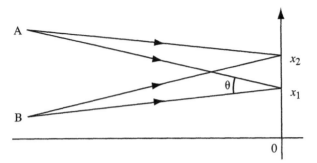

Figure 17.1. Geometry of a fourth-order interference experiment.

However, if we use the two photodetectors at x_1 and x_2 to measure the joint probability, $\mathcal{P}_{12}(x_1, x_2)\Delta x_1 \Delta x_2$, of detecting a photon within Δx_1 and another photon within Δx_2, we have [Ghosh *et al.*, 1986]

$$\mathcal{P}_{12}(x_1, x_2)\Delta x_1 \Delta x_2 = 2K_1 K_2 \Delta x_1 \Delta x_2 \left[1 + \cos\left(\frac{2\pi(x_1 - x_2)}{L}\right) \right], \qquad (17.2)$$

where $L = \lambda/\theta$ is the spacing of the second-order interference fringes corresponding to the geometry of Fig. 17.1. We can regard the effects observed as interference between two, different, two-photon probability amplitudes, because the system cannot distinguish between photons from A and B being detected at x_1 and x_2, respectively, or vice versa. These fourth-order interference fringes have a visibility of unity. However, with classical fields, it can be shown [Mandel, 1983] that the visibility of the fourth-order interference fringes cannot exceed 0.5, and the joint probability $\mathcal{P}_{12}(x_1, x_2)\Delta x_1 \Delta x_2$ never drops to zero.

A similar experiment has been described by Ribeiro *et al.* [1994] using a simple two-slit arrangement. Observations of coincidences between one beam (the signal beam), which was transmitted through the two slits, and its conjugate (the idler beam), which was transmitted through a pinhole, showed that the visibility of the Young's fringes formed by the signal beam depended on the diameter of the pinhole in the idler beam. This result can be regarded as showing that, because of the entanglement of the down-converted beams, the "effective" source area for the signal beam varies with a change in the diameter of the pinhole in the idler beam [Řeháček and Peřina, 1996].

In another version of this experiment, two correlated photons were made to pass through a double slit. Since, in principle, it was possible to identify which slit one of the photons passed through, from the position of the other photon before it reached the double slit, no interference patterns due to the individual photons were observed. However, the joint-detection counting rate of the photon pairs revealed a spatial interference pattern [Hong and Noh, 1998].

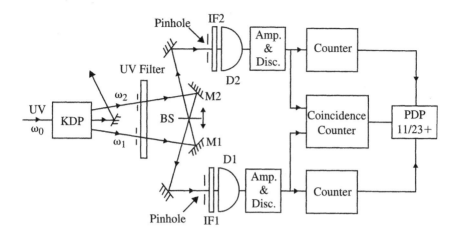

Figure 17.2. Experimental arrangement used to demonstrate fourth-order interference effects with a variable optical path difference [Hong, Ou, and Mandel, 1987].

A more striking example of a nonclassical fourth-order interference effect is when the inputs to the two ports of a beamsplitter are one-photon Fock states [Fearn and Loudon, 1987, 1989; Ou, Hong, and Mandel, 1987]. In this case, a sharp reduction in the coincidence count rate can be observed [Hong, Ou, and Mandel, 1987; Rarity and Tapster, 1989]. As shown in Fig. 17.2, the superimposed beams leaving the beamsplitter go to two photodetectors, D1 and D2, and measurements are made of the rate at which photons are detected in coincidence as the beamsplitter is displaced in small steps $c\Delta\tau$ from the point at which the two optical paths are equal.

We label the field modes on the input side of the beamsplitter as 01,02 and on the output side as 1,2 and assume that the light is perfectly monochromatic. If the input state resulting from degenerate down-conversion is the two-photon Fock state $|1\rangle_{01}|1\rangle_{02}$, then the state on the output side of the beamsplitter is

$$|\psi_{\text{out}}\rangle = (R - T)|1\rangle_1|1\rangle_2 + i(2RT)^{1/2}[|2\rangle_1|0\rangle_2 + |0\rangle_1|2\rangle_2], \qquad (17.3)$$

where R and T are the reflectance and transmittance of the beamsplitter, with $R+T = 1$. If $R = T = 0.5$, the first term on the right-hand side of Eq. (17.3) is zero, corresponding to destructive interference of the two-photon probability amplitudes, and the entangled state $|2; 0\rangle_{1,2}^{\pi/2,0}$ is obtained. Since this corresponds to the photon pair being in a superposition state where it adopts either path 1 or path 2, no coincidences should be recorded.

However, the down-converted photons are never monochromatic, so that the two-photon state can be represented more correctly by the linear superposition

$$|\psi\rangle = \int d\omega f(\omega, \omega_0 - \omega)|1\rangle_\omega|1\rangle_{\omega_0-\omega}, \qquad (17.4)$$

where $f(\omega_1, \omega_2)$ is a weight function that has its maximum when $\omega_1 = \omega_2 = (1/2)\omega_0$. The joint probability for the detection of photons at D1 and D2 at times t and $t + \tau$, respectively, is then

$$
\mathcal{P}_{12}(\tau) = K|G(0)|^2 \{ T^2|G_0(\tau)|^2 + R^2|G_0(2\Delta\tau - \tau)|^2
$$
$$
- RT[G_0^*(\tau)G_0(2\Delta\tau - \tau) + c.c.] \}, \tag{17.5}
$$

where $G(\tau)$ is the Fourier transform of the weight function, $G_0(\tau) \equiv G(\tau)/G(0)$, and K is a constant, characteristic of the detectors.

The coincidence measurement corresponds to an integration of the probability $\mathcal{P}_{12}(\tau)$ over a time that is much longer than the correlation time. Accordingly, if we assume that the weighting function is a Gaussian with bandwidth $\Delta\omega$, the expected number of coincidences N_c is

$$
N_c = C(T^2 + R^2)\left[1 - \frac{2RT}{R^2 + T^2} \exp(-\Delta\omega\Delta\tau)^2 \right]. \tag{17.6}
$$

When $\Delta\tau \gg G_0(\tau)$, $N_c = C(T^2 + R^2)$, but when $\Delta\tau = 0$, $N_c = C(R - T)^2$, which vanishes when $R = T = 0.5$.

As shown in Fig. 17.3, the rate of coincidences drops to a few percent of its normal value when the two optical paths are equal, because of destructive interference of the two-photon probability amplitudes. The width of the dip in the coincidence rate gives a measure of the length of the photon wave packet, which agrees with the value derived from the width of the pass band of the interference filters, F1 and F2.

The occurrence of an almost complete null at the center of the dip confirms that this is a nonclassical effect, since, according to classical theory, the visibility cannot exceed 0.5 [Mandel, 1983]. The drop in the number of coincidences is associated with an increase in the number of photon pairs leaving the beamsplitter in the same direction and arises from the Bose–Einstein commutation properties of the photon-creation and -annihilation operators [Fearn and Loudon, 1989].

An extension of this experiment [Ou and Mandel, 1988] involved the use of interference filters with pass bands centered on different, nonoverlapping frequencies, ω_1 and ω_2, chosen so that $\omega_1 + \omega_2 = \omega_0$, and corresponding to a frequency difference $(\omega_1 - \omega_2)/2\pi = 27 \times 10^{12}$ Hz. If the frequency responses of the filters are Gaussian functions with an rms width σ, the expected coincidence probability is

$$
\mathcal{P}_{12} \propto 2^{1/2}\pi\sigma^3 \left[T^2 + R^2 - 2TR \exp\left(\frac{-\sigma^2\Delta\tau^2}{2} \right) \cos(\omega_1 - \omega_2)\Delta\tau \right], \tag{17.7}
$$

which corresponds to an interference pattern whose visibility is unity when $\Delta\tau = 0$, but falls off exponentially on either side.

Figure 17.4 shows the observed two-photon coincidence rate as a function of the position of the beamsplitter and the corresponding time delay $\Delta\tau$ between the signal and idler photons. The coincidence rate exhibits interference fringes with a

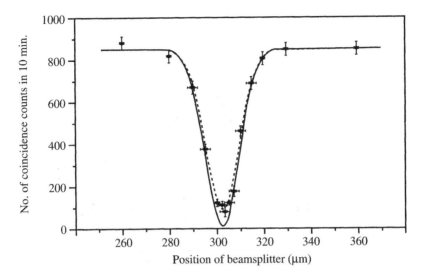

Figure 17.3. Measured number of coincidences as a function of the position of the beamsplitter, superimposed on the theoretical (solid) curve obtained from Eq. (17.6) with $R/T = 0.95$ and $\Delta\omega = 3 \times 10^{13}$ rad/s. The dashed curve was obtained by multiplying the factor $2RT/(R^2 + T^2)$ in Eq. (17.6) by 0.9 [Hong, Ou, and Mandel, 1987].

spatial period of 5.5 μm, corresponding to a temporal period of 37 fs, the period of the beat frequency. This beat is observed in the coincidence rate even though the two photodetectors, individually, do not register a beat.

This experiment demonstrates that even though a fundamental limit exists to the localization of a photon, this limit does not apply to the average time interval between photons.

17.2 Interference in Separated Interferometers

Fourth-order interference effects can also be observed when pairs of photons enter one or more interferometers, and the coincidence rate is monitored at the output ports [Kwiat *et al.*, 1990; Ou *et al.*, 1990a; Rarity *et al.*, 1990]. In the arrangement used by Ou *et al.* [1990a] (see Fig. 17.5), the two photons traveled to two photodetectors via two unbalanced Michelson interferometers, which were adjusted so that the difference in the propagation time between the longer and shorter paths was much greater than the coherence time of the individual photons, but was the same in both channels. Under these conditions, the count rate registered by the two detectors showed no

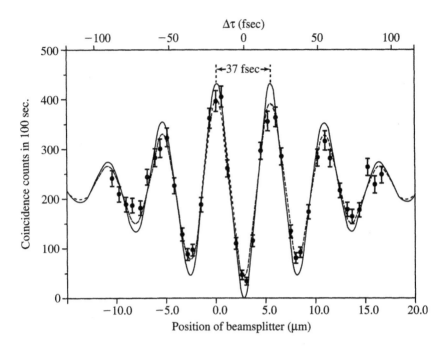

Figure 17.4. Number of coincidences as a function of the time delay $\Delta\tau$ between the signal and idler photons, along with the theoretical (solid) curve obtained from Eq. (17.7). The dotted curve was obtained by multiplying the interference term by 0.8 [Ou and Mandel, 1988].

dependence on the optical path difference in either of the interferometers. However, measurements of the two-photon coincidence rate, as a function of the position of one of the mirrors, revealed interference fringes with a spatial period corresponding to the wavelength of the pump beam. The visibility obtained in this experiment was only 0.5, due to the limited time resolution of the detector system; however, subsequent measurements with better time resolution gave coincidence fringes with a visibility of 0.87 [Brendel, Mohler, and Martienssen, 1991].

Unusual interference patterns have also been observed with nondegenerate photon pairs [Larchuk *et al.*, 1993]. In their experiments, pairs of photons whose center wavelengths differed by approximately 40 nm were used as the inputs to single and dual Mach–Zehnder interferometers (MZIs). In the single MZI configuration, the paths of both down-converted beams overlapped completely within the interferometer. In the dual MZI configuration, their paths did not overlap; this is equivalent to sending each beam into a separate interferometer. Second-order interference was observed by counting the number of photons at each of the output ports, while observations of fourth-order interference were made by recording coincidences as the optical path difference was varied from zero to very large values.

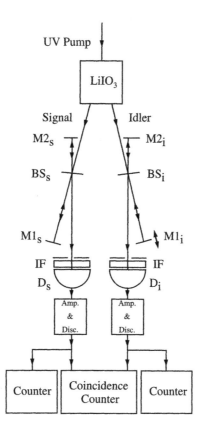

Figure 17.5. Experimental arrangement used to observe fourth-order interference effects with photon pairs in two separated interferometers [Ou *et al.*, 1990a].

The second-order interference patterns at both the output ports were the same for photon pairs as that predicted by classical theory for the center frequencies, ω_1 and ω_2, and, as expected, disappeared when the optical path difference exceeded the second-order coherence length. However, the behavior of the fourth-order interference patterns was quite different. For small path differences, the coincidence rates exhibited interference fringes corresponding to the difference frequency $\omega_d = |\omega_1 - \omega_2|$, as well as the sum frequency $\omega_s = |\omega_1 + \omega_2|$, when the beams overlapped, and also when they did not overlap. When the beams did not overlap, interference fringes were also obtained corresponding to the center frequencies, ω_1 and ω_2. A striking observation was the existence of interference effects at the sum (pump) frequency at path-length differences that were greater than the second-order coherence length.

In another fourth-order interference experiment [Ou *et al.*, 1990b], two photons produced simultaneously provided the two inputs to a Mach–Zehnder interferometer, as shown in Fig. 17.6, and the photons emerging at the two outputs were counted. Even though the average intensities of both outputs remained constant, the rate of

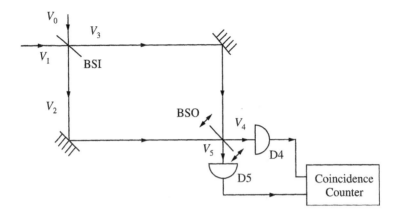

Figure 17.6. Experimental arrangement used to study interference by two-photon superposition states formed by pairs of photons at a beamsplitter [Ou *et al.*, 1990b].

coincidences was found to exhibit interference fringes with high visibility when the optical path difference was varied. This behavior can be attributed to the fact that when two similar photons enter a beamsplitter simultaneously at ports 0 and 1, two photons always emerge together, either at port 2 or at port 3 [Hong, Ou, and Mandel, 1987]. The resulting fourth-order interference fringes, which have a visibility of unity, are a consequence of the interference of photon pairs, rather than single photons.

17.3   The Geometric Phase

In both the experiments described above, there is often no difference in the effects predicted by a classical treatment or a quantum-mechanical treatment, since the measurements involve optical path differences. However, the effects produced by the geometric (Pancharatnam) phase in fourth-order interference have been shown to be quite different by Brendel, Dultz, and Martienssen [1995] using the experimental system shown in Fig. 17.7. In this arrangement, photon pairs generated by downconversion of blue light ($\lambda = 458$ nm) from an Ar^+ laser in a beta barium borate (BBO) crystal traversed a Michelson interferometer. One arm of this interferometer contained two quarter-wave plates, one of which was fixed at an azimuth of $45°$, while the other could be rotated. Rotation of the second quarter-wave plate through an angle θ introduced geometric (Pancharatnam) phases $\Delta\phi = \pm 2\theta$, respectively, for the two orthogonal polarizations.

Experiments were carried out using two BBO crystals cut, respectively, for type-I and type-II phase matching, so that the photons of a pair could be prepared either in the same state of polarization (type-I) or in orthogonal states of polarization (type-II). The photon pairs emerging from the interferometer were incident on a second beamsplitter BS_2 which directed them to two photodetectors D_1 and D_2. With type-I

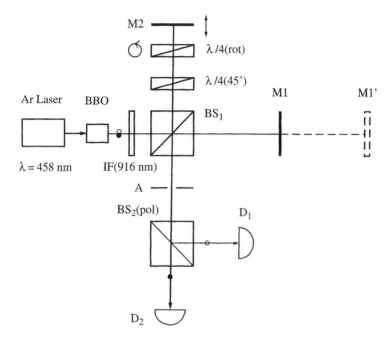

Figure 17.7. Experimental setup used to demonstrate the effects of the geometric phase in fourth-order interference [Brendel, Dultz, and Martienssen, 1995].

phase matching, BS_2 was a normal beamsplitter, while with type-II phase matching, BS_2 was a polarizing beamsplitter.

Measurements with this system showed that the effects observed depended on the initial states of polarization of the two photons in a pair and the optical path difference. With near-zero optical path differences, second-order interference fringes were observed. Under these conditions, the effects of the dynamic phase and the geometric phase were equivalent. However, with large optical path differences and coincidence detection, no changes in the coincidence rate were observed as the second quarter-wave plate was rotated, in the case of type-II phase matching, whereas with type-I phase matching, interference fringes with a visibility of 0.78 were obtained with a period half that expected with a classical light field.

These results imply that pairs of photons with parallel polarizations acquire twice the geometric phase of single photons and behave like single particles with spin 2. On the other hand, pairs of photons with orthogonal polarizations acquire geometric phases with opposite signs and behave like a single particle with total spin 0.

Chapter 18

Two-Photon Interferometry

The incompatibility of quantum theory and local realism was first pointed out in a classic paper by Einstein, Podolsky, and Rosen [1935]. They concluded that the quantum-mechanical description of a system was incomplete and postulated the existence of "hidden variables," which predetermined the results of any measurements. Subsequently, Bell [1965] formulated an inequality based on a "thought experiment" of Bohm [1951] involving spin-half particles, that could distinguish between quantum theory and theories based on local realism.

18.1 Tests of Bell's Inequality

The first tests of Bell's inequality [Aspect, Dalibard, and Roger, 1982] used measurements of the correlation of the polarizations of the two photons in a pair produced by an atomic cascade (see Section 16.5). However, the generation of correlated photon pairs by parametric down-conversion (see Section 16.3.1) has made possible more precise tests of Bell's inequality using two-photon interferometry [Horne, Shimony, and Zeilinger, 1989].

In the arrangement for two-photon interferometry shown in Fig. 18.1, pairs of photons, having wavelengths λ_1 and λ_2, are selected by four pinholes to produce four beams, A, B, C, D, with wave vectors $\mathbf{k}_A$, $\mathbf{k}_B$, $\mathbf{k}_C$, and $\mathbf{k}_D$, where

$$|\mathbf{k}_A| = |\mathbf{k}_D|, \ |\mathbf{k}_B| = |\mathbf{k}_C|, \text{ but } |\mathbf{k}_B| \neq |k_A|, \tag{18.1}$$

and

$$\mathbf{k}_A + \mathbf{k}_C = \mathbf{k}_B + \mathbf{k}_D = \mathbf{k}, \tag{18.2}$$

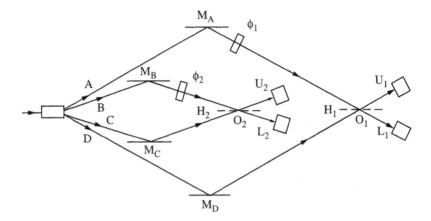

Figure 18.1. Arrangement used to test Bell's inequality using two-photon interference [Horne, Shimony, and Zeilinger, 1989].

where $\mathbf{k}$ is the wave vector of the beam incident on the crystal. Each pair in the ensemble of photon pairs is in the quantum state

$$|\psi\rangle = 2^{1/2}[|A\rangle_1|C\rangle_2 + |D\rangle_1|B\rangle_2], \tag{18.3}$$

which is a coherent superposition of the probability amplitudes corresponding to two distinct pairs of correlated paths. In one case, a photon with wavelength λ_1 goes into beam A, and a photon with wavelength λ_2 goes into beam C; in the other, a photon with wavelength λ_1 goes into beam D, and a photon with wavelength λ_2 goes into beam B. A variable phase difference ϕ_1 can be introduced between the beams A and D before they are recombined by the 50:50 beamsplitter H_1 and reach the photodetectors U_1 and L_1. Similarly, a variable phase difference ϕ_2 can be introduced between the beams B and C before they are recombined by the 50:50 beamsplitter H_2 and reach the photodetectors U_2 and L_2.

The quantum mechanical probabilities for the joint detection of both photons by the detector pairs (U_1, U_2), (L_1, L_2), (U_1, L_2), and (U_2, L_1), are then

$$\mathcal{P}(U_1, U_2|\phi_1, \phi_2) = \mathcal{P}(L_1, L_2|\phi_1, \phi_2)$$
$$= (\eta^2/4)[1 + \cos(\phi_2 - \phi_1 + \phi_0)] \tag{18.4}$$

and

$$\mathcal{P}(U_1, L_2|\phi_1, \phi_2) = \mathcal{P}(L_1, U_2|\phi_1, \phi_2)$$
$$= (\eta^2/4)[1 - \cos(\phi_2 - \phi_1 + \phi_0)], \tag{18.5}$$

where η is the quantum efficiency of the photodetectors, and ϕ_0 is a constant phase factor.

On the other hand, the probabilities for detecting single photons by the four detectors are

$$\mathcal{P}(U_1|\phi_1, \phi_2) = \mathcal{P}(L_1|\phi_1, \phi_2) = \mathcal{P}(U_2|\phi_1, \phi_2) = \mathcal{P}(L_2|\phi_1, \phi_2)$$
$$= \eta/2. \tag{18.6}$$

It follows that while the count rate for single photons, which is defined by Eq. (18.6), is constant and independent of ϕ_1 and ϕ_2, the count rates for coincidences, which are defined by Eqs. (18.4) to (18.5), will vary sinusoidally with the phase shifts ϕ_1 and ϕ_2.

The correlation coefficient

$$\mathcal{E}(\phi_1, \phi_2)$$
$$= \frac{\mathcal{P}(U_1, U_2|\phi_1, \phi_2) + \mathcal{P}(L_1, L_2|\phi_1, \phi_2) - \mathcal{P}(U_1, L_2|\phi_1, \phi_2) - \mathcal{P}(L_1, U_2|\phi_1, \phi_2)}{\mathcal{P}(U_1, U_2|\phi_1, \phi_2) + \mathcal{P}(L_1, L_2|\phi_1, \phi_2) + \mathcal{P}(U_1, L_2|\phi_1, \phi_2) + \mathcal{P}(L_1, U_2|\phi_1, \phi_2)}$$
$$\tag{18.7}$$

gives the distribution of coincidences between detectors on the same side and opposite sides of the beamsplitters for these values of the phase. A generalization of Bell's inequality [Clauser and Shimony, 1978] states that a combination of four such measurements, given by the relation

$$\mathcal{S} = \mathcal{E}(\phi_1, \phi_2) - \mathcal{E}(\phi_1, \phi_2') + \mathcal{E}(\phi_1', \phi_2) + \mathcal{E}(\phi_1', \phi_2'), \tag{18.8}$$

should always lie within the bounds

$$-2 \leq \mathcal{S} \leq 2, \tag{18.9}$$

if we assume local realism. However, quantum theory indicates that, for appropriately chosen values of the phase angles ($\phi_1 = 0$, $\phi_1' = \pi/2$, $\phi_2 = \pi/4$, $\phi_2' = 3\pi/4$),

$$\mathcal{S} = 2\sqrt{2}. \tag{18.10}$$

In actual measurements [Rarity and Tapster, 1990], a value of $\mathcal{S} = 2.21 \pm 0.022$ was obtained. This lower value could be attributed to a reduced visibility $\mathcal{V} = 0.78$ of the interference fringes due to misalignment of the apparatus, but corresponded to a violation of Bell's inequality by 10 standard deviations.

18.1.1 Tests with Unbalanced Interferometers

Another experimental test of Bell's inequality was proposed by Franson [1989] and carried out by several groups [Franson, 1991a; Brendel, Mohler, and Martienssen,

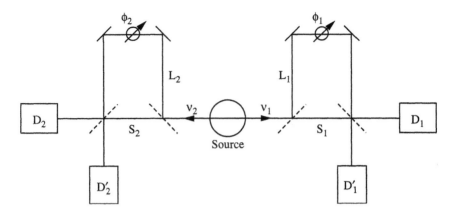

Figure 18.2. Schematic of an experiment to test Bell's inequality with unbalanced interferometers [Franson, 1989].

1992; Kwiat, Steinberg, and Chiao, 1993; Shih, Sergienko, and Rubin, 1993]. As shown in Fig. 18.2, each of the photons from a down-converted pair was sent into an unbalanced interferometer, presenting a short (S) and a long (L) path to the final output. Examination of the singles count rates, when the differences in the optical paths in each interferometer were greater than the coherence length of the down-converted photons, revealed no interference effects. However, when the difference of the imbalances in the two interferometers was less than the coherence length of the down-converted photons, observations of the coincidence rates revealed interference effects arising from the impossibility of distinguishing between the two processes that led to coincidences. These interference effects could be observed even when the extra optical path traversed by one of the photons was quite long [Franson, 1991b; Rarity and Tapster, 1992].

With detectors fast enough to exclude the possibility of one photon taking the short path and the other taking the long path, fringes could be obtained corresponding to observations of the quantum state

$$|\psi\rangle = \frac{1}{2}[|S_1, S_2\rangle - \exp(\mathrm{i}\phi)|L_1, L_2\rangle], \qquad (18.11)$$

where ϕ is proportional to the sum of the relative phases in the two interferometers. The visibility of the fringes was greater than 0.8 [Kwiat, Steinberg, and Chiao, 1993], whereas the maximum possible without violating Bell's inequality would be 0.71.

18.1.2   Tests with Perfect Correlations

A shortcoming of experiments with a pair of particles is that they reveal contradictions between the predictions of quantum mechanics and the EPR postulates only in situations involving imperfect correlations. A way to overcome this limitation, proposed by Greenberger, Horne, and Zeilinger [1989], was to use three or more particles in an entangled state. Greenberger *et al.* [1990] showed that, in this case, contradictions emerge even at the level of perfect correlations. They also described a "thought experiment" using three entangled photons to illustrate this point.

While no method for generating three entangled photons exists, Tewari and Hariharan [1997] have shown how four entangled photons can be generated by parametric down-conversion of two pump beams, and a design for an experiment using four-photon interference has been presented by Hariharan, Samuel, and Sinha [1999]. In this case, it should be possible, with just two measurements, to demonstrate that quantum mechanics contradicts any theory based on local realism, even at the level of perfect correlations.

18.2 Quantum Cryptography

A promising application for two-photon interferometry is in quantum cryptography.

Completely secure encryption can be ensured if a key shared between a sender (A) and a receiver (B) is (1) as long as the message, (2) is purely random, and (3) is used only once. The major drawback of this technique of encryption (the one-time pad) is that, to establish the key for each message, A needs a perfectly secure channel to transmit the key to B. This problem can be overcome with a quantum channel using correlated photons, since intrusion by an eavesdropper (E) results in a loss of photon correlations.

A scheme for quantum key distribution [Ekert *et al.*, 1992] uses pairs of photons that travel along separate optical fibers to identical unbalanced interferometers (see Fig. 18.2), one with A and the other with B. Each interferometer contains a shorter path and a longer path with a difference in transit time greater than the coherence time of the down-converted photons [Franson, 1989]. If the phase differences in the two interferometers are, respectively, $(\phi_A + \theta)$ and $(\phi_B + \theta)$, with $\theta = 2m\pi$, where m is an integer, the coefficient of correlation of the four outputs is

$$J(\phi_A, \phi_B) = \cos(\phi_A + \phi_B). \tag{18.12}$$

To establish a key, A chooses, at random, values of ϕ_A of 0 or $\pi/2$, while B chooses, at random, values of ϕ_B of 0 or $-\pi/2$. After a sufficiently large number of photons have been recorded, A and B reveal the settings of ϕ_A and ϕ_B that they have used, but not which detectors recorded a photon. They then discard all measurements in which $\phi_A + \phi_B \neq 0$, as well as those in which one or both of them failed to detect a photon. Finally, A and B compare the results of the photocounts on a randomly chosen subset of the measurements. If the results exhibit a sufficiently high degree of correlation, the key is considered secure and can be used.

A drawback of this system is the need to exchange information on the phase shifts on an open channel. In addition, errors due to transmission losses, imperfect detector efficiency, and dark counts can corrupt the key and conceal the activities of an eavesdropper. Various proposals have been made to overcome these limitations, including the use of three-state systems [Phoenix, Barnett, and Chefles, 2000; Bechmann-Pasquinicci and Peres, 2000] and four-photon interference [Hariharan and Sanders, 2002].

18.3 Beams from Two Down-Converters

Another class of two-photon interference experiments makes use of the down-converted light beams from two nonlinear crystals that are optically pumped by mutually coherent beams from the same laser. In one arrangement (see Fig. 18.3), the signal beams s_1 and s_2 from the two down-converters were combined by one beamsplitter (BS_A) and allowed to fall on one photodetector (D_A), while the two idler beams i_1 and i_2 were combined by another beamsplitter (BS_B) and taken to another photodetector (D_B) [Ou *et al.*, 1990c]. Measurements of the counting rates of the individual photodetectors showed no change as the optical path difference was varied, confirming that the mutual coherence of the pump beams did not produce any mutual coherence, either between the two signal beams s_1 and s_2 from the two down-converters, or between the two idler beams i_1 and i_2. However, measurements of the coincidence rate for simultaneous detection of photons by both D_A and D_B, as a function of the optical path difference, revealed interference effects.

A modification of this arrangement, shown in Fig. 18.4, used a single nonlinear crystal traversed by the pump beam in opposite senses [Herzog *et al.*, 1994]. In this case, down-converted photons could be generated on either of the two passes, and it was possible to make the idler modes from the two processes overlap at one photodetector, while the signal modes overlapped at the other. Since the two production processes were indistinguishable, interference effects were observed in the count rates at the individual detectors, as well as the coincidence rates, when any one of the mirrors was moved, even when the distances to the mirrors were much greater than the coherence lengths of the down-converted beams.

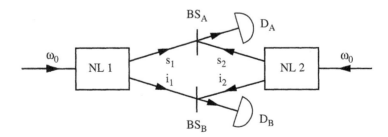

Figure 18.3. Optical system used to observe interference effects produced by down-converted light beams from two nonlinear crystals [Ou *et al.*, 1990c].

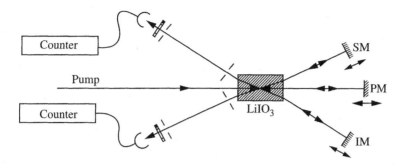

Figure 18.4. Arrangement using a single nonlinear crystal to generate two sets of down-converted light beams [Herzog *et al.*, 1994].

Nonclassical effects can also be seen in some second-order interference experiments in which only one photon is detected [Mandel, 1982]. Figure 18.5 is a schematic of the optical system for such an interference experiment with beams from two parametric down-converters [Zou, Wang, and Mandel, 1991]. In this case also, both the nonlinear crystals, NL1 and NL2, were optically pumped by mutually coherent beams derived from the same laser by means of a beamsplitter. However, while the two signal beams s_1 and s_2 were combined by means of another beamsplitter and taken to a photodetector (D_s), the idler beam i_1 was allowed to pass through the nonlinear crystal NL2 and fall, along with the other idler beam i_2, directly on the other photodetector D_i.

In this case, the photon counting rate at D_s was found to oscillate as the optical path difference was varied by translating the beamsplitter BS_0, indicating that s_1 and s_2 were mutually coherent. These oscillations could be observed as long as i_1 and i_2

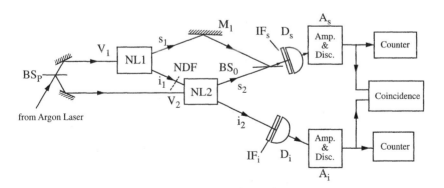

Figure 18.5. Experimental arrangement used to demonstrate interference effects produced by the signal beams from two parametric down-converters [Zou, Wang, and Mandel, 1991].

were collinear, but if either i_1 or i_2 was misaligned, or if i_1 was blocked so that it could not reach NL2, they disappeared. If, instead of blocking i_1, an attenuator with a complex amplitude transmittance t was placed between NL1 and NL2, the visibility of the interference pattern registered by D_s was found to be proportional to $|t|$. The average rate of photon counts was the same in both cases, implying that the degree of mutual coherence of the two beams could be controlled without affecting their intensities.

In addition, the introduction of a delay τ, by varying the length of the path of the idler i_1 between the two nonlinear crystals NL1 and NL2, was found to affect the visibility of the interference effects produced by the signal beams, s_1 and s_2, exactly as if the delay had been introduced in one of the signal paths [Zou et al., 1993]. When $\tau > \tau_c$, where τ_c was the coherence time, the interference effects disappeared. A phase shift of the idler beam i_1 introduced through the geometric (Pancharatnam) phase (see Section 16.6) also had the same effect on the interference pattern produced by the signal beams [Grayson, Torgerson, and Barbosa, 1994].

As is well known, even when $\tau \gg \tau_c$, interference effects can still be seen in the spectral domain [Mandel, 1962]. Such effects were observed in this experiment by inserting a scanning Fabry–Perot interferometer before the detector D_s [Zou, Grayson, and Mandel, 1992]. The modulation of the spectrum could then be observed even with a differential delay $\tau \approx 5$ ps $\approx 5\tau_c$, but disappeared when the idler beam i_1 was blocked.

All these effects can be attributed to the indistinguishability of the paths taken by the beams through the interferometer. In the arrangement shown in Fig. 18.3, when a coincidence is registered, there is no way to determine whether the pair of photons involved originated in NL1 or NL2. Similarly, in the arrangement shown in Fig. 18.5, it is not possible to determine the origins of the photons reaching the photodetector D_s, as long as both i_1 and i_2 are incident on the detector D_i.

18.4   The Quantum Eraser

A consequence of the uncertainty principle is that whenever "which path" information is available, interference effects cannot be observed. However, interference effects may reappear if the information can somehow be "erased." This procedure is the basis of what is now known as the "quantum eraser" [Scully and Druhl, 1982; Scully, Englert, and Walther, 1991; Zajonc et al., 1991].

One demonstration of a quantum eraser [Kwiat, Steinberg, and Chiao, 1992] used the interferometer shown in Fig. 18.6 in which the plane of polarization of one of the beams could be rotated by a half-wave plate inserted in one of the paths before the beamsplitter. When the polarization of this beam was orthogonal to that of the other beam, the coincidence null disappeared, since it became possible to identify the paths taken by each of the photons. However, this information could be erased by inserting two polarizers just in front of the photodetectors, after the photons had left the beamsplitter. Interference could not be restored with a single polarizer in front of

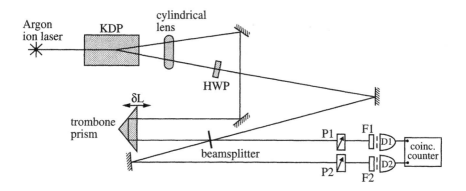

Figure 18.6. Two-photon interferometer used as a quantum eraser [Kwiat, Steinberg, and Chiao, 1992].

one detector, since "which-path" information was available from the photon reaching the other photodetector.

In particular, if the initial polarization of the down-converted photons was horizontal, and the half-wave plate rotated one polarization to vertical, polarizers at 45° before each detector restored the original coincidence null. However, as shown in Fig. 18.7, if one polarizer was set at 45° and the other at −45°, an interference peak was observed instead of a dip.

A demonstration of the quantum eraser is also possible with the interferometer shown in Fig. 18.3 [Ou *et al.*, 1990c]. In this case, removal of the beamsplitter BS_B, which at first sight should not affect the results, destroys the interference. The reason is that since the signal and idler photons are produced simultaneously, it then becomes possible, from the output of the photodetector D_B, to decide whether the corresponding signal photon comes from NL1 or NL2. Insertion of BS_B mixes the idlers and erases the information on the paths taken by the photons [Zajonc *et al.*, 1991].

The experimental arrangement shown in Fig. 18.5 [Zou, Wang, and Mandel, 1991] could also be modified to demonstrate this concept by using a half-wave plate between the two crystals to rotate the polarization of the idler photons from NL1 so that it was orthogonal to the polarization of the idlers from NL2. In this case, interference could be recovered by using a polarizer in front of D_i and correlating the counts of the two detectors.

18.5 Single-Photon Tunneling

If two right-angle prisms are placed with their hypotenuse faces opposite each other, but separated by an air gap, a beam of light incident on the interface at an angle greater than the critical angle is totally internally reflected. However, if the air gap is reduced to a fraction of a wavelength, some of the light is transmitted. The fact that light tunnels through such a gap confirms the wave-like behavior of light.

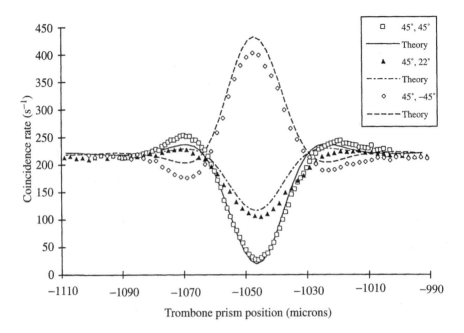

Figure 18.7. Experimental data obtained with a two-photon interferometer with the polarizer in one path set at 45° and the polarizer in the other path set at various angles [Kwiat, Steinberg, and Chiao, 1992].

On the other hand, if the same experiment is repeated with single-photon states, quantum mechanics predicts that photons will be detected in perfect anticoincidence in the transmitted and reflected beams [Ghose, Home, and Agarwal, 1991]. This prediction has been verified experimentally by Mizobuchi and Ohtake [1992]. Accordingly, we have a situation where single-photon states simultaneously display wave-like properties (tunneling) and particle-like properties (anticoincidence).

While all quantum particles can, in principle, tunnel through normally forbidden regions of space, the question of how much time it takes for a particle to tunnel through a barrier is quite controversial. Interferometric experiments have made it possible to study this aspect of photon tunneling.

18.5.1 Dispersion Cancelation

It follows from the uncertainty principle that, to make measurements of transit times with high resolution, it is necessary to make the energy uncertainty or spectral bandwidth quite large. With such large spectral bandwidths, dispersion can result in significant broadening of a pulse, and a consequent decrease in time resolution [Franson, 1992]. This problem can be solved by making measurements with correlated photon pairs. It is then possible to take advantage of quantum-mechanical effects to obtain an effective cancelation of dispersion [Steinberg, Kwiat, and Chiao,

1992a, 1992b, 1993]. If we use an interferometer similar to that described by Hong, Ou, and Mandel [1987] (see Fig. 17.2), with a dispersive medium (say, a glass plate) in one beam, one photon of each pair travels through the dispersive medium while its conjugate travels through a path containing only air. However, after the photons are recombined at the beamsplitter, it is impossible to determine which one of them traveled through the glass plate. This indistinguishability leads to a cancelation of first-order dispersion effects, so that there is no broadening of the coincidence minimum. As a result, the shift in the position of the minimum in the rate of coincidences can be used to make high-resolution measurements of the propagation delay produced by the glass plate.

18.5.2 Measurements of Tunneling Time

In the arrangement used for measurements of tunneling time shown in Fig. 18.8 [Steinberg, Kwiat, and Chiao, 1993], the tunnel barrier was a dielectric mirror consisting of 11 alternating layers of low- and high-index material, each a quarter of a wavelength thick at the wavelength used (700 nm in air), coated on one half of the surface of a high-quality optical substrate. In such a periodic structure, the multiple reflections interfere so as to damp the incident wave exponentially, resulting in the equivalent of a photonic bandgap [Yablonovitch, 1993].

To make measurements of the tunneling time, the multilayer structure was moved into and out of the beam periodically, while the optical path difference between the beams was slowly varied by translating the reflecting prism. A Gaussian curve was then fitted to each of the two dips in the coincidence rate, and the distance between their

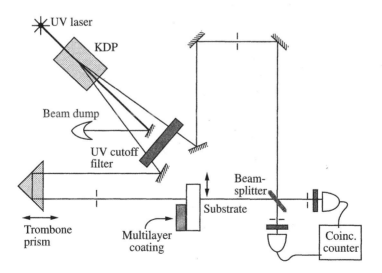

Figure 18.8. Apparatus used for measurements of the single-photon tunneling time [Steinberg, Kwiat, and Chiao, 1993].

centers was calculated. The average of several such measurements showed that the peak arrived 1.47 ± 0.21 fs earlier when the multilayer was in the path. Measurements of the tunneling time as a function of the angle of incidence made it possible to study its energy dependence [Steinberg and Chiao, 1995].

18.6   Conclusions

The complementarity of wave and particle behavior appears to be the reason for the puzzling features of quantum theory. However, in most cases the "photon picture" gives a good description of interference phenomena, provided that we apply Dirac's dictum that "a photon interferes only with itself." This rule can cover higher-order interference effects, if we generalize it to read "each pair of photons ... interferes only with itself."

An alternative explanation is to regard optical interference as due to the existence of indistinguishable paths [Mandel, 1991]. We can then apply Feynman's rules for interference, after taking into account the fact that outcomes which are distinguishable, even in principle, do not interfere. In all other cases, the coherent addition of the probability amplitudes associated with each path, and the evaluation of the squared modulus of this sum, yields the probability of detection of a photon.

Appendix A

Two-Dimensional Linear Systems

The distribution of the complex amplitude in an image can be expressed by means of the two-dimensional Fourier transform as a function of two orthogonal spatial frequencies [Goodman, 1996]. As a result, the original light wave is decomposed into component plane waves whose direction cosines correspond to the various spatial frequencies in the image. The propagation of these components through an optical system can then be analyzed using the Fresnel–Kirchhoff integral (see Appendix B).

A.1 The Fourier Transform

The two-dimensional Fourier transform of $g(x, y)$ is defined as

$$\mathcal{F}\{g(x, y)\} = \int_{-\infty}^{\infty} \int_{-\infty}^{\infty} g(x, y) \exp[-i2\pi(\xi x + \eta y)]dxdy$$

$$= G(\xi, \eta). \tag{A.1}$$

Similarly, the inverse Fourier transform of $G(\xi, \eta)$ is defined as

$$\mathcal{F}^{-1}\{G(\xi, \eta)\} = \int_{-\infty}^{\infty} \int_{-\infty}^{\infty} G(\xi, \eta) \exp[i2\pi(\xi x + \eta y)]d\xi d\eta$$

$$= g(x, y). \tag{A.2}$$

Equations (A.1) and (A.2) can be written symbolically as

$$g(x, y) \leftrightarrow G(\xi, \eta). \tag{A.3}$$

Some important theorems are:

a. The linearity theorem

$$\mathcal{F}\{ag(x, y) + bh(x, y)\} = aG(\xi, \eta) + bH(\xi, \eta), \qquad (A.4)$$

where a and b are constants, and $g(x, y) \leftrightarrow G(\xi, \eta), h(x, y) \leftrightarrow H(\xi, \eta)$.

b. The shift theorem

$$\mathcal{F}\{g(x - a, y - b) = G(\xi, \eta) \exp[-i2\pi(\xi a + \eta b)]. \qquad (A.5)$$

c. The similarity theorem

$$\mathcal{F}\{g(ax, by) = (1/|ab|)G(\xi/a, \eta/b). \qquad (A.6)$$

d. Plancherel's theorem (also called Rayleigh's theorem)

$$\int_{-\infty}^{\infty} \int_{-\infty}^{\infty} |g(x, y)|^2 dx dy = \int_{-\infty}^{\infty} \int_{-\infty}^{\infty} |G(\xi, \eta)|^2 d\xi d\eta. \qquad (A.7)$$

A.2   Convolution and Correlation

The output from any linear system is the convolution of the input and the instrument function. In two dimensions, the convolution of two functions, $g(x, y)$ and $h(x, y)$, is

$$f(x, y) = \int_{-\infty}^{\infty} \int_{-\infty}^{\infty} g(u, v)h(x - u, y - v)du dv, \qquad (A.8)$$

which can also be written as

$$f(x, y) = g(x, y) * h(x, y), \qquad (A.9)$$

where the symbol $*$ denotes the convolution operation.

The cross-correlation of two functions, $g(x, y)$ and $h(x, y)$, is

$$c(x, y) = \int_{-\infty}^{\infty} \int_{-\infty}^{\infty} g^*(u, v)h(x + u, y + v)du dv, \qquad (A.10)$$

where $g^*(u, v)$ is the complex conjugate of $g(u, v)$. Equation (A1.10) can be written as

$$c(x, y) = g(x, y) \star h(x, y), \qquad (A.11)$$

where the symbol $\star$ denotes the correlation operation. A comparison of Eq. (A.10) with Eq. (A.8) shows that the cross-correlation can also be expressed as a convolution

$$c(x, y) = g^*(x, y) * h(-x, -y).$$ (A.12)

The autocorrelation of a function $g(x, y)$ is then

$$a(x, y) = \int_{-\infty}^{\infty} \int_{-\infty}^{\infty} g^*(u, v)g(x + u, y + v)\mathrm{d}u\mathrm{d}v$$

$$= g(x, y) \star g(x, y).$$ (A.13)

Two useful results which follow are:

a. The convolution theorem
  If $g(x, y) \leftrightarrow G(\xi, \eta)$ and $h(x, y) \leftrightarrow H(\xi, \eta)$,

$$\mathcal{F}\{g(x, y) * h(x, y)\} = G(\xi, \eta)H(\xi, \eta).$$ (A.14)

b. The autocorrelation (Wiener–Khinchin) theorem

$$\mathcal{F}\{g(x, y) \star g(x, y)\} = |G(\xi, \eta)|^2.$$ (A.15)

A.3 The Dirac Delta Function

The two-dimensional Dirac delta function is a convenient representation of a point source. The delta function takes the values

$$\delta(x, y) = \infty, \ (x = 0, \ \text{and} \ y = 0),$$

$$\delta(x, y) = 0, \ (x \neq 0, \ \text{or} \ y \neq 0),$$ (A.16)

and its integral is equal to unity.
  By definition, convolution of a function with the Dirac delta function yields the original function, so that, in two dimensions,

$$\int_{-\infty}^{\infty} \int_{-\infty}^{\infty} f(u, v)\delta(x - u, y - v)\mathrm{d}u\mathrm{d}v = f(x, y).$$ (A.17)

A.4 Random Functions

The correlation function, as normally defined, cannot be used with randomly varying quantities, since the correlation integral does not have a finite value. The cross-correlation of two stationary random functions, $g(t)$ and $h(t)$, is therefore written

[Papoulis, 1965] as

$$R_{gh}(\tau) = \lim_{T \to \infty} \frac{1}{2T} \int_{-T}^{T} g^*(t)h(t + \tau)\mathrm{d}t$$
$$= \langle g^*(t)h(t + \tau) \rangle. \tag{A.18}$$

The autocorrelation function of a random function, say $g(t)$, is therefore

$$R_{gg}(\tau) = \langle g^*(t)g(t + \tau) \rangle. \tag{A.19}$$

The power spectrum $S(\omega)$ of $g(t)$ is then defined as the Fourier transform of its autocorrelation function

$$S(\omega) \leftrightarrow R_{gg}(\tau). \tag{A.20}$$

Appendix B

The Fresnel–Kirchhoff Integral

If, as shown in Fig. B.1, a plane wave with an amplitude a is incident on an object with an amplitude transmittance $t(x_1, y_1)$ located in the plane $z = 0$, the complex amplitude at a point $P(x, y, z)$ is given by the Fresnel–Kirchhoff integral [Born and Wolf, 1999]

$$a(x, y, z) = (ia/\lambda) \int_{-\infty}^{\infty} \int_{-\infty}^{\infty} t(x_1, y_1)$$

$$\times \frac{\exp\{(-i2\pi/\lambda)[(x - x_1)^2 + (y - y_1)^2 + z^2]^{1/2}\}}{[(x - x_1)^2 + (y - y_1)^2 + z^2]^{1/2}}$$

$$\times \cos\theta \, dx_1 dy_1. \tag{B.1}$$

When $z \gg (x - x_1)$ and $(y - y_1)$, $\cos\theta \approx 1$, and a binomial approximation to the terms in square brackets allows us to omit a factor $\exp(-i2\pi/\lambda z)$, which only affects the overall phase, and write Eq. (B.1) as

$$a(x, y, z) = (ia/\lambda z) \int_{-\infty}^{\infty} \int_{-\infty}^{\infty} t(x_1, y_1)$$

$$\times \exp\{(-i\pi/\lambda z)[(x - x_1)^2 + (y - y_1)^2]\} dx_1 dy_1$$

$$= (ia/\lambda z) \int_{-\infty}^{\infty} \int_{-\infty}^{\infty} t(x_1, y_1) \exp[(-i\pi/\lambda z)(x^2 + y^2)] \exp[(-i\pi/\lambda z)$$

$$\times (x_1^2 + y_1^2)] \exp\{i2\pi[x_1(x/\lambda z) + y_1(y/\lambda z)]\} dx_1 dy_1. \tag{B.2}$$

If, in addition, z is in the far field of the object, in which case

$$z \gg (x_1^2 + y_1^2)/\lambda, \tag{B.3}$$

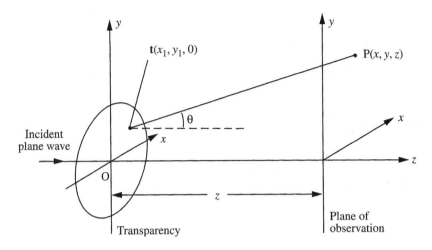

Figure B.1. Derivation of the Fresnel–Kirchhoff integral.

and if we set

$$\xi = x/\lambda z, \quad \eta = y/\lambda z, \tag{B.4}$$

Eq. (B.2) reduces to

$$
\begin{aligned}
a(x, y, z) &= (ia/\lambda z)\exp[(-i\pi/\lambda z)(x^2 + y^2)] \\
&\quad \times \int_{-\infty}^{\infty}\int_{-\infty}^{\infty} \mathbf{t}(x_1, y_1)\exp[i2\pi(\xi x_1 + \eta y_1)]dx_1 dy_1 \\
&= (ia/\lambda z)\exp[(-i\pi/\lambda z)(x^2 + y^2)]\mathbf{T}(\xi, \eta),
\end{aligned}
\tag{B.5}
$$

where

$$\mathbf{t}(x_1, y_1) \leftrightarrow \mathbf{T}(\xi, \eta). \tag{B.6}$$

Accordingly, the complex amplitude in the plane of observation is given by the Fourier transform of the amplitude transmittance of the object, multiplied by a spherical phase factor. If the object transparency is placed in the front focal plane of a lens, and the plane of observation is located in its back focal plane, the exact Fourier transform is obtained. In this case, z tends to infinity, and the phase factor $\exp[(-i\pi/\lambda z)(x^2 + y^2)]$ reduces to unity.

Appendix C

Reflection and Transmission

C.1 The Fresnel Formulas

Consider a plane monochromatic wave incident, as shown in Fig. C.1, on the interface between two isotropic transparent media with refractive indices n_1 and n_2.

If θ_1 and θ_2 are the angles of incidence and refraction, the coefficients of reflection and transmission for amplitude are given by the Fresnel formulas, which can be written as

$$r_\parallel = \frac{\tan(\theta_1 - \theta_2)}{\tan(\theta_1 + \theta_2)}, \tag{C.1}$$

$$t_\parallel = \frac{2\sin\theta_2\cos\theta_1}{\sin(\theta_1 + \theta_2)\cos(\theta_1 - \theta_2)}, \tag{C.2}$$

for light polarized with its electric vector in the plane of incidence (the p-component), and

$$r_\perp = -\frac{\sin(\theta_1 - \theta_2)}{\sin(\theta_1 + \theta_2)}, \tag{C.3}$$

$$t_\perp = \frac{2\sin\theta_2\cos\theta_1}{\sin(\theta_1 + \theta_2)}, \tag{C.4}$$

for light polarized with the electric vector perpendicular to the plane of incidence (the s-component). At the Brewster angle, which is defined by the condition $r_\parallel = 0$,

$$\theta_1 = \pi/2 - \theta_2 \tag{C.5}$$

and

$$\tan\theta_1 = n_2/n_1. \tag{C.6}$$

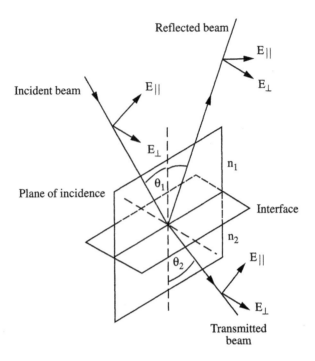

Figure C.1. Incident, transmitted, and reflected fields at an interface ($n_1 > n_2$) when $\theta_1 + \theta_2 < \pi/2$.

Since, below the critical angle, $t_\parallel$ and $t_\perp$ are always positive, there is no change in the phase of the transmitted wave. On the other hand, $r_\perp$, which is positive when $n_1 > n_2$, is negative when $n_1 < n_2$, indicating, in the latter case, a phase shift of π. In the case of $r_\parallel$, the phase shifts at small angles of incidence are the same as those for $r_\perp$ (the difference in their signs arises from the choice of the coordinate axes), but there is a change in sign at the Brewster angle.

C.2 The Stokes Relations

Let r and t be the reflected and transmitted amplitudes, respectively, for a wave of unit amplitude incident, as shown in Fig. C.2a, on the interface between two transparent media. If, then, as shown in Fig. C.2b, the direction of the reflected ray is reversed, it will give rise to a reflected component with an amplitude r^2 and a transmitted component with an amplitude rt. Similarly, if the direction of the transmitted ray is reversed, it will give rise to a reflected component with an amplitude tr' and a transmitted component with an amplitude tt', where r' and t' are, respectively, the reflectance and transmittance for amplitude for a ray incident on the interface from below.

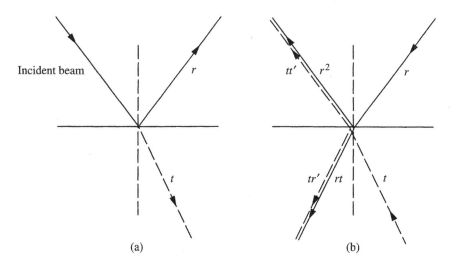

Figure C.2. Derivation of the Stokes relations.

If there are no losses at the interface,

$$r^2 + tt' = 1 \tag{C.7}$$

and

$$rt + tr' = 0. \tag{C.8}$$

Equations (C.7) and (C.8) lead to the Stokes relations

$$tt' = 1 - r^2 \tag{C.9}$$

and

$$r' = -r. \tag{C.10}$$

From Eq. (C.10), it follows that the phase shifts on reflection at the two sides of the interface differ by π.

Appendix D

The Jones Calculus

In the Jones calculus [Jones, 1941], the characteristics of a polarized beam are described by means of a two-element column vector, known as the Jones vector. For a plane monochromatic light wave propagating along the z axis, which has components $E_x = |a_x| \cos(\omega t + \phi_x)$ and $E_y = |a_y| \cos(\omega t + \phi_y)$ along the x and y directions, respectively, the Jones vector takes the form

$$\mathbf{A} = \begin{bmatrix} E_x \\ E_y \end{bmatrix}. \tag{D.1}$$

With monochromatic light, the variation with time of E_x and E_y is usually not of interest. Equation (D.1) is, therefore, often written in the form

$$\mathbf{A} = \begin{bmatrix} a_x \\ a_y \end{bmatrix} = \begin{bmatrix} |a_x| \exp(i\phi_x) \\ |a_y| \exp(i\phi_y) \end{bmatrix}, \tag{D.2}$$

which is known as the full Jones vector.

The intensity at a point is then

$$I = a_x a_x^* + a_y a_y^*$$
$$= \mathbf{A}^\dagger \mathbf{A}. \tag{D.3}$$

The intensity is, therefore, the product of the Jones vector and its Hermitian conjugate.

For convenience in calculations, the normalized form of the Jones vector is commonly used. This is obtained by multiplying the full Jones vector by a complex scalar that reduces the intensity to unity.

For example, for light that is linearly polarized at 45°, the amplitudes of the x and y components are the same, and the full Jones vector can be written as

$$\begin{bmatrix} a \\ a \end{bmatrix}. \tag{D.4}$$

The corresponding normalized vector is

$$2^{-1/2} \begin{bmatrix} 1 \\ 1 \end{bmatrix}. \tag{D.5}$$

Optical elements which modify the state of polarization of the beam are described by matrices containing four elements, known as Jones matrices. The Jones vector of the incident beam is multiplied by the Jones matrix of the optical element to obtain the Jones vector of the emerging beam.

For example, the Jones matrix for a half-wave plate with its axis vertical or horizontal is

$$\begin{bmatrix} 1 & 0 \\ 0 & -1 \end{bmatrix}. \tag{D.6}$$

If a beam linearly polarized at 45° is incident on this half-wave plate, the Jones vector for the emerging beam is

$$\begin{bmatrix} 1 & 0 \\ 0 & -1 \end{bmatrix} 2^{-1/2} \begin{bmatrix} 1 \\ 1 \end{bmatrix} = 2^{-1/2} \begin{bmatrix} 1 \\ -1 \end{bmatrix}, \tag{D.7}$$

which is a beam linearly polarized at $-45°$.

An important element in interferometry is a metal film (a mirror or a beamsplitter). Oblique reflection at such a film introduces a phase shift between the components of the incident beam polarized parallel and perpendicular to the plane of incidence. As a result, the amplitude reflection coefficients $\mathbf{r}_{\parallel}$ and $\mathbf{r}_{\perp}$ are complex, and the film can be represented by the Jones matrix

$$\mathbf{M} = \begin{bmatrix} \mathbf{r}_{\parallel} & 0 \\ 0 & \mathbf{r}_{\perp} \end{bmatrix}. \tag{D.8}$$

Note that account must be taken of the change from right-hand to left-hand coordinates on reflection.

In an optical system in which the beam passes through several elements in succession, the Jones vector of the beam leaving the system can be obtained by writing down the Jones matrices of the various elements and carrying out the multiplication in the proper sequence. Alternatively, where the state of polarization of the incident beam is to be varied, it is possible to compute the overall matrix for the system, and then multiply the Jones vectors for the various states of the incident beam by this matrix.

The use of the Jones calculus has been described in detail by Shurcliff [1962], Shurcliff and Ballard [1964], Jerrard [1982], and Collet [1992].

Appendix E

The Pancharatnam Phase

E.1 The Poincaré Sphere

The Poincaré sphere is a convenient way of representing the state of polarization of a beam of light; it also makes it very easy to visualize the effects of retarders [Ramachandran and Ramaseshan, 1952; Jerrard, 1954, 1982; Born and Wolf, 1999].

As shown in Fig. E.1, right- and left-circular polarized states are represented by the north and south poles of the sphere, while linearly polarized states lie on the equator, with the plane of polarization rotating by 180° for a change in longitude of 360°. If we consider a point on the equator at longitude 0° as representing a vertical linearly polarized state, any other point on the surface of the sphere with a latitude 2ω and a longitude $2l$ represents an elliptical vibration with an ellipticity $|\tan \omega|$ whose major axis makes an angle l with the vertical. Two states of polarization at the opposite ends of a diameter can be regarded as orthogonally polarized states.

A birefringent plate with a retardation δ, whose fast and slow axes are at angles θ and $\theta + 90°$ with the vertical, is represented by the points O and O' at longitudes of 2θ and $180° + 2\theta$. The effect of passage through such a retarder is equivalent to the operation of rotating the sphere about the diameter OO' by an angle δ and moves the linearly polarized state represented by the point P to P'.

E.2 The Pancharatnam Phase

Whereas it is a simple matter to evaluate the phase difference between two beams in the same state of polarization, difficulties arise when the two beams are in different states of polarization. Pancharatnam [1956] was able to solve this problem by considering the intensity produced when the two beams were made to interfere. He regarded the two beams as being "in phase" when the resultant intensity was a maximum.

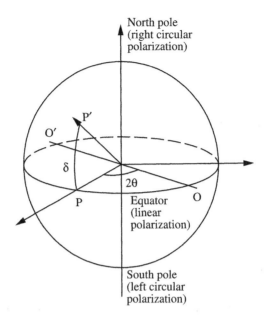

Figure E.1. The Poincaré sphere: effect of a birefringent plate on the state of polarization of a beam of light.

This approach made it possible to define how a beam changed its phase when its state of polarization was altered. It also led to the observation that if a beam was taken from one polarization state, without introducing any phase changes, through two other polarization states back to its original state, it could exhibit a phase shift. The magnitude of this phase shift (the Pancharatnam phase) was equal to half the solid angle subtended at the center of the Poincaré sphere by the circuit traversed by the point representing the state of polarization of the beam.

Appendix F

Holography

Holography permits complete reconstruction of an image, that is to say, reproduction of the relative phases as well as the relative amplitudes of the light waves scattered by an object. Conventional imaging techniques cannot do this, since all available recording materials respond only to the intensity. Gabor [1948] solved this problem by adding a coherent background. The phase variations across the wavefront are then encoded as variations in the intensity and the position of the resulting interference fringes.

F.1   The Off-Axis Hologram

We shall assume that, as shown in Fig. F.1, in addition to the object beam, a reference beam (a collimated beam of uniform intensity) is incident at an angle θ on the photographic plate [Leith and Upatnieks, 1964]. The complex amplitude due to the reference beam at any point $P(x, y)$ on the plate can be written [see Hariharan, 1996d] as

$$r(x, y) = r \exp(\mathrm{i}2\pi \xi_r x), \tag{F.1}$$

where $\xi_r = (\sin \theta)/\lambda$, while that due to the object beam is

$$o(x, y) = |o(x, y)| \exp[-\mathrm{i}\phi(x, y)]. \tag{F.2}$$

The resultant intensity in the interference pattern formed by these two waves is, therefore,

$$I(x, y) = |r(x, y) + o(x, y)|^2$$
$$= r^2 + |o(x, y)|^2$$

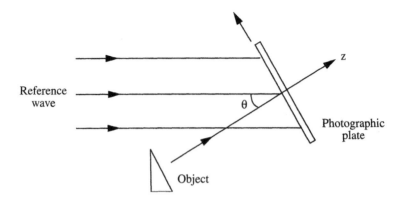

Figure F.1. Hologram recording with an off-axis reference beam.

$$+ r|o(x, y)| \exp\{-i[2\pi\xi_r x + \phi(x, y)]\}$$
$$+ r|o(x, y)| \exp\{i[2\pi\xi_r x + \phi(x, y)]\}$$
$$= r^2 + |o(x, y)|^2 + 2r|o(x, y)| \cos[2\pi\xi_r x + \phi(x, y)]. \qquad \text{(F.3)}$$

Equation (F.3) shows that the intensity distribution consists of a set of fine fringes, constituting a carrier with a spatial frequency ξ_r, whose visibility and spacing are modulated by the amplitude and phase of the object wave.

If the recording material has a linear response to exposure, its amplitude transmittance can be written as

$$\mathbf{t} = \mathbf{t}_0 + \beta T I, \qquad \text{(F.4)}$$

where $\mathbf{t}_0$ is a constant corresponding to the transmittance of the unexposed material, T is the exposure time, and β is the slope (negative) of the amplitude transmittance vs exposure characteristic of the material. The amplitude transmittance of the hologram is then

$$\mathbf{t}(x, y) = \mathbf{t}_0 + \beta T[r^2 + |o(x, y)|^2$$
$$+ r|o(x, y)| \exp\{-i[2\pi\xi_r x + \phi(x, y)]\}$$
$$= r|o(x, y)| \exp\{i[2\pi\xi_r x + \phi(x, y)]\}]. \qquad \text{(F.5)}$$

To reconstruct the image, the hologram is illuminated, as shown in Fig. F.2, with the reference beam. The complex amplitude of the transmitted wave is then

$$u(x, y) = r(x, y)\mathbf{t}(x, y)$$
$$= (\mathbf{t}_0 + \beta T r^2)r \exp(i2\pi\xi_r x)$$

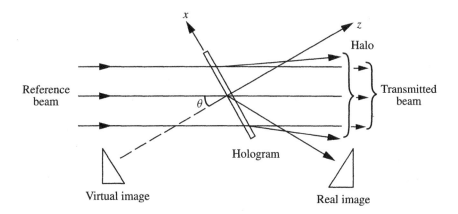

Figure F.2. Image reconstruction by a hologram recorded with an off-axis reference beam.

$$+ \beta T r |o(x, y)|^2 \exp(\mathrm{i}2\pi \xi_r x)$$
$$+ \beta T r^2 o(x, y)$$
$$+ \beta T r^2 o^*(x, y) \exp(\mathrm{i}4\pi \xi_r x). \tag{F.6}$$

Inspection of Eq. (F.6) shows that the first term on the right-hand side is merely the directly transmitted reference beam, while the second term corresponds to a halo around it. The third term, which is the one of interest, is the same as the original object wave, except for a constant factor, and generates a virtual image of the object in its original position. The fourth term corresponds to a conjugate wave which is propagated at an angle to the axis which is twice that made by the reference beam with it, and produces a real image.

If the angle between the object beam and the reference beam is made large enough, it is possible to view the virtual image without interference from the other transmitted and diffracted beams.

F.2 Computer-Generated Holograms

The production of a computer-generated hologram (CGH) involves two basic steps [Lee, 1978; Yaroslavskii and Merzlyakov, 1980]. The first step is to calculate the complex amplitude of the object wave at the hologram plane which, for convenience, is usually taken to be the Fourier transform of the complex amplitude in the object plane. Since the transform has to be evaluated digitally, it is necessary to sample the wavefront in the object plane. If the object wave is sampled at $N \times N$ equally spaced points, the $N \times N$ coefficients of its discrete Fourier transform can then be calculated using the fast Fourier-transform algorithm [Cochran *et al.*, 1967].

The next step is to produce, using the calculated values of the discrete Fourier transform, a transparency (the hologram) which, when properly illuminated, reconstructs the object wave. This process can be simplified considerably if the hologram has only two levels of amplitude transmittance—zero and one.

The best known hologram of this type is the binary detour-phase hologram [Brown and Lohmann, 1966, 1969]. In this type of hologram, the area of the hologram is divided into $N \times N$ cells, and each complex coefficient of the discrete Fourier transform is represented by a single transparent window within the corresponding cell. If such a hologram is illuminated at an angle by a collimated beam, it is apparent that a shift of the window within any cell would result in the light diffracted by it traveling by a longer or shorter path to the reconstructed image. Accordingly, it is possible to encode the complex coefficients by making the areas of the windows proportional to the moduli of these coefficients, while their phases are represented by the positions of the windows within the cells.

A problem is noise due to quantization of the modulus and phase of the Fourier coefficients. A number of encoding techniques have been described to minimize this problem [Lee, 1974, 1979].

Appendix G

Speckle

G.1 Speckle Statistics

If the scattered light from a rough surface illuminated with coherent light falls on a screen, a granular pattern, commonly called a speckle pattern, is seen [Dainty, 1976]. The complex amplitude at any point in this speckle pattern is the sum of the complex amplitudes of the diffracted waves from N microscopic elements on the scatterer and is given by the relation

$$A(x, y) = N^{-1/2} \sum_{m=1}^{N} |A_m| \exp(-i\phi_m). \tag{G.1}$$

The second-order statistics of the intensity distribution in the speckle pattern can be evaluated [Goldfischer, 1965; Goodman, 1984] by making use of the fact that the complex amplitude $A(x_1, y_1)$ at the scattering plane and the complex amplitude $A(x, y)$ at the observation plane are related by the Fresnel–Kirchhoff integral (see Appendix B). If an image of the scattering surface is formed by a lens, the dimensions of the speckles correspond to the width of the autocorrelation peak and are determined by the aperture of the lens, which acts as a low-pass spatial filter [Lowenthal and Arsenault, 1970]. For a circular pupil of radius ρ, the average dimensions of the speckles are

$$\Delta x = \Delta y = 0.61 \lambda f / \rho, \tag{G.2}$$

where f is the focal length of the lens, and correspond to those of a diffraction-limited image of a point.

The other quantity of interest is the power spectral density of the intensity distribution, which is given by the Fourier transform of the autocorrelation function and is made up of two components, each containing half the total power. One is a delta function at the origin, while the other is the normalized autocorrelation function of the intensity distribution over the scattering surface or, in the case of image speckle, the imaging lens.

G.2 Young's Fringes

We assume that two identical speckle patterns are recorded with a small mutual displacement on a photographic film whose amplitude transmittance, after processing, can be written as

$$\mathbf{t} = \mathbf{t}_0 + \beta T I, \tag{G.3}$$

where $\mathbf{t}_0$ is the transmittance of the unexposed material, T is the exposure time, and β is the slope (negative) of the amplitude transmittance vs exposure characteristic of the material. The amplitude transmittance of the resulting transparency is then

$$\mathbf{t}(x, y) = \mathbf{t}_0 + \beta T [I(x, y) + I(x, y - y_0)]$$
$$= \mathbf{t}_0 + \beta T I(x, y) * [\delta(x, y) + \delta(x, y - y_0)], \tag{G.4}$$

where $I(x, y)$ is the intensity distribution in the individual speckle patterns, and y_0 is the amount by which they are displaced.

If this transparency is placed in the front focal plane of a lens and illuminated by a collimated beam of monochromatic light, the amplitude in the back focal plane of the lens is proportional to the Fourier transform of $\mathbf{t}(x, y)$ (see Appendix A.1), which is

$$g(u, v) = \mathbf{t}_0 \delta(u, v) + \beta T h(u, v)[1 + \exp(ikvy_0)], \tag{G.5}$$

where $h(u, v) \leftrightarrow I(x, y)$ and $k = 2\pi/\lambda$. The first term on the right-hand side of Eq. (G.5) is merely the directly transmitted beam, which is brought to a point focus on the axis, while the intensity due to the second term is

$$I(u, v) = \beta^2 T^2 |h(u, v)|^2 (1 + \cos kvy_0), \tag{G.6}$$

which corresponds to the power spectrum of the speckle pattern modulated by a set of equally spaced interference fringes (Young's fringes).

Bibliography

Adrian, R. J. (ed.) (1993). "Laser Doppler Velocimetry," Vol. MS78. SPIE, Bellingham, WA.

Agarwal, G. S. (ed.) (1995). "Fundamentals of Quantum Optics," Vol. MS103. SPIE, Bellingham, WA.

Bell, R. J. (1972). "Introductory Fourier Transform Spectroscopy." Academic Press, New York.

Born, M., and Wolf, E. (1999). "Principles of Optics." Cambridge University Press, Cambridge, UK.

Boyd, R. W. (1992). "Nonlinear Optics." Academic Press, Boston.

Brown, G. M. (ed.) (2000). "Modern Interferometry," Selected SPIE Papers on CD-ROM, Vol. 15. SPIE, Bellingham, WA.

Brown, R. Hanbury (1974). "The Intensity Interferometer." Taylor and Francis, London.

Butcher, P. N., and Cotter, D. (1990). "The Elements of Nonlinear Optics." Cambridge University Press, Cambridge, UK.

Candler, C. (1951). "Modern Interferometers." Hilger and Watts, London.

Chamberlain, J. (1979). "The Principles of Interferometric Spectroscopy." Wiley, Chichester, UK.

Collet, E. (1992). "Polarized Light." Marcel Dekker, New York.

Culshaw, B. (1984). "Optical Fibre Sensing and Signal Processing." Peregrinus, London.

Culshaw, B., and Dakin, J. P. (eds.) (1989). "Optical Fibre Sensors," Vol. II. Academic Press, San Diego.

Davis, S. P., Abrams, M. C., and Brault, J. W. (2001). "Fourier Transform Spectrometry." Academic Press, San Diego.

Durst, F., Melling, A., and Whitelaw, J. H. (1976). "Principles and Practice of Laser-Doppler Anemometry." Academic Press, London.

Françon, M., and Mallick, S. (1971). "Polarization Interferometers: Applications in Microscopy and Macroscopy." Wiley-Interscience, London.

Ghiglia, D. C., and Pritt, M. (1998). "Two-Dimensional Phase Unwrapping: Theory, Algorithms and Software." Wiley, New York.

Goodman, J. W. (1996). "Introduction to Fourier Optics." McGraw-Hill, New York.

Hariharan, P. (ed.) (1991). "Interferometry," Vol. MS28. SPIE, Bellingham, WA.

Hariharan, P. (1992). "Basics of Interferometry." Academic Press, San Diego.

Hariharan, P., and Malacara-Hernandez, D. (eds.) (1995). "Interference, Interferometry and Interferometric Metrology," Vol. MS110. SPIE, Bellingham, WA.

Hernandez, G. (1986). "Fabry–Perot Interferometers." Cambridge University Press, Cambridge, UK.

Krug, W., Rienitz, J., and Schulz, G. (1964). "Contributions to Interference Microscopy." Hilger and Watts, London.

Lawson, P. R. (ed.) (1997). "Long Baseline Stellar Interferometry," Vol. MS139. SPIE, Bellingham, WA.

Leonhardt, K. (1981). "Optische Interferenzen." Wissenschaftliche Vergesellschaft, Stuttgart.

Malacara, D. (ed.) (1992). "Optical Shop Testing." John Wiley, New York.

Malacara, D., Servin, M., and Malacara, Z. (1998). "Interferogram Analysis for Optical Testing." Marcel Dekker, New York.

Mandel, L., and Wolf, E. (1995). "Optical Coherence and Quantum Optics." Cambridge University Press, Cambridge, UK.

Peřina, J. (1971). "Coherence of Light." Van Nostrand Reinhold, New York.

Robinson, D. W., and Reid, G. T. (1993). "Interferogram Analysis: Digital Processing Techniques for Fringe Pattern Measurement." IOP Publishing, London.

Silvfast, W. T. (1996). "Laser Fundamentals." Cambridge University Press, Cambridge, UK.

Steel, W. H. (1983). "Interferometry." Cambridge University Press, Cambridge, UK.

Svelto, O. (1989). "Principles of Lasers." Plenum Press, New York.

Tolansky, S. (1955). "An Introduction to Interferometry." Longmans, London.

Tolansky, S. (1961). "Surface Microtopography." Longmans, London.

Udd, E. (ed.) (1991). "Fiber Optic Sensors: An Introduction for Engineers and Scientists." John Wiley, New York.

Vaughan, J. M. (1989). "The Fabry–Perot Interferometer." Adam Hilger, Bristol.

References

Abitbol, C., Gallion, P., Nakajima, H., and Chabran, C. (1984). *J. Opt. (Paris)* **15**, 411–418.

Abramovici, A., Althouse, W. E., Camp, J., Durance, D., Giaime, J. A., Gillespie, A., Kawamura, S., Kuhnert, A., Lyons, T., Raab, F. J., Savage Jr., R. L., Shoemaker, D., Sievers, L., Spero, R. E., Vogt, R. E., Weiss, R., Whitcomb, S., and Zucker, M. (1996). *Phys. Lett. A* **218**, 157–163.

Abramovici, A., Althouse, W. E., Drever, R. W. P., Gürsel, Y., Kawamura, S., Raab, F. J., Shoemaker, D., Sievers, L., Spero, R. E., Thorne, K. S., Vogt, R. E., Weiss, R., Whitcomb, S. E., and Zucker, M. E. (1992). *Science* **256**, 325–333.

Abramovici, A., Vager, Z., and Weksler, M. (1986). *J. Phys. E: Sci. Instrum.* **19**, 182–188.

Abrams, M. C., Goldman, A., Gunson, M. R., Rinsland, C. P., and Zander, R. (1996a). *Appl. Opt.* **35**, 2747–2751.

Abrams, M. C., Gunson, M. R., Chang, A. Y., Rinsland, C. P., and Zander, R. (1996b). *Appl. Opt.* **35**, 2774–2786.

Abramson, N. (1969). *Optik* **30**, 56–71.

Adrian, R. J. (ed.) (1993). "Laser Doppler Velocimetry," Vol. MS78. SPIE, Bellingham, WA.

Agarwal, G. S. (ed.) (1995). "Fundamentals of Quantum Optics," Vol. MS103. SPIE, Bellingham, WA.

Agarwal, G. S., and Hariharan, P. (1993). *Opt. Commun.* **103**, 111–115.

Agarwal, G. S., and James, D. F. V. (1993). *J. Mod. Opt.* **40**, 1431–1436.

Agarwal, G. S., and Wolf, E. (1993). *J. Mod. Opt.* **40**, 1489–1496.

Ai, C. (1997). *Appl. Opt.* **36**, 8135–8138.

Ai, C., and Wyant, J. C. (1993). *Appl. Opt.* **32**, 4698–4705.

Aiki, K., Nakamura, M., Kuroda, T., Umeda, J., Ito, R., Chinone, N., and Maeda, M. (1978). *IEEE J. Quantum Electron.* **QE-14**, 89–94.

Alford, W. P., and Gold, A. (1958). *Am. J. Phys.* **26**, 481–484.

Alum, Kh. P., Koval'chuk, Yu. V., and Ostrovskaya, G. V. (1984). *Sov. Phys. Tech. Phys.* **29**, 534–538.

Ammann, E. O., and Chang, I. C. (1965). *J. Opt. Soc. Am.* **55**, 835–841.

Anderson, D. J., and Jones, J. D. C. (1992). *J. Mod. Opt.* **39**, 1837–1847.

Anderson, D. Z., Lininger, D. M., and Feinberg, J. (1987). *Opt. Lett.* **12**, 123–125.

Anderson, R., Bilger, H. R., and Stedman, G. E. (1994). *Am. J. Phys.* **62**, 975–985.

ANSI (1993). *American National Standard for the Safe Use of Lasers* **Z.136**.1–1993. American Standards Institute, New York.

Araya, A., Telada, S., Tochikubo, K., Taniguchi, S., Takahashi, R., Kawabe, K., Tatsumi, D., Yamazaki, T., Kawamura, S., Miyoki, S., Moriwaki, S., Musha, M., Nagano, S., Fujimoto, M.-K., Horikoshi, K., Mio, N., Naito, Y., Takamori, A., and Yamamoto, K. (1999). *Appl. Opt.* **38**, 2848–2856.

Armitage, J. D., Jr., and Lohmann, A. (1965). *Opt. Acta* **12**, 185–192.

Armstrong, J. T., Hutter, D. J., Johnston, K. J., and Mozurkewich, D. (May 1995). *Phys. Today*, 42–49.

Armstrong, J. T., Mozurkewich, D., Rickard, L. J., Hutter, D. J., Benson, J. A., Bowers, P. F., Elias II, N. M., Hummel, C. A., Johnston, K. J., Buscher, D. F., Clark III, J. H., Ha, L., Ling, L. C., White, N. M., and Simon, R. S. (1998). *Astrophys. J.* **496**, 550–571.

Arnautov, G. P., Boulanger, Yu. D., Kalish, E. N., Koronkevitch, V. P., Stua, Yu. F., and Tarasyuk, V. G. (1983). *Metrologia* **19**, 49–55.

Arnold, S. J., Boksenberg, A., and Sargent, W. L. W. (1979). *Astrophys. J.* **234**, L159–L163.

Arnold, S. M. (1985). *Opt. Eng.* **24**, 803–807.

Arnold, S. M. (1988). *Proc. SPIE* **884**, 23–26.

Arnold, S. M. (1989). *Proc. SPIE* **1052**, 191–197.

Aronowitz, F. (1971). *In* "Laser Applications," Vol. I (M. Ross, ed.), pp. 133–200. Academic Press, New York.

Ashby, D. E. T. F., and Jephcott, D. F. (1963). *Appl. Phys. Lett.* **3**, 13–16.

Ashby, D. E. T. F., Jephcott, D. F., Malein, A., and Raynor, F. A. (1965). *J. Appl. Phys.* **36**, 29–34.

Aspect, A., Dalibard, J., and Roger, G. (1982). *Phys. Rev. Lett.* **49**, 1804–1807.

Aspect, A., Grangier, P., and Roger, G. (1981). *Phys. Rev. Lett.* **47**, 460–463.

Atherton, P. D., Reay, N. K., Ring, J., and Hicks, T. R. (1981). *Opt. Eng.* **20**, 806–814.

Baba, N., Murakami, N., and Ishigaki, T. (2001). *Opt. Lett.* **26**, 1167–1169.

Baird, K. M. (1954). *J. Opt. Soc. Am.* **44**, 11–13.

Baird, K. M. (1963). *Appl. Opt.* **2**, 471–479.

Baird, K. M. (1981). *J. Phys. Colloq. (France)* **42(C-8)**, 485–494.

Baird, K. M. (1983a). *Phys. Today* **36(1)**, 52–57.

Baird, K. M. (1983b). *In* "Quantum Metrology and Fundamental Physical Constants" (P. H. Cutler and A. A. Lucas, eds.), pp. 143–162. Plenum, New York.

Baird, K. M., and Howlett, L. E. (1963). *Appl. Opt.* **2**, 455–463.

Baker, D. J., Steed, A. J., and Stair, A. T., Jr. (1973). *J. Geophys. Res.* **78**, 8859–8863.

Baldwin, J. E., Beckett, M. G., Boysen, R. C., Burns, D., Buscher, D. F., Cox, G. C., Haniff, C. A., Mackay, C. D., Nightingale, N. S., Rogers, J., Scheuer, P. A. G., Scott, T. R., Tuthill, P. G., Warner, P. J., Wilson, D. M. A., and Wilson, R. W. (1996). *Astron. Astrophys.* **306**, L13–L16.

Baldwin, J. E., Boysen, R. C., Cox, G. C., Haniff, C. A., Rogers, J., Warner, P. J., Wilson, D. M. A., and Mackay, C. D. (1994). *Proc. SPIE* **2200**, 118–128.

Baldwin, J. E., Haniff, C. A., Mackay, C. D., and Warner, P. J. (1986). *Nature* **320**, 595–597.

Balhorn, R., Kunzmann, H., and Lebowsky, F. (1972). *Appl. Opt.* **11**, 742–744.

Bar-Joseph, I., Hardy, A., Katzir, Y., and Silberberg, Y. (1981). *Opt. Lett.* **6**, 414–416.

Barker, L. M. (1971). *Rev. Sci. Instrum.* **42**, 276–278.

Barker, L. M., and Hollenbach, R. E. (1965). *Rev. Sci. Instrum.* **36**, 1617–1620.

Barnes, T. H., Eiju, T., and Matsuda, K. (1996). *Opt. Commun.* **132**, 494–501.

Barrell, H., and Sears, J. E. (1939). *Phil. Trans. Roy. Soc. A* **238**, 1–64.

Bartelt, H. O., and Jahns, J. (1979). *Opt. Commun.* **30**, 268–274.

Bates, J. B. (1976). *Science* **191**, 31–37.

Bates, R. H. T., Gough, P. T., and Napier, P. J. (1973). *Astron. Astrophys.* **22**, 319–320.

Bates, W. J. (1947). *Proc. Phys. Soc.* **59**, 940–950.

Bay, Z., Luther, G. G., and White, J. A. (1972). *Phys. Rev. Lett.* **29**, 189–192.

Beard, P. C., and Mills, T. N. (1996). *Appl. Opt.* **35**, 663–675.

Bearden, A., O'Neill, M. P., Osborne, L. C., and Wong, T. L. (1993). *Opt. Lett.* **18**, 238–240.

Bechmann-Pasquinicci, H., and Peres, A. (2000). *Phys. Rev. Lett.* **85**, 3313–3316.

Beddoes, D. R., Dainty, J. C., Morgan, B. L., and Scadden, R. J. (1976). *J. Opt. Soc. Am.* **66**, 1247–1251.

Bell, J. S. (1965). *Physics (USA)* **1**, 195–200.

Bell, R. J. (1972). "Introductory Fourier Transform Spectroscopy." Academic Press, New York.

Benedict, L. H., Nobach, H., and Tropea, C. (2000). *Meas. Sci. Technol.* **4**, 1089–1104.

Bennett, H. E., and Bennett, J. M. (1967). *In* "Physics of Thin Films" (G. Hass and R. E. Thun, eds.), Vol. IV, pp. 1–96. Academic Press, New York.

Bennett, J. M., and Mattson, L. (1989). "Introduction to Surface Roughness and Scattering." Optical Society of America, Washington, DC.

Bennett, S. J., Ward, R. E., and Wilson, D. C. (1973). *Appl. Opt.* **12**, 1406.

Beran, M. J., and Parrent, G. B., Jr. (1964). "Theory of Partial Coherence." Prentice-Hall, Englewood Cliffs, NJ.

Berger, J., and Lovberg, R. H. (1969). *Rev. Sci. Instrum.* **40**, 1569–1575.

Bergh, R. A., Lefevre, H. C., and Shaw, H. J. (1981a). *Opt. Lett.* **6**, 198–200.

Bergh, R. A., Lefevre, H. C., and Shaw, H. J. (1981b). *Opt. Lett.* **6**, 502–504.

Bergh, R. A., Lefevre, H. C., and Shaw, H. J. (1984). *IEEE J. Lightwave Technol.* **LT-2**, 91–107.

Bernard, J. E., Madej, A. A., Marmet, L., Whitford, B. G., Siemsen, K. J., and Cundy, S. (1999). *Phys. Rev. Lett.* **82**, 3228–3231.

Berry, M. V. (1984). *Proc. Roy. Soc. (London) A* **392**, 45–57.

Berry, M. V. (1987). *J. Mod. Opt.* **34**, 1401–1407.

Bester, M., Danchi, W. C., and Townes, C. H. (1990). *Proc. SPIE* **1237**, 40–48.

Bhandari, R., and Samuel, J. (1988). *Phys. Rev. Lett.* **60**, 1211–1213.

Bhushan, B., Wyant, J. C., and Koliopoulos, C. L. (1985). *Appl. Opt.* **24**, 1489–1497.

Biddles, B. J. (1969). *Opt. Acta* **16**, 137–157.

Billing, H., Maischberger, K., Rudiger, A., Schilling, R., Schnupp, L., and Winkler, W. (1979). *J. Phys. E: Sci. Instrum.* **12**, 1043–1050.

Birch, K. G. (1973). *J. Phys. E: Sci. Instrum.* **6**, 1045–1048.

Bjelkhagen, H. (1999). *Opt. Eng.* **38**, 55–61.

Blazit, A., Bonneau, D., Koechlin, L., and Labeyrie, A. (1977). *Astrophys. J.* **214**, L79–L84.

Bliss, E. S., Speck, D. R., and Simmons, W. W. (1974). *Appl. Phys. Lett.* **25**, 728–730.

Bloom, A. L. (1974). *J. Opt. Soc. Am.* **64**, 447–452.

Bobroff, N. (1993). *Meas. Sci. Technol.* **4**, 907–926.

Boggess, N. W., Mather, J. C., Weiss, R., Bennett, C. L., Cheng, E. S., Dwek, E., Gulkis, S., Hauser, M. G., Janssen, M., Kelsall, T., Meyer, S. S., Mosely, S. H., Murdock, T. L., Shafer, R. A., Silverberg, R. F., Smoot, G. F., Wilkinson, D. T., and Wright, E. L. (1992). *Astrophys. J. (USA)* **397**, 420–429.

Bohm, D. (1951). "Quantum Theory." Prentice-Hall, Englewood Cliffs, NJ.

Bold, G. T., Barnes, T. H., Gourlay, J., Sharples, R. M., and Haskell, T. G. (1998). *Opt. Commun.* **148**, 323–330.

Bondurant, R. S., and Shapiro, J. H. (1984). *Phys. Rev. D* **30**, 2548–2556.

Bone, D. J. (1991). *Appl. Opt.* **30**, 3627–3632.

Bone, D. J., Bachor, H.-A., and Sandemann, R. J. (1986). *Appl. Opt.* **25**, 1653–1660.

Bonneau, D., and Labeyrie, A. (1973). *Astrophys. J.* **181**, L1–L4.

Born, M., and Wolf, E. (1999). "Principles of Optics." Cambridge University Press, Cambridge, UK.

Bouchareine, P., and Connes, P. (1963). *J. de Phys.* **24**, 134–138.

Bourdet, G. L., and Orszag, A. G. (1979). *Appl. Opt.* **18**, 225–227.

Boyd, G. D., and Gordon, J. P. (1961). *Bell Syst. Tech. J.* **40**, 489–508.

Boyd, G. D., and Kogelnik, H. (1962). *Bell Syst. Tech. J.* **41**, 1347–1369.

Bracewell, R. N. (1978). *Nature* **274**, 780–781.

Bradley, D. J., and Mitchell, C. J. (1968). *Phil Trans. Roy. Soc. A* **263**, 209–223.

Brault, J. W. (1996). *Appl. Opt.* **35**, 2891–2896.

Braun, M., Maier, J., and Liening, H. (1987). *J. Phys. E: Sci. Instrum.* **20**, 1247–1249.

Breckenridge, J. B. (ed.) (1994). "Amplitude and Intensity Spatial Interferometry II." *Proc. SPIE* **2200**. SPIE, Bellingham, WA.

Brendel, J., Dultz, W., and Martienssen, W. (1995). *Phys. Rev. A* **52**, 2551–2556.

Brendel, J., Mohler, E., and Martienssen, W. (1991). *Phys. Rev. Lett.* **66**, 1142–1145.

Brendel, J., Mohler, E., and Martienssen, W. (1992). *Europhys. Lett.* **20**, 575–580.

Briers, J. D. (1972). *Opt. Laser Technol.* **4**, 28–41.

Brillet, A., and Cerez, P. (1981). *J. Phys. Colloq. (France)* **42** **(C-8)**, 73–82.

Brillet, A., and Hall, J. L. (1979). *Phys. Rev. Lett.* **42**, 549–552.

Brooks, J. L., Moslehi, B., Kim, B. Y., and Shaw, H. J. (1987). *J. Lightwave Technol.* **LT-5**, 1014–1023.

Brooks, J. L., Wentworth, R. H., Youngquist, R. C., Tur, M., Kim, B. Y., and Shaw, H. J. (1985). *J. Lightwave Technol.* **LT-3**, 1062–1071.

Brossel, J. (1947). *Proc. Phys. Soc.* **59**, 224–234.

Brown, B. R., and Lohmann, A. W. (1966). *Appl. Opt.* **5**, 967–969.

Brown, B. R., and Lohmann, A. W. (1969). *IBM J. Res. Dev.* **13**, 160–167.

Brown, D. S. (1959). *In* "Interferometry: N.P.L. Symposium No. 11," pp. 253–256. Her Majesty's Stationery Office, London.

Brown, N. (1981). *Appl. Opt.* **20**, 3711–3714.

Brown, R. Hanbury (1964). *Sky Telesc.* **28**, 64–69.

Brown, R. Hanbury (1974). "The Intensity Interferometer." Taylor and Francis, London.

Brown, R. Hanbury, Hazard, C., Davis, J., and Allen, L. R. (1964). *Nature* **201**, 1111–1112.

Brown, R. Hanbury, and Twiss, R. Q. (1954). *Phil. Mag.* **45**, 663–682.

Brown, R. Hanbury, and Twiss, R. Q. (1956). *Nature* **177**, 27–29.

Brown, R. Hanbury, and Twiss, R. Q. (1957a). *Proc. Roy. Soc. A* **242**, 300–324.

Brown, R. Hanbury, and Twiss, R. Q. (1957b). *Proc. Roy. Soc. A* **243**, 291–319.

Brown, R. Hanbury, and Twiss, R. Q. (1958a). *Proc. Roy. Soc. A* **248**, 199–221.

Brown, R. Hanbury, and Twiss, R. Q. (1958b). *Proc. Roy. Soc. A* **248**, 222–237.

Bruce, C. F., and Hill, R. M. (1961). *Aust. J. Phys.* **14**, 64–88.

Bruning, J. H., Heriott, D. R., Gallagher, J. W., Rosenfeld, D. P., White, A. D., and Brangaccio, D. J. (1974). *Appl. Opt.* **13**, 2693–2703.

Bryngdahl, O. (1965). *In* "Progress in Optics" (E. Wolf, ed.), Vol. IV, pp. 37–83. North-Holland, Amsterdam.

Bucaro, J. A., Dardy, H. D., and Carome, E. F. (1977). *Appl. Opt.* **16**, 1761–1762.

Bucholtz, F., Kersey, A. D., and Dandridge, A. (1986). *Electron. Lett.* **22**, 451–453.

Bünnagel, R. von (1956). *Z Angew. Phys.* **8**, 342–350.

Bünnagel, R. von. (1965). *Z. Instrumkde.* **73**, 214–215.

Burch, J. M. (1953). *Nature* **171**, 889–890.

Burch, J. M. (1972). *In* "Optical Instruments and Techniques" (J. Home-Dickson, ed.), pp. 213–229. Oriel Press, Newcastle-upon-Tyne, UK.

Burnham, D. C., and Weinberg, D. L. (1970). *Phys. Rev. Lett.* **25**, 84–87.

Burns, W. K., Moeller, R. P., Villaruel, C. A., and Abebe, M. (1984). *Opt. Lett.* **9**, 570–572.

Burton, C. H., Leistner, A. J., and Rust, D. M. (1987). *Appl. Opt.* **26**, 2637–2642.

Butter, C. D., and Hocker, G. R. (1978). *Appl. Opt.* **17**, 2867–2869.

Byer, R. L., Paul, J., and Duncan, M. D. (1977). *In* "Laser Spectroscopy III" (J. L. Hall and J. L. Carlsten, eds.), pp. 414–416. Springer-Verlag, Berlin.

Caber, P. J. (1993). *Appl. Opt.* **32**, 3438–3441.

Cageao, R. P., Blavier, J.-F., McGuire, J. P., Jiang, Y., Nemtchinov, V., Mills, F. P., and Sander, S. P. (2001). *Appl. Opt.* **40**, 2024–2030.

Cagnet, M. (1954). *Rev. Opt.* **33**, 1–25, 113–124, 229–241.

Cahill, R. F., and Udd, E. (1979). *Opt. Lett.* **4**, 93–95.

Calatroni, J., Guerrero, A. L., Sainz, C., and Escalona, R. (1996). *Opt. Laser Technol.* **28**, 485–489.

Campos, R. A., Saleh, B. E. A., and Teich, M. C. (1989). *Phys. Rev. A* **40**, 1371–1384.

Candler, C. (1951). "Modern Interferometers." Hilger and Watts, London.

Carere, C. A., Neil, W. S., and Sloan, J. J. (1996). *Appl. Opt.* **35**, 2857–2866.

Carré, P. (1966). *Metrologia* **2**, 13–23.

Caves, C. M. (1980). *Phys. Rev. Lett.* **45**, 75–79.

Caves, C. M. (1981). *Phys. Rev. D* **23**, 1693–1708.

Chabbal, R. (1958). *Rev. Opt.* **37**, 49–103, 336–370, 501–551.

Chakraborty, A. K. (1970). *Ind. J. Pure Appl. Phys.* **8**, 814–816.

Chakraborty, A. K. (1973). *Nouv. Rev. Optique* **4**, 331–335.

Chamberlain J. (1971). *Infrared Phys.* **11**, 25–55.

Chamberlain, J. (1979). "The Principles of Interferometric Spectroscopy." Wiley, Chichester, UK.

Chamberlain, J., and Gebbie, H. A. (1971). *Infrared Phys.* **11**, 56–73.

Chan, R. K. Y., Jones, J. D. C., and Jackson, D. A. (1985). *Opt. Acta* **32**, 241–246.

Chen, J., Ishii, Y., and Murata, K. (1988). *Appl. Opt.* **27**, 124–128.

Cheng, Y.-Y., and Wyant, J. C. (1984). *Appl. Opt.* **23**, 4539–4543.

Cheng, Y.-Y., and Wyant, J. C. (1985a). *Appl. Opt.* **24**, 804–807.

Cheng, Y.-Y., and Wyant, J. C. (1985b). *Appl. Opt.* **24**, 3049–3052.

Chien, P.-Y., Chang, Y.-S., and Chang, M.-W. (1997). *Opt. Commun.* **135**, 198–202.

Chien, P.-Y., Pan, C.-L., and Chang, L.-W. (1991). *Opt. Lett.* **16**, 1701–1703.

Chim, S. S. C., and Kino, G. S. (1992). *Appl. Opt.* **31**, 2550–2553.

Chow, W. W., Gea-Banacloche, J., Pedrotti, L. M., Sanders, V. E., Schleich, W., and Scully, M. O. (1985). *Rev. Mod. Phys.* **57**, 61–104.

Chow, W. W., Hambenne, J. B., Hutchings, T. J., Sanders, V. E., Sargent, M., and Scully, M. O. (1980). *IEEE J. Quantum Electron.* **QE-16**, 918–935.

Churnside, J. H. (1984). *Appl. Opt.* **23**, 61–66.

Chyba, T. H., Wang, L. J., Mandel, L., and Simon, R. (1988). *Opt. Lett.* **13**, 562–564.

Ciddor, P. E. (1973). *Aust. J. Phys.* **26**, 783–796.

Ciddor, P. E. (1996). *Appl. Opt.* **35**, 1566–1572.

Ciddor, P. E., and Bruce, C. F. (1967). *Metrologia* **3**, 109–118.

Ciddor, P. E., and Duffy, R. M. (1983). *J. Phys. E: Sci. Instrum.* **16**, 1223–1227.

Ciddor, P. E., and Hariharan, P. (2001). *J. Mod. Opt.* **117**, 15–19.

Ciddor, P. E., and Hill, R. J. (1999). *Appl. Opt.* **38**, 1663–1667.

Clark, R. J., Hause, C. D., and Bennett, G. S. (1953). *J. Opt. Soc. Am.* **43**, 408–409.

Clauser, J. F. (1974). *Phys. Rev. D* **9**, 853–860.

Clauser, J. F., and Shimony, A. (1978). *Rep. Prog. Phys.* **41**, 1881–1927.

Clunie, D. M., and Rock, N. H. (1964). *J. Sci. Instrum.* **41**, 489–492.

Cochran, W. T., Cooley, J. W., Favin, D. L., Helms, H. D., Kaenel, R. A., Lang, W. W., Maling, G. C., Jr., Nelson, D. E., Radder, C. M., and Welch, P. D. (1967). *Proc. IEEE* **55**, 1664–1674.

Cogswell, C. J., Smith, N. I., Larkin, K. G., and Hariharan, P. (1997). *Proc. SPIE* **2984**, 72–81.

Colavita, M. M., Shao, M., Hines, B. E., Wallace, J. K., Gursel, Y., Malbet, F., Yu, J. W., Singh, H., Beichman, C. A., Pan, X. P., Nakajima, T., and Kulkarni, S. R. (1994). *Proc. SPIE* **2200**, 89–97.

Colavita, M. M., Shao, M., and Rayman, M. D. (1993). *Appl. Opt.* **32**, 1789–1797.

Colavita, M. M., Wallace, J. K., Hines, B. E., Gursel, Y., Malbet, F., Palmer, D. L., Pan, X. P., Shao, M., Yu, J. W., Boden, A. F., Dumont, P. J., Gubler, J.,

Koresko, C. D., Kulkarni, S. R., Lane, B. F., Mobley, D. W., and van Belle, G. T. (1999). *Astrophys. J.* **510**, 505–521.

Collet, E. (1992). "Polarized Light." Marcel Dekker, New York.

Collins, R. J., Nelson, D. F., Schawlow, A. L., Bond, W., Garrett, C. G. B., and Kaiser, W. (1960). *Phys. Rev. Lett.* **5**, 303–305.

Connes, J. (1961). *Rev. Opt.* **40**, 45–79, 116–140, 171–190, 231–265.

Connes, J., and Connes, P. (1966). *J. Opt. Soc. Am.* **56**, 896–910.

Connes, J., and Nozal, V. (1961). *J. Phys. Radium* **22**, 359–366.

Connes, P. (1956). *Rev. Opt.* **35**, 37–43.

Connes, P. (1958). *J. Phys. Radium* **19**, 262–269.

Cooley, J. W., and Tukey, J. W. (1965). *Math. Comput.* **19**, 297–301.

Crane, R. (1969). *Appl. Opt.* **8**, 538–542.

Creath, K. (1985). *Appl. Opt.* **24**, 1291–1293.

Creath, K. (1987). *Appl. Opt.* **26**, 2810–2816.

Creath, K. (1988). *In* "Progress in Optics" (E. Wolf, ed.), Vol. XXVI, pp. 349–393. Elsevier, Amsterdam.

Creath, K. (1991). *Proc. SPIE* **1553**, 213–220.

Creath, K., and Hariharan, P. (1994). *Appl. Opt.* **33**, 24–25.

Culshaw, B. (1984). "Optical Fibre Sensing and Signal Processing." Peregrinus, London.

Culshaw, B., and Dakin, J. P. (eds.) (1989). "Optical Fibre Sensors," Vol. II. Academic Press, San Diego.

Dahlquist, J. A., Peterson, D. G., and Culshaw, W. (1966). *Appl. Phys. Lett.* **9**, 181–183.

Dai, X., and Seta, K. (1998). *Meas. Sci. Technol.* **9**, 1031–1035.

Dainty, J. C. (1976). *In* "Progress in Optics" (E. Wolf, ed.), Vol. XIV, pp. 3–46. North-Holland, Amsterdam.

Dandridge, A., Miles, R. O., and Giallorenzi, T. G. (1980). *Electron. Lett.* **16**, 948–949.

Dandridge, A., and Tveten, A. B. (1981). *Appl. Opt.* **20**, 2337–2339.

Dandridge, A., and Tveten, A. B. (1982). *Opt. Lett.* **7**, 279–281.

Dandridge, A., Tveten, A. B., Sigel, G. H., Jr., West, E. J., and Giallorenzi, T. G. (1980). *Electron. Lett.* **16**, 408–409.

Daria, V. R., and Saloma, C. (2000). *Appl. Opt.* **39**, 108–113.

Darlin, J. S., Kothiyal, M. P., and Sirohi, R. S. (1998). *J. Mod. Opt.* **45**, 2371–2378.

Davis, J. (1979). *In* "High Angular Resolution Stellar Interferometry" (I.A.U. Colloquim No. 50), (J. Davis and W. J. Tango, eds.), pp. 14.1–14.12. Chatterton Astronomy Department, University of Sydney.

Davis, J. (1984). *In* "Proc. International Symposium on Measurement and Processing for Indirect Imaging, Sydney, 1983" (J. Roberts, ed.), pp. 125–141. Cambridge University Press, Cambridge, UK.

Davis, J., and Tango, W. J. (1986). *Nature* **323**, 234–235.

Davis, J., Tango, W. J., Booth, A. J., Minard, R. A., Owens, S. M., and Shobbrook, R. R. (1994). *Proc. SPIE* **2200**, 231–241.

Davis, J., Tango, W. J., Booth, A. J., ten Brummelaar, A., Minard, R. A., and Owens, S. M. (1999). *Mon. Not. R. Astron. Soc.* **303**, 773–782.

Davis, J. L., and Ezekiel, S. (1981). *Opt. Lett.* **6**, 505–507.

Deferrari, H. A., Darby, R. A., and Andrews, F. A. (1967). *J. Acoust. Soc. Am.* **42**, 982–990.

de Groot, P. (1995). *Appl. Opt.* **34**, 4723–4730.

de Groot, P. (2000a). *Appl. Opt.* **39**, 1527–1530.

de Groot, P. (2000b). *Appl. Opt.* **39**, 2658–2663.

de Groot, P. (2001). *Opt. Eng.* **40**, 28–32.

de Groot, P., and Deck, L. (1993). *Opt. Lett.* **18**, 1462–1464.

de Groot, P., and Deck, L. (1995). *J. Mod. Opt.* **42**, 389–401.

de Groot, P., de Lega, X. C., Kramer, J., and Turzhitsky, M. (2002). *Appl. Opt.* **41**, 4571–4578.

de Groot, P., de Lega, X. C., and Stephenson, D. (2000). *Opt. Eng.* **39**, 86–90.

de Groot, P., Gallatin, G., Gardopee, G., and Dixon, R. (1989). *Appl. Opt.* **28**, 2462–2464.

de Groot, P. J., and Gallatin, G. M. (1989). *Opt. Lett.* **14**, 165–167.

de Groot, P., Gallatin, G. M., and Macomber, S. H. (1988). *Appl. Opt.* **27**, 4475–4480.

Delage, L., Reynaud, F., and Lannes, A. (2000). *Appl. Opt.* **39**, 6406–6420.

Delaunay, G. (1953). *Rev. Opt.* **32**, 610–614.

DeSouza, P. D., and Mermelstein, M. D. (1982). *Appl. Opt.* **21**, 4214–4218.

Diddams, S. A., Jones, D. J., Ma, L.-S., Cundiff, S. T., and Hall, J. L. (2000). *Opt. Lett.* **25**, 186–188.

Diedrich, F., and Walther, H. (1987). *Phys. Rev. Lett.* **58**, 203–206.

Dil, J. G., van Hijningen, N. C. J. A., van Dorst, F., and Aarts, R. M. (1981). *Appl. Opt.* **20**, 1374–1381.

Dirac, P. A. M. (1958). "The Principles of Quantum Mechanics." Oxford University Press, Oxford.

Dobrowolski, J. A., and Traub, W. A. (1996). *Appl. Opt.* **35**, 2934–2946.

Doran, N. J., and Wood, D. (1988). *Opt. Lett.* **13**, 56–58.

Dörband, B. (1982). *Optik* **60**, 161–174.

Dörband, B., and Tiziani, H. J. (1985). *Appl. Opt.* **24**, 2604–2611.

Dorrio, B. V., and Fernández, J. L. (1999). *Meas. Sci. Technol.* **10**, R33–R55.

Dorschner, T. A., Haus, H. A., Holz, M., Smith, I. E., and Statz, H. (1980). *IEEE J. Quantum Electron.* **QE-16**, 1376–1379.

Downs, M. J., McGivern, W. H., and Ferguson, H. J. (1985). *Precis. Eng.* **7**, 211–215.

Dresel, T., Hausler, G., and Venzke, H. (1992). *Appl. Opt.* **31**, 919–925.

Drever, R. W. P. (1983). *In* "Gravitational Radiation" (N. Deruelle and T. Piran, eds.), pp. 321–338. North-Holland, Amsterdam.

Drever, R. W. P., Hoggan, S., Hough, J., Meers, B. J., Munley, A. J., Newton, G. P., Ward, H., Anderson, D. Z., Gursel, Y., Hereld, M., Spero, R. E., and Whitcomb, S. E. (1983). *In* "Proceedings of the Third Marcel Grossmann Meeting on General Relativity," Vol. I. (H. Ning, ed.), pp. 739–753. Science Press, Beijing and North-Holland, Amsterdam.

Drever, R. W. P., Hough, J., Munley, A. J., Lee, S. A., Spero, R., Whitcomb, S. E., Ward, H., Ford, G. M., Hereld, M., Robertson, N. A., Kerr, I., Pugh, J. R., Newton, G. P., Meers, B., Brooks, E. D., III, and Gursel, Y. (1981). *In* "Laser Spectroscopy V" (A. R. W. McKellar, T. Oka, and B. P. Stoicheff, eds.), pp. 33–40. Springer-Verlag, Berlin.

Dubois, A., Selb, J., Vabre, L., and Boccara, A-C. (2000). *Appl. Opt.* **39**, 2326–2331.

Dufour, C. (1951). *Ann. Phys.* **6**, 5–107.

Dukes, J. N., and Gordon, G. B. (1970). *Hewlett-Packard J.* **21**(12), 2–8.

Durst, F., Melling, A., and Whitelaw, J. H. (1976). "Principles and Practice of Laser-Doppler Anemometry." Academic Press, London.

Durst, F., and Whitelaw, J. H. (1971). *J. Phys. E: Sci. Instrum.* **4**, 804–808.

Dyer, S. D., and Christensen, D. A. (1997). *Opt. Eng.* **36**, 2440–2447.

Dyson, J. (1957). *J. Opt. Soc. Am.* **47**, 386–390.

Dyson, J. (1963). *J. Opt. Soc. Am.* **53**, 690–694.

Dyson, J., Flude, M. J. C., Middleton, S. P., and Palmer, E. W. (1972). *In* "Optical Instruments and Techniques" (J. Home Dickson, ed.), pp. 93–103. Oriel Press, Newcastle-upon-Tyne, UK.

Ebberg, A., and Schiffner, G. (1985). *Opt. Lett.* **10**, 300–302.

Eberhardt, F. J., and Andrews, F. A. (1970). *J. Acoust. Soc. Am.* **48**, 603–609.

Eckstein, J. N., Ferguson, A. I., and Hänsch, T. W. (1978). *Phys. Rev. Lett.* **40**, 847–850.

Edelstein, W. A., Hough, J., Pugh, J. R., and Martin, W. (1978). *J. Phys. E: Sci. Instrum.* **11**, 710–712.

Edlen, B. (1966). *Metrologia* **2**, 71–80.

Einstein, A., Podolsky, B., and Rosen, N. (1935). *Phys. Rev.* **47**, 777–780.

Ekert, A. K., Rarity, J. G., Tapster, P. R., and Palma, G. M. (1992). *Phys. Rev. Lett.* **69**, 1293–1295.

Elssner, K.-E., Vogel, A., Grzanna, J., and Schulz, G. (1994). *Appl. Opt.* **33**, 2437–2446.

Elster, C. (2000). *Appl. Opt.* **39**, 5353–5359.

Elster, C., and Weingärtner, I. (1999a). *J. Opt. Soc. Am. A* **16**, 2281–2285.

Elster, C., and Weingärtner, I. (1999b). *Appl. Opt.* **39**, 5024–5031.

Evans, C. J. (1998). *Opt. Eng.* **37**, 1880–1882.

Evans, C. J., and Kestner, R. N. (1996). *Appl. Opt.* **35**, 1015–1021.

Evenson, K. M., Day, G. W., Wells, J. S., and Mullen, L. O. (1972). *Appl. Phys. Lett.* **20**, 133–134.

Evenson, K. M., Jennings, D. A., and Petersen, F. R. (1981). *J. Phys. Colloq. (France)* **42**(C-8), 473–483.

Evenson, K. M., Wells, J. S., Petersen, F. R., Danielson, B. L., and Day, G. W. (1973). *Appl. Phys. Lett.* **22**, 192–195.

Ezbiri, A., and Tatam, R. P. (1995). *Opt. Lett.* **20**, 1818–1820.

Ezekiel, S., and Arditty, H. J. (eds.) (1982). "Fiber-Optic Rotation Sensors and Related Technologies." Springer-Verlag, New York.

Ezekiel, S., and Balsamo, S. R. (1977). *Appl. Phys. Lett.* **30**, 478–480.

Ezekiel, S., and Knausenberger, G. E. (eds.) (1978). "Laser Inertial Rotation Sensors." *Proc. SPIE* **157**. SPIE, Bellingham, WA.

Fairman, P. S., Ward, B. K., Oreb, B. F., Farrant, D. I., Gilliand, Y., Freund, C. H., Leistner, A. J., Seckold, J. A., and Walsh, C. J. (1999). *Opt. Eng.* **38**, 1371–1379.

Farahi, F., Gerges, A. S., Jones, J. D. C., and Jackson, D. A. (1988a). *Electron. Lett.* **24**, 54–55.

Farahi, F., Jones, J. D. C., and Jackson, D. A. (1988). *Electron. Lett.* **24**, 409–410.

Farahi, F., Newson, T. P., Jones, J. D. C., and Jackson, D. A. (1988b). *Opt. Commun.* **65**, 319–321.

Farahi, F., Takahashi, N., Jones, J. D. C., and Jackson, D. A. (1989). *J. Mod. Opt.* **36**, 337–348.

Farr, K. B., and George, N. (1992). *Opt. Eng.* **31**, 2191–2196.

Faulde, M., Fercher, A. F., Torge, R., and Wilson, R. N. (1973). *Opt. Commun.* **7**, 363–365.

Fearn, H., and Loudon, R. (1987). *Opt. Commun.* **64**, 485–490.

Fearn, H., and Loudon, R. (1989). *J. Opt. Soc. Am. B* **6**, 917–927.

Feinberg, J. (1983). *Opt. Lett.* **8**, 569–571.

Feinberg, J., and Bacher, G. D. (1984). *Opt. Lett.* **9**, 420–422.

Fellgett, P. (1958). *J. Phys. Radium* **19**, 237–240.

Fellgett, P. B. (1951). Thesis, University of Cambridge, England.

Ferguson, J. B., and Morris, R. H. (1978). *Appl. Opt.* **17**, 2924–2929.

Feynman, R. P., Leighton, R. B., and Sands, M. (1963). "Lectures on Physics," Vol. 3. Addison-Wesley, London.

Fischer, A., Kullmer, R., and Demtroder, W. (1981). *Opt. Commun.* **39**, 277–282.

Fisher, A. D., and Warde, C. (1979). *Opt. Lett.* **4**, 131–133.

Fisher, A. D., and Warde, C. (1983). *Opt. Lett.* **8**, 353–355.

Fork, R. L., Herriott, D. R., and Kogelnik, H. (1964). *Appl. Opt.* **3**, 1471–1484.

Forman, M., Steel, W. H., and Vanasse, G. A. (1966). *J. Opt. Soc. Am.* **56**, 59–63.

Forman, M. L. (1966). *J. Opt. Soc. Am.* **56**, 978–979.

Fornaro, G., Franceschetti, G., Lanari, R., Sansosti, E., and Tesauro, M. (1997). *J. Opt. Soc. Am. A* **14**, 2702–2708.

Forward, R. L. (1978). *Phys. Rev. D* **17**, 379–390.

Fournier, J. M. (1991). *J. Optics* **22**, 259–266.

Fournier, J. M. (1994). *J. Imag. Sci. Tech.* **38**, 507–512.

Françon, M. (1957). *J. Opt. Soc. Am.* **47**, 528–535.

Françon, M. (1961). "Progress in Microscopy." Pergamon, Oxford.

Françon, M., and Mallick, S. (1971). "Polarization Interferometers: Applications in Microscopy and Macroscopy." Wiley-Interscience, London.

Franson, J. D. (1989). *Phys. Rev. Lett.* **62**, 2205–2208.

Franson, J. D. (1991a). *Phys. Rev. Lett.* **67**, 290–293.

Franson, J. D. (1991b). *Phys. Rev. A* **44**, 4552–4555.

Franson, J. D. (1992). *Phys. Rev. A* **45**, 3126–3132.

Frantz, L. M., Sawchuk, A. A., and von der Ohe, W. (1979). *Appl. Opt.* **18**, 3301–3306.

Franze, B., and Tiziani, H. J. (1998). *J. Mod. Opt.* **45**, 861–872.

Freed, C., and Javan, A. (1970). *Appl. Phys. Lett.* **17**, 53–56.

Freischlad, K., and Koliopoulos, C. L. (1990). *J. Opt. Soc. Am. A* **7**, 542–551.

Freischlad, K. R. (2001). *Appl. Opt.* **40**, 1637–1648.

Freniere, E. R., Toler, O. E., and Race, R. (1981). *Opt. Eng.* **20**, 253–255.

Friberg, S., Hong, C. K., and Mandel, L. (1985). *Phys. Rev. Lett.* **54**, 2011–2013.

Frins, E. M., and Dultz, W. (1997). *Opt. Commun.* **136**, 354–356.

Fritschel, P., Bork, R., González, G., Mavalvala, N., Ouimette, D., Rong, H., Sigg, D., and Zucker, M. (2001). *Appl. Opt.* **40**, 4988–4998.

Fritschel, P., González, G., Mavalvala, N., Shoemaker, D., Sigg, D., and Zucker, M. (1998). *Appl. Opt.* **37**, 6734–6747.

Fritschel, P., Shoemaker, D., and Weiss, R. (1992). *Appl. Opt.* **31**, 1412–1418.

Fritz, B. S. (1984). *Opt. Eng.* **23**, 379–383.

Froehly, C. (1982). *In* "Proceedings of the European Southern Observatory (ESO) Conference" (M. H. Ulrich and K. Kjär, eds.), pp. 285–293. European Southern Observatory, Garching.

Gabor, D. (1948). *Nature* **161**, 777–778.

Garcia-Marquez, J., Malacara-Hernandez, D., and Servin, M. (1998). *Appl. Opt.* **37**, 7977–7982.

Gardner, J. L. (1983). *Opt. Lett.* **8**, 91–93.

Garvey, D. W., Li, Q., Kuzyk, M. G., Dirk, C. W., and Martinez, S. (1996). *Opt. Lett.* **21**, 104–106.

Gates, J. W. (1955). *Nature* **176**, 359–360.

Gauthier, D. J., Boyd, R. W., Jungquist, R. K., Lisson, J. B., and Voci, L. L. (1989). *Opt. Lett.* **14**, 323–325.

Gay, J., and Journet, A. (1973). *Nature Phys. Sci.* **241**, 32–33.

Ge, Z., Kobayashi, F., Matsuda, S., and Takeda, M. (2001). *Appl. Opt.* **40**, 1649–1657.

Gerardo, J. B., and Verdeyen, J. T. (1963). *Appl. Phys. Lett.* **3**, 121–123.

Gerges, A. S., Newson, T. P., Jones, J. D. C., and Jackson, D. A. (1989). *Opt. Lett.* **14**, 251–253.

Gezari, D. Y., Labeyrie, A., and Stachnik, V. (1972). *Astrophys. J.* **173**, L1–L5.

Ghiglia, D. C., and Pritt, M. (1998). "Two-Dimensional Phase Unwrapping: Theory, Algorithms and Software." Wiley, New York.

Ghiglia, D. G., Mastin, G. A., and Romero, L. A. (1987). *J. Opt. Soc. Am. A* **4**, 267–280.

Ghiglia, D. G., and Romero, L. A. (1994). *J. Opt. Soc. Am. A* **11**, 107–117.

Ghose, P., Home, D., and Agarwal, G. S. (1991). *Phys. Lett. A* **153**, 403–406.

Ghosh, R., Hong, C. K., Ou, Z. Y., and Mandel, L. (1986). *Phys. Rev. A* **34**, 3962–3968.

Ghosh, R., and Mandel, L. (1987). *Phys. Rev. Lett.* **59**, 1903–1905.

Giallorenzi, T. G., Bucaro, J. A., Dandridge, A., Sigel, G. H., Jr., Cole, J. H., Rashleigh, S. C., and Priest, R. G. (1982). *IEEE J. Quant. Electron.* **QE-18**, 626–665.

Gillard, C. W., and Buholz, N. E. (1983). *Opt. Eng.* **22**, 348–353.

Gillard, C. W., Buholz, N. E., and Ridder, D. W. (1981). *Opt. Eng.* **20**, 129–134.

Gilles, I. P., Uttam, D., Culshaw, B., and Davies, D. E. N. (1983). *Electron. Lett.* **19**, 14–15.

Gilliland, K. E., Cook, H. D., Mielenz, K. D., and Stephens, R. B. (1966). *Metrologia* **2**, 95–98.

Givens, R. B., Wickenden, D. K., Oursler, D. A., Oslander, R., Champion, J. L., and Kistenmacher, T. J. (1999). *Appl. Phys. Lett.* **74**, 1472–1474.

Goldfischer, L. I. (1965). *J. Opt. Soc. Am.* **55**, 247–253.

Goldmeer, J. S., Urban, D. L., and Yuan, Z. (2001). *Appl. Opt.* **40**, 4816–5253.

Goodman, J. W. (1984). *In* "Laser Speckle and Related Phenomena" (J. C. Dainty, ed.), pp. 9–75. Springer-Verlag, Berlin.

Goodman, J. W. (1996). "Introduction to Fourier Optics." McGraw-Hill, New York.

Gordon, S. K., and Jacobs, S. F. (1974). *Appl. Opt.* **13**, 231.

Gori, F., Guattari, G., Palma, C., and Padovani, C. (1988). *Opt. Commun.* **67**, 1–4.

Gough, P. T., and Bates, R. H. T. (1974). *Opt. Acta* **21**, 243–254.

Grangier, P., Roger, G., and Aspect, A. (1986). *Europhys. Lett.* **1**, 173–179.

Grangier, P., Slusher, R. E., Yurke, B., and LaPorta, A. (1987). *Phys. Rev. Lett.* **59**, 2153–2156.

Grayson, T. P., Torgerson, J. R., and Barbosa, G. A. (1994). *Phys. Rev. A* **49**, 626–628.

Greenberger, D. M., Horne, M., and Zeilinger, A. (1989). *In* "Bell's Theorem, Quantum Theory and Conceptions of the Universe" (M. Kafatos, ed.), pp. 73–76. Kluwer, Dordrecht.

Greenberger, D. M., Horne, M. A., Shimony, A., and Zeilinger, A. (1990). *Am. J. Phys.* **58**, 1131–1143.

Greivenkamp, J. E. (1987). *Appl. Opt.* **26**, 5245–5258.

Greivenkamp, J. E., and Bruning, J. H. (1992). *In* "Optical Shop Testing" (D. Malacara, ed.), pp. 501–598. John Wiley, New York.

Greivenkamp, J. E., Lowman, A. E., and Palum, R. J. (1996). *Opt. Eng.* **35**, 2962–2969.

Griffin, D. W. (2001). *Opt. Lett.* **26**, 140–141.

Grohe, O., Gottschling, H., Jennewein, H., and Tschudi, T. (2001). *Opt. Eng.* **40**, 529–532.

Grzanna, J. (1994). *Appl. Opt.* **33**, 6654–6661.

Grzanna, J., and Schultz, G. (1990). *Opt. Commun.* **77**, 107–112.

Grzanna, J., and Schultz, G. (1992). *Appl. Opt.* **31**, 3767–3780.

Guardalben, M. J., Ning, L., Jain, N., Battaglia, J., and Marshall, K. L. (2002). *Appl. Opt.* **41**, 1353–1365.

Guelachvili, G. (1978). *Appl. Opt.* **17**, 1322–1326.

Hale, P. D. S., Bester, M., Danchi, W. C., Fitelson, W., Hoss, S., Lipman, E. A., Monnier, J. D., Tuthill, P. G., and Townes, C. H. (2000). *Astrophys. J.* **537**, 998–1012.

Hall, J. L. (1968). *IEEE J. Quantum Electron.* **QE-4**, 638–641.

Hall, J. L., and Lee, S. A. (1976). *Appl. Phys. Lett.* **29**, 367–369.

Halmos, M. J., and Shamir, J. (1982). *Appl. Opt.* **21**, 265–273.

Hanel, R., and Kunde, V. G. (1975). *Space Sci. Rev.* **18**, 201–256.

Hanes, G. R. (1963). *Appl. Opt.* **2**, 465–470.

Hanes, G. R., Baird, K. M., and De Remigis, J. (1973). *Appl. Opt.* **12**, 1600–1605.

Hanes, G. R., and Dahlstrom, C. E. (1969). *Appl. Phys. Lett.* **14**, 362–364.

Haniff, C. A., Mackay, C. D., Titterington, D. J., Sivia, D., Baldwin, J. E., and Warner, P. J. (1987). *Nature* **328**, 694–696.

Hansen, G. (1955). *Optik* **12**, 5–16.

Hansen, G., and Kinder, W. (1958). *Optik* **15**, 560–564.

Harasaki, A., Schmit, J., and Wyant, J. C. (2000). *Appl. Opt.* **39**, 2107–2115.

Harasaki, A., Schmit, J., and Wyant, J. C. (2001). *Appl. Opt.* **40**, 2102–2106.

Harasaki, A., and Wyant, J. C. (2000). *Appl. Opt.* **39**, 2101–2106.

Harbers, G., Kunst, P. J., and Leibbrandt, G. W. R. (1996). *Appl. Opt.* **35**, 6162–6172.

Hariharan, P. (1969a). *Appl. Opt.* **8**, 1925–1926.

Hariharan, P. (1969b). *J. Opt. Soc. Am.* **59**, 1384.

Hariharan, P. (1975a). *Opt. Eng.* **14**, 257–258.

Hariharan, P. (1975b). *Appl. Opt.* **14**, 1056–1057.

Hariharan, P. (1975c). *Appl. Opt.* **14**, 2319–2321.

Hariharan, P. (1982). *Opt. Lett.* **7**, 274–275.

Hariharan, P. (1987a). *In* "Progress in Optics" (E. Wolf, ed.), Vol. XXIV, pp. 105–164. Elsevier, Amsterdam.

Hariharan, P. (1987b). *Appl. Opt.* **26**, 2506–2507.

Hariharan, P. (1989). *Appl. Opt.* **28**, 27–28.

Hariharan, P. (1996a). *Opt. Eng.* **35**, 484–485.

Hariharan, P. (1996b). *Opt. Eng.* **35**, 3265–3266.

Hariharan, P. (1996c). *J. Mod. Opt.* **43**, 1305–1306.

Hariharan, P. (1996d). "Optical Holography: Theory, Techniques and Applications," 2nd ed. Cambridge University Press, Cambridge, UK.

Hariharan, P. (1997a). *Opt. Eng.* **36**, 2330–2334.

Hariharan, P. (1997b). *Opt. Eng.* **36**, 2478–2481.

Hariharan, P. (1998). *Opt. Eng.* **37**, 2751–2753.

Hariharan, P. (2000). *Opt. Eng.* **39**, 967–969.

Hariharan, P., Brown, N., Fujima, I., and Sanders, B. C. (1993b). *J. Mod. Opt.* **40**, 1477–1488.

Hariharan, P., Brown, N., Fujima, I., and Sanders, B. C. (1995). *J. Mod. Opt.* **42**, 565–567.

Hariharan, P., Brown, N., and Sanders, B. C. (1993). *J. Mod. Opt.* **40**, 113–122.

Hariharan, P., and Ciddor, P. E. (1994). *Opt. Commun.* **110**, 13–17.

Hariharan, P., and Ciddor, P. E. (1995). *Opt. Commun.* **117**, 13–15.

Hariharan, P., and Ciddor, P. E. (1999). *Opt. Eng.* **38**, 1078–1080.

Hariharan, P., and Hegedus, Z. S. (1974). *Opt. Commun.* **14**, 148–151.

Hariharan, P., Larkin, K. G., and Roy, M. (1994). *J. Mod. Opt.* **41**, 663–667.

Hariharan, P., Mujumdar, S., and Ramachandran, H. (1999). *J. Mod. Opt.* **46**, 1443–1446.

Hariharan, P., Oreb, B. F., and Brown, N. (1982). *Opt. Commun.* **41**, 393–396.

Hariharan, P., Oreb, B. F., and Eiju, T. (1987). *Appl. Opt.* **26**, 2504–2506.

Hariharan, P., Oreb, B. F., and Leistner, A. J. (1984). *Opt. Eng.* **23**, 294–297.

Hariharan, P., Oreb, B. F., and Wanzhi, Z. (1984). *Opt. Acta* **31**, 989–999.

Hariharan, P., and Roy, M. (1992). *J. Mod. Opt.* **39**, 1811–1815.

Hariharan, P., and Roy, M. (1994). *J. Mod. Opt.* **41**, 2197–2201.

Hariharan, P., and Roy, M. (1995). *J. Mod. Opt.* **42**, 2357–2360.

Hariharan, P., and Roy, M. (1996a). *Opt. Commun.* **126**, 220–222.

Hariharan, P., and Roy, M. (1996b). *J. Mod. Opt.* **43**, 1797–1800.

Hariharan, P., Roy, M., Robinson, P. A., and O'Byrne, J. W. (1993a). *J. Mod. Opt.* **40**, 871–877.

Hariharan, P., Samuel, J., and Sinha, S. (1999). *J. Opt. B. Quantum Semiclass.* **1**, 199–205.

Hariharan, P., and Sanders, B. C. (1996). *In* "Progress in Optics" (E. Wolf, ed.), Vol. XXXVI, pp. 49–128. Elsevier, Amsterdam.

Hariharan, P., and Sanders, B. C. (2000). *J. Mod. Opt.* **47**, 1739–1744.

Hariharan, P., and Sanders, B. C. (2002). *Optics Express* **10**, 1222–1226.

Hariharan, P., and Sen, D. (1959a). *J. Sci. Instrum.* **36**, 70–72.

Hariharan, P., and Sen, D. (1959b). *J. Opt. Soc. Am.* **49**, 232–234.

Hariharan, P., and Sen, D. (1960a). *J. Opt. Soc. Am.* **50**, 1026–1027.

Hariharan, P., and Sen, D. (1960b). *J. Opt. Soc. Am.* **50**, 357–361.

Hariharan, P., and Sen, D. (1960c). *J. Opt. Soc. Am.* **50**, 999–1001.

Hariharan, P., and Sen, D. (1960d). *J. Sci. Instrum.* **37**, 374–376.

Hariharan, P., and Sen, D. (1960e). *Proc. Phys. Soc.* **75**, 434–438.

Hariharan, P., and Sen, D. (1961a). *J. Opt. Soc. Am.* **51**, 400–404.

Hariharan, P., and Sen, D. (1961b). *J. Opt. Soc. Am.* **51**, 1212–1218.

Hariharan, P., and Sen, D. (1961c). *J. Sci. Instrum.* **38**, 428–432.

Hariharan, P., and Sen, D. (1961d). *J. Opt. Soc. Am.* **51**, 398–399.

Hariharan, P., and Singh, R. G. (1959a). *J. Sci. Instrum.* **36**, 323–324.

Hariharan, P., and Singh, R. G. (1959b). *J. Opt. Soc. Am.* **49**, 732–733.

Hariharan, P., and Singh, R. G. (1961). *J. Opt. Soc. Am.* **51**, 1307.

Hariharan, P., Steel, W. H., and Wyant, J. C. (1974). *Opt. Commun.* **11**, 317–320.

Harris, S. E., Ammann, E. O., and Chang, I. C. (1964). *J. Opt. Soc. Am.* **54**, 1267–1279.

Harris, S. E., Oshman, M. K., and Byer, R. L. (1967). *Phys. Rev. Lett.* **18**, 732–734.

Heckenberg, N. R., and Smith, W. I. B. (1971). *Rev. Sci. Instrum.* **42**, 977–980.

Hegeman, P., Christmann, C., Visser, M., and Braat, J. (2001). *Appl. Opt.* **40**, 4526–4533.

Heintze, L. R. (1967). *Appl. Opt.* **6**, 1924–1929.

Helen, S. S., Kothiyal, M. P., and Sirohi, R. S. (1998). *Opt. Commun.* **154**, 249–254.

Helen, S. S., Kothiyal, M. P., and Sirohi, R. S. (1999). *J. Mod. Opt.* **46**, 993–1001.

Helen, S. S., Kothiyal, M. P., and Sirohi, R. S. (2001). *Opt. Eng.* **40**, 1329–1336.

Henry, C. H. (1991). *In* "Semiconductor Diode Lasers," Vol. 1, pp. 36–41. IEEE, New York.

Herbst, T. M., and Beckwith, S. V. W. (1989). *Appl. Opt.* **28**, 5275–5277.

Hercher, M. (1968). *Appl. Opt.* **7**, 951–966.

Hercher, M. (1969). *Appl. Opt.* **8**, 1103–1106.

Hernandez, G. (1986). "Fabry-Perot Interferometers." Cambridge University Press, Cambridge, UK.

Herold, H., and Jahoda, F. C. (1969). *Rev. Sci. Instrum.* **40**, 145–147.

Herriot, D. R. (1962). *J. Opt. Soc. Am.* **52**, 31–37.

Herriott, D. R. (1963). *Appl. Opt.* **2**, 865–866.

Herzog, T., Rarity, J. G., Weinfurter, H., and Zeilinger, A. (1994). *Phys. Rev. Lett.* **72**, 629–632.

Hettwer, A., Kranz, J., and Schwider, J. (2000). *Opt. Eng.* **39**, 960–966.

Hibino, K. (1997). *Appl. Opt.* **36**, 2084–2093.

Hibino, K., Oreb, B. F., and Farrant, D. I. (1995). *J. Opt. Soc. Am. A* **12**, 761–768.

Hibino, K., Oreb, B. F., Farrant, D. I., and Larkin, K. G. (1997). *J. Opt. Soc. Am. A* **14**, 918–930.

Hicks, T. R., Reay, N. K., and Scaddan, R. J. (1974). *J. Phys E: Sci. Instrum.* **7**, 27–30.

Hill, R. M., and Bruce, C. F. (1962). *Aust. J. Phys.* **15**, 194–222.

Hill, R. M., and Bruce, C. F. (1963). *Aust. J. Phys.* **16**, 282–285.

Hocker, G. B. (1979). *Appl. Opt.* **18**, 1445–1448.

Hodgkinson, I. J., and Vukusic, J. I. (1978). *Appl. Opt.* **17**, 1944–1948.

Holland, M. J., and Burnett, K. (1993). *Phys. Rev. Lett.* **71**, 1355–1358.

Holtom, G., and Teschke, O. (1974). *IEEE J. Quantum Electron.* **QE-10**, 577–579.

Holzwarth, R., Udem, Th., Hänsch, T. W., Knight, J. C., Wadsworth, W. J., and Russell, P. St. J. (2000). *Phys. Rev. Lett.* **85**, 2264–2267.

Hong, C. K., and Mandel, L. (1985). *Phys. Rev. A* **31**, 2409–2418.

Hong, C. K., and Mandel, L. (1986). *Phys. Rev. Lett.* **56**, 58–60.

Hong, C. K., and Noh, T. G. (1998). *J. Opt. Soc. Am. A* **15**, 1192–1197.

Hong, C. K., Ou, Z. Y., and Mandel, L. (1987). *Phys. Rev. Lett.* **59**, 2044–2046.

Hooper, E. B., Jr., and Bekefi, G. (1966). *J. Appl. Phys.* **37**, 4083–4094.

Hopf, F. A. (1980). *J. Opt. Soc. Am.* **70**, 1320–1322.

Hopf, F. A., and Cervantes, M. (1982). *Appl. Opt.* **21**, 668–677.

Hopf, F. A., Tomita, A., and Al-Jumaily, G. (1980). *Opt. Lett.* **5**, 386–388.

Hopf, F. A., Tomita, A., Al-Jumaily, G., Cervantes, M., and Liepmann, T. (1981). *Opt. Commun.* **36**, 487–490.

Hopkins, H. H. (1951). *Proc. Roy. Soc. A* **208**, 263–277.

Hopkins, H. H. (1953). *Proc. Roy. Soc. A* **217**, 408–432.

Hopkins, H. H. (1955). *Opt. Acta* **2**, 23–29.

Hopkinson, G. R. (1978). *J. Optics (Paris)* **9**, 151–155.

Horne, M. A., Shimony, A., and Zeilinger, A. (1989). *Phys. Rev. Lett.* **62**, 2209–2212.

Houston J. B., Jr., Buccini, C. J., and O'Neill, P. K. (1967). *Appl. Opt.* **6**, 1237–1242.

Howes, W. L. (1986a). *Appl. Opt.* **25**, 473–474.

Howes, W. L. (1986b). *Appl. Opt.* **25**, 3167–3170.

Huang, C.-C. (1984). *Opt. Eng.* **23**, 365–370.

Huntley, J. M. (1998). *J. Opt. Soc. Am. A* **8**, 2233–2241.

Ikegami, T., Sudo, S., and Sakai, Y. (1995). "Frequency Stabilization of Semiconductor Laser Diodes." Artech House, Norwood, UK.

Ikram, M., and Hussain, G. (1999). *Appl. Opt.* **38**, 113–120.

Imai, M., Ohashi, T., and Ohtsuka, Y. (1981). *Opt. Commun.* **39**, 7–10.

Inci, M. N., Kidd, S. R., Barton, J. S., and Jones, J. D. C. (1992). *Meas. Sci. Technol.* **3**, 678–684.

Ishii, Y., Chen, J., and Murata, K. (1987). *Opt. Lett.* **12**, 233–235.

Ishii, Y., and Onodera, R. (1991). *Opt. Lett.* **16**, 1523–1525.

Jackson, D. A., Dandridge, A., and Sheem, S. K. (1980). *Opt. Lett.* **5**, 139–141.

Jackson, D. A., and Jones, J. D. C. (1987). *Opt. Acta* **33**, 1469–1503.

Jackson, D. A., Kersey, A. D., Corke, M., and Jones, J. D. C. (1982). *Electron. Lett.* **18**, 1081–1082.

Jackson, D. A., Priest, R., Dandridge, A., and Tveten, A. B. (1980). *Appl. Opt.* **19**, 2926–2929.

Jacobs, S. F., and Shough, D. (1981). *Appl. Opt.* **20**, 3461–3463.

Jacobs, S. F., Shough, D., and Connors, C. (1984). *Appl. Opt.* **23**, 4237–4244.

Jacquinot, P. (1960). *Rep. Progr. Phys.* **23**, 267–312.

Jacquinot, P., and Dufour, C. (1948). *J. Rech. C.N.R.S.* **6**, 91–103.

Jacquinot, P., and Roizen-Dossier, B. (1964). *In* "Progress in Optics" (E. Wolf, ed.), Vol. III, pp. 29–186. North-Holland, Amsterdam.

Jahns, J., and Lohmann, A. W. (1979). *Opt. Commun.* **28**, 263–267.

James, D. F. V., and Wolf, E. (1991a). *Opt. Commun.* **81**, 151–154.

James, D. F. V., and Wolf, E. (1991b). *Phys. Lett. A* **157**, 6–10.

James, D. F. V., and Wolf, E. (1998). *Opt. Commun.* **145**, 1–4.

Jarrett, S. M., and Young, J. F. (1979). *Opt. Lett.* **4**, 176–178.

Jaseja, T. S., Javan, A., Murray, J., and Townes, C. H. (1964). *Phys. Rev. A* **133**, 1221–1225.

Javan, A., Ballik, E. A., and Bond, W. L. (1962). *J. Opt. Soc. Am.* **52**, 96–98.

Javan, A., Bennett, W. R., and Herriott, D. R. (1961). *Phys. Rev. Lett.* **6**, 106–110.

Jensen, S. C., Chow, W. W., and Lawrence, G. N. (1984). *Appl. Opt.* **23**, 740–745.

Jentink, H. W., de Mul, F. F. M., Suichies, H. E., Aarnoudse, J. G., and Greve, J. (1988). *Appl. Opt.* **27**, 379–385.

Jerrard, H. G. (1954). *J. Opt. Soc. Am.* **44**, 634–640.

Jerrard, H. G. (1982). *Opt. Laser Technol.* **14**, 309–319.

Jiao, H., Wilkinson, S. R., Chiao, R. Y., and Nathel, H. (1989). *Phys. Rev. A* **39**, 3475–3486.

Jinno, M., and Matsumoto, T. (1991). *Opt. Lett.* **16**, 220–222.

Joenathan, C. (1994). *Appl. Opt.* **33**, 4147–4155.

Johnson, G. W., Leiner, D. C., and Moore, D. T. (1977). *Proc. SPIE* **126**, 152–160.

Johnson, G. W., Leiner, D. C., and Moore, D. T. (1979). *Opt. Eng.* **18**, 46–52.

Johnson, M. A., Betz, A. L., and Townes, C. H. (1974). *Phys. Rev. Lett.* **33**, 1617–1620.

Jones, F. E. (1981). *J. Res. NBS* **86**, 27–32.

Jones, J. D. C., Corke, M., Kersey, A. D., and Jackson, D. A. (1982). *Electron. Lett.* **18**, 967–969.

Jones, R. C. (1941). *J. Opt. Soc. Am.* **31**, 488–493.

Jordan, T. F., and Ghielmetti, F. (1964). *Phys. Rev. Lett.* **12**, 607–609.

Judge, Th. R., Quan, C. H., and Bryanston-Cross, P. J. (1992). *Opt. Eng.* **31**, 533–543.

Juncar, P., and Pinard, J. (1975). *Opt. Commun.* **14**, 438–441.

Juncar, P., and Pinard, J. (1982). *Rev. Sci. Instrum.* **53**, 939–948.

Kahane, A., O'Sullivan, M. S., Sanford, N. M., and Stoicheff, B. P. (1983). *Rev. Sci. Instrum.* **54**, 1138–1142.

Kajimura, T., Koroda, T., Yamashita, S., Nakamura, M., and Umeda, J. (1979). *Appl. Opt.* **18**, 1812–1815.

Kamegai, M. (1974). *Appl. Opt.* **13**, 1997–1998.

Kandpal, H. C., Vaishya, J. S., and Joshi, K. C. (1989). *Opt. Commun.* **73**, 169–172.

Kelsall, D. (1959). *Proc. Phys. Soc.* **73**, 465–479.

Kennedy, R. J. (1926). *Proc. Natl. Acad. Sci. USA* **12**, 621–629.

Kersey, A. D., Berkoff, T. A., and Morey, W. W. (1993). *Opt. Lett.* **18**, 72–74.

Kersey, A. D., Jackson, D. A., and Corke, M. (1982). *Electron. Lett.* **18**, 559–561.

Kersey, A. D., Jackson, D. A., and Corke, M. (1983a). *Electron. Lett.* **19**, 102–103.

Kersey, A. D., Jackson, D. A., and Corke, M. (1983b). *Opt. Commun.* **45**, 71–74.

Kersey, A. D., Lewin, A. C., and Jackson, D. A. (1984). *Electron. Lett.* **20**, 368–370.

Khavinson, V. M. (1999). *Appl. Opt.* **38**, 126–135.

Kikuta, H., Iwata, K., and Nagata, R. (1986). *Appl. Opt.* **25**, 2976–2980.

Killpatrick, J. (1967). *IEEE Spectrum* **4**, 44–55.

Kim, B. Y., and Shaw, H. J. (1984a). *Opt. Lett.* **9**, 375–377.

Kim, B. Y., and Shaw, H. J. (1984b). *Opt. Lett.* **9**, 378–380.

Kim, M.-S., and Kim, S.-W. (2002). *Appl. Opt.* **41**, 5938–5942.

Kim, S.-W., Kang, M.-G., and Han, G.-S. (1997). *Opt. Eng.* **36**, 3101–3106.

Kim, S.-W., and Kim, G.-H. (1999). *Appl. Opt.* **38**, 5968–5973.

Kimble, H. J., and Walls, D. F. (1987). *J. Opt. Soc. Am. B* **4**, 1450.

Kingslake, R. (1925–26). *Trans. Opt. Soc.* **27**, 94–105.

Kinnstaetter, K., Lohmann, A. W., Schwider, J., and Streibl, N. (1988). *Appl. Opt.* **27**, 5082–5089.

Kino, G. S., and Chim, S. S. C. (1990). *Appl. Opt.* **29**, 3775–3783.

Kinosita, K. (1953). *J. Phys. Soc. Japan* **8**, 219–225.

Klimcak, C. M., and Camparo, J. C. (1988). *J. Opt. Soc. Am. B* **5**, 211–214.

Knight, D. J. E., Edwards, G. J., Pearce, P. R., and Cross, N. R. (1980). *IEEE Trans. Instrum. Measurement* **IM-29**, 257–264.

Knox, J. D., and Pao, Y. (1970). *Appl. Phys. Lett.* **16**, 129–131.

Koch, C. (1999). *Appl. Opt.* **38**, 2812–2819.

Koch, G. J., Cook, A. L., Fitzgerald, C. M., and Dharamsi, A. N. (2001). *Opt. Eng.* **40**, 525–528.

Koechner, W. (1976). "Solid-State Laser Engineering." Springer-Verlag, New York.

Kogelnik, H., and Li, T. (1966). *Appl. Opt.* **5**, 1550–1567.

Kohno, T., Matsumoto, D., Yazawa, T., and Uda, Y. (2000). *Opt. Eng.* **39**, 2696–2699.

Koo, K. P., and Sigel, J. H., Jr. (1982). *IEEE J. Quantum Electron.* **QE-18**, 670–675.

Koo, K. P., and Sigel J. H., Jr. (1984). *Opt. Lett.* **9**, 257–259.

Koppelmann, G., and Krebs, K. (1961a). *Optik* **18**, 349–357.

Koppelmann, G., and Krebs, K. (1961b). *Optik* **18**, 358–372.

Kothiyal, M. P., and Delisle, C. (1985). *Appl. Opt.* **24**, 4439–4442.

Kowalski, F. V., Hawkins, R. T., and Schawlow, A. L. (1976). *J. Opt. Soc. Am.* **66**, 965–966.

Kowalski, F. V., Teets, R. E., Demtröder, W., and Schawlow, A. L. (1978). *J. Opt. Soc. Am.* **68**, 1611–1613.

Kowalski, T., Neumann, R., Noehte, S., Schwarzwald, R., Suhr, H., and zu Putlitz, G. (1985). *Opt. Commun.* **53**, 141–146.

Kristal, R., and Peterson, R. W. (1976). *Rev. Sci. Instrum.* **47**, 1357–1359.

Krug, W., Rienitz, J., and Schulz, G. (1964). "Contributions to Interference Microscopy." Hilger and Watts, London.

Kubota, H. (1950). *J. Opt. Soc. Am.* **40**, 146–149.

Kubota, H. (1961). *In* "Progress in Optics" (E. Wolf, ed.), Vol. I, pp. 213–251. North-Holland, Amsterdam.

Kubota, T., Nara, M., and Yoshino, T. (1987). *Opt. Lett.* **12**, 310–312.

Küchel, M. F. (2001). *Optik* **112**, 381–391.

Kujawinska, M., and Wojciak, J. (1991). *Opt. Lasers Eng.* **14**, 325–339.

Kuwamura, S., and Yamaguchi, I. (1997). *Appl. Opt.* **37**, 4473–4482.

Kwiat, P. G., and Chiao, R. Y. (1991). *Phys. Rev. Lett.* **66**, 588–591.

Kwiat, P. G., Steinberg, A. M., and Chiao, R. Y. (1992). *Phys. Rev. A* **45**, 7729–7739.

Kwiat, P. G., Steinberg, A. M., and Chiao, R. Y. (1993). *Phys. Rev. A* **47**, R2472–R2475.

Kwiat, P. G., Vareka, W. A., Hong, C. K., Nathel, H., and Chiao, R. Y. (1990). *Phys. Rev. A* **41**, 2910–2913.

Kwon, O. (1984). *Opt. Lett.* **9**, 59–61.

Kwon, O., Wyant, J. C., and Hayslett, C. R. (1980). *Appl. Opt.* **19**, 1862–1869.

Kwong, S.-K., Yariv, A., Cronin-Golomb, M., and Ury, I. (1985). *Appl. Phys. Lett.* **47**, 460–462.

Kyuma, K., Tai, S., and Nunoshita, M. (1982). *Opt. Lasers Eng.* **3**, 155–182.

Labeyrie, A. (1970). *Astron. Astrophys.* **6**, 85–87.

Labeyrie, A. (1974). *Nouv. Rev. Optique* **5**, 141–151.

Labeyrie, A. (1976). *In* "Progress in Optics" (E. Wolf, ed.), Vol. XIV, pp. 49–87. North-Holland, Amsterdam.

Labeyrie, A. (1978). *Ann. Rev. Astron. Astrophys.* **16**, 77–102.

Labeyrie, A., Bonneau, D., Stachnik, R. V., and Gezari, D. Y. (1974). *Astrophys. J.* **194**, L147–L151.

Lacroix, S., Bures, J., Parent, M., and Lapierre, J. (1984). *Opt. Commun.* **51**, 65–67.

LaGasse, M. J., Liu-Wong, D., Fujimoto, J. G., and Haus, H. A. (1989). *Opt. Lett.* **14**, 311–313.

Laming, R., Gold, M. P., Payne, D. N., and Halliwell, N. A. (1985). *Proc. SPIE* **586**, 38–44.

Larchuk, T. S., Campos, R. A., Rarity, J. G., Tapster, P. R., Jakeman, E., Saleh, B. E. A., and Teich, M. C. (1993). *Phys. Rev. Lett.* **70**, 1603–1606.

Larkin, K. G. (1996). *J. Opt. Soc. Am. A* **13**, 832–843.

Larkin, K. G., and Oreb, B. F. (1992). *J. Opt. Soc. Am. A* **9**, 1740–1748.

Lattes, A., Haus, H. A., Leonberger, F. J., and Ippen, E. P. (1983). *IEEE J. Quantum Electron.* **19**, 1718–1723.

Lavan, M. J., Cadwallender, W. K., and De Young, T. F. (1975). *Rev. Sci. Instrum.* **46**, 525–527.

Lavan, M. J., Cadwallender, W. K., De Young, T. F., and Van Damme, G. E. (1976). *Appl. Opt.* **15**, 2627–2628.

Lawson, P. R. (ed.) (1997). "Long Baseline Stellar Interferometry," Vol. MS139. SPIE, Bellingham, WA.

Layer, H. P. (1980). *IEEE Trans. Instrum. Measurement* **IM-29**, 358–361.

Lee, B. S., and Strand, T. C. (1990). *Appl. Opt.* **29**, 3784–3788.

Lee, P. H., and Skolnick, M. L. (1967). *Appl. Phys. Lett.* **10**, 303–305.

Lee, S. H., Naulleau, P., Goldberg, K. A., Piao, F., Oldham, W., and Bokar, J. (2000). *Appl. Opt.* **31**, 5768–5772.

Lee, W. H. (1974). *Appl. Opt.* **13**, 1677–1682.

Lee, W. H. (1978). *In* "Progress in Optics" (E. Wolf, ed.), Vol. XVI, pp. 121–232. North-Holland, Amsterdam.

Lee, W. H. (1979). *Appl. Opt.* **18**, 3661–3669.

Leeb, W. R., Schiffner, G., and Scheiterer, E. (1979). *Appl. Opt.* **18**, 1293–1295.

Leibbrandt, G. W. R., Harbers, G., and Kunst, P. J. (1996). *Appl. Opt.* **35**, 6151–6161.

Leilabady, P. A., Jones, J. D. C., Corke, M., and Jackson, D. A. (1986). *J. Phys. E: Sci. Instrum.* **19**, 143–146.

Leith, E. N., and Upatnieks, J. (1964). *J. Opt. Soc. Am.* **54**, 1295–1301.

Lena, P. J., and Quirrenbach, A. (eds.) (2000). "Interferometry in Optical Astronomy." *Proc. SPIE* **4006**. SPIE, Bellingham, WA.

Lenouvel, L., and Lenouvel, F. (1938). *Rev. Opt.* **17**, 350–361.

Leonhardt, K. (1972). *Optik* **35**, 509–523.

Leonhardt, K. (1974). *Opt. Commun.* **11**, 312–316.

Leonhardt, K. (1981). "Optische Interferenzen." Wissenschaftliche Vergesellschaft, Stuttgart.

Lewin, A. C., Kersey, A. D., and Jackson, D. A. (1985). *J. Phys. E: Sci. Instrum.* **18**, 604–608.

Li, K., Xiong, Z., Peng, G. D., and Chu, P. L. (1997). *Opt. Commun.* **136**, 223–226.

Liepmann, T. W., and Hopf, F. A. (1985). *Appl. Opt.* **24**, 1485–1488.

Liewer, K. M. (1979). *In* "High Angular Resolution Stellar Interferometry" (I.A.U. Colloquium No. 50) (J. Davis and W. J. Tango, eds.), pp. 8-1–8-14. Chatterton Astronomy Department, University of Sydney.

Lin, S. C., and Giallorenzi, T. G. (1979). *Appl. Opt.* **18**, 915–931.

Lindermeir, E., Haschberger, P., Tank, V., and Dietl, H. (1992). *Appl. Opt.* **31**, 4527–4533.

Lindlein, N. (2001). *Appl. Opt.* **40**, 2698–2708.

Lindsay, S. M., and Shepherd, I. W. (1977). *J. Phys. E: Sci. Instrum.* **10**, 150–154.

Linnik, V. P. (1933). *C. R. Acad. Sci. URSS* **1**, 18.

Linnik, V. P. (1942). *Comptes Rendus (Doklady) URSS* **35**, 16–19.

Liu, C. Y. C., and Lohmann, A. W. (1973). *Opt. Commun.* **8**, 372–377.

Liu, L. S., and Klinger, J. H. (1979). *Proc. SPIE* **192**, 17–26.

Löfdahl, M. G., and Eriksson, H. (2001). *Opt. Eng.* **40**, 984–990.

Loheide, S. (1997). *Opt. Commun.* **141**, 254–258.

Lohmann, A. W., and Silva, D. E. (1971). *Opt. Commun.* **2**, 413–415.

Loomis, J. S. (1980). *Opt. Eng.* **19**, 679–685.

Loos, G. C. (1992). *Appl. Opt.* **31**, 6632–6636.

Loudon, R. (1980). *Rep. Progr. Phys.* **43**, 913–949.

Love, G. D. (1993). *Appl. Opt.* **32**, 2222–2223.

Lowenthal, S., and Arsenault, H. H. (1970). *J. Opt. Soc. Am.* **60**, 1478–1483.

Luc, P., and Gerstenkorn, S. (1978). *Appl. Opt.* **17**, 1327–1331.

Lyot, B. (1944). *Ann. Astrophys.* **7**, 31–79.

McAlister, H. A., Hartkopf, W. I., Bagnuolo W. G., Jr., Sowell, J. R., Franz, O. G., and Evans, D. S. (1988). *Astron. J. (USA)* **96**, 1431–1438.

McAlister, H. A., Hartkopf, W. I., and Franz, O. G. (1990). *Astron. J. (USA)* **99**, 965–978.

McAlister, H. A., Hartkopf, W. I., Sowell, J. R., Dombrowski, E. G., and Franz, O. G. (1989). *Astron. J. (USA)* **97**, 510–531.

Macek, W., and Davis, D. (1963). *Appl. Phys. Lett.* **2**, 67–68.

MacGovern, A. J., and Wyant, J. C. (1971). *Appl. Opt.* **10**, 619–624.

Mack, J. E., McNutt, D. P., Rossler, F. L., and Chabbal, R. (1963). *Appl. Opt.* **2**, 873–885.

Macy, W. W., Jr. (1983). *Appl. Opt.* **22**, 3898–3901.

Maeda, M. W., Kumar, P., and Shapiro, J. H. (1987). *Opt. Lett.* **12**, 161–163.

Magyar, G., and Mandel, L. (1963). *Nature* **198**, 255–256.

Mahal, V., and Arie, A. (1996). *Appl. Opt.* **35**, 3010–3015.

Maillard, J. P. (1996). *Appl. Opt.* **35**, 2734–2746.

Maiman, T. H. (1960). *Nature* **187**, 493–494.

Maischberger, K., Rudiger, A., Schilling, R., Schnupp, L., Winkler, W., and Billing, H. (1981). *In* "Laser Spectroscopy V" (A. R. W. McKellar, T. Oka, and B. P. Stoicheff, eds.), pp. 25–32. Springer-Verlag, Berlin.

Malacara, D. (1974). *Appl. Opt.* **13**, 1781–1784.

Malacara, D. (1992a). *In* "Optical Shop Testing" (D. Malacara, ed.), pp. 51–94. John Wiley, New York.

Malacara, D. (1992b). *In* "Optical Shop Testing" (D. Malacara, ed.), pp. 173–206. John Wiley, New York.

Malacara, D., Carpio-Valadez, J. M., and Sanchez Mondragon, J. J. (1987). *Proc. SPIE* **813**, 1273–1274.

Malacara, D., Carpio-Valadez, J. M., and Sanchez Mondragon, J. J. (1990). *Opt. Eng.* **29**, 672–675.

Malacara, D., and DeVore, S. L. (1992). *In* "Optical Shop Testing" (D. Malacara, ed.), pp. 455–499. John Wiley, New York.

Malacara, D., and Harris, O. (1970). *Appl. Opt.* **9**, 1630–1633.

Malacara, D., and Mendez, M. (1968). *Opt. Acta* **15**, 59–63.

Malacara, D., Servin, M., and Malacara, Z. (1998). "Interferogram Analysis for Optical Testing." Marcel Dekker, New York.

Malyshev, Yu., Ovchimikov, S. N., Rastorguev, Yu. G., Tatarenkov, V. M., and Titov, A. N. (1980). *Sov. J. Quant. Electron.* **10**, 376–377.

Mandel, L. (1962). *J. Opt. Soc. Am.* **52**, 1335–1340.

Mandel, L. (1963). *In* "Progress in Optics" (E. Wolf, ed.), Vol. II, pp. 181–248. North-Holland, Amsterdam.

Mandel, L. (1964). *Phys. Rev.* **134**, A10–A15.

Mandel, L. (1976). *In* "Progress in Optics" (E. Wolf, ed.), Vol. XIII, pp. 27–68. North-Holland, Amsterdam.

Mandel, L. (1982). *Phys. Lett. A* **89**, 325–326.

Mandel, L. (1983). *Phys. Rev. A* **28**, 929–943.

Mandel, L. (1991). *Opt. Lett.* **16**, 1882–1883.

Mandel, L., and Wolf, E. (1965). *Rev. Mod. Phys.* **37**, 231–287.

Mandel, L., and Wolf, E. (1976). *J. Opt. Soc. Am.* **66**, 529–535.

Mandel, L., and Wolf, E. (1995). "Optical Coherence and Quantum Optics." Cambridge University Press, Cambridge, UK.

Mantravadi, M. V. (1992a). *In* "Optical Shop Testing" (D. Malacara, ed.), pp. 1–49. John Wiley, New York.

Mantravadi, M. V. (1992b). *In* "Optical Shop Testing" (D. Malacara, ed.), pp. 123–172. John Wiley, New York.

Maréchal, A., Lostis, P., and Simon, J. (1967). *In* "Advanced Optical Techniques" (A. C. S. van Heel, ed.), pp. 435–446. North-Holland, Amsterdam.

Maron, Y. (1977). *Rev. Sci. Instrum.* **48**, 1668–1672.

Maron, Y. (1978). *Rev. Sci. Instrum.* **49**, 1598–1599.

Marrakchi, A., Huignard, J. P., and Herriau, J. P. (1980). *Opt. Commun.* **34**, 15–18.

Massie, N. A. (1980). *Appl. Opt.* **19**, 154–160.

Massie, N. A., Nelson, R. D., and Holly, S. (1979). *Appl. Opt.* **18**, 1797–1803.

Mather, J. C., Cheng, E. S., Eplee, R. E., Jr., Isaacman, R. B., Meyer, S. S., Shafer, R. A., Weiss, R., Wright, E. L., Bennett, C. I., Boggess, N. W., Dwek, E., Gulkis, S., Hauser, M. G., Janssen, M., Kelsall, T., Lubin, P. M., Mosely, S. H., Jr., Murdock, T. L., Silverberg, R. F., Smoot, G. F., and Wilkinson, D. T. (1990). *Astrophys. J. Lett. (USA)* **354**, L37–L40.

Matsubara, K., Stork, W., Wagner, A., Drescher, J., and Müller-Glaser, K. D. (1997). *Appl. Opt.* **36**, 4516–4520.

Matsumoto, H. (1986). *Appl. Opt.* **25**, 493–498.

Matsumoto, H., Seino, S., and Sakurai, Y. (1980). *Metrologia* **16**, 169–175.

Matthews, H. J., Hamilton, D. K., and Sheppard, J. R. (1986). *Appl. Opt.* **25**, 2372–2374.

Medecki, H., Tejnil, E., Goldberg, K. A., and Bokor, J. (1996). *Opt. Lett.* **21**, 1526–1528.

Meers, B. J. (1988). *Phys. Rev. D* **38**, 2317–2326.

Mehta, D. S., Sugai, M., Hinosugi, H., Saito, S., Takeda, M., Kurokawa, T., Takahashi, H., Ando, M., Shishido, M., and Yoshizawa, T. (2002). *Appl. Opt.* **41**, 3874–3885.

Meisner, J. A. (1996). *Opt. Eng.* **35**, 1927–1935.

Mercer, C. R., and Creath, K. (1996). *Appl. Opt.* **35**, 1633–1642.

Michelson, A. A., and Pease, F. G. (1921). *Astrophys. J.* **53**, 249–259.

Mizobuchi, Y., and Ohtake, Y. (1992). *Phys. Lett. A* **168**, 1–5.

Mizuno, J., Strain, K. A., Nelson, P. G., Chen, J. M., Schilling, R., Rüdiger, A., Winkler, W., and Danzmann, K. (1993). *Phys. Lett. A* **175**, 273–276.

Monchalin, J. P., Kelly, M. J., Thomas, J. E., Kurnit, N. A., Szöke, A., Zernike, F., Lee, P. H., and Javan, A. (1981). *Appl. Opt.* **20**, 736–757.

Montgomery, A. J. (1964). *J. Opt. Soc. Am.* **54**, 191–198.

Moran, M. J., She, C., and Carman, R. L. (1975). *IEEE J. Quantum Electron.* **QE-11**, 259–263.

Morris, R. H., Ferguson, J. B., and Warniak, J. S. (1975). *Appl. Opt.* **14**, 2808.

Moss, G. E., Miller, L. R., and Forward, R. L. (1971). *Appl. Opt.* **10**, 2495–2498.

Mottier, F. M. (1979). *Opt. Eng.* **18**, 464–468.

Mourat, G., Servagent, N., and Bosch, T. (2000). *Opt. Eng.* **39**, 738–743.

Munnerlyn, C. R. (1969). *Appl. Opt.* **8**, 827–829.

Munnerlyn, C. R., and Latta, M. (1968). *Appl. Opt.* **7**, 1858–1859.

Murphy, K. A., Gunther, M. F., Vengsarkar, A. M., and Claus, R. O. (1991). *Opt. Lett.* **16**, 273–275.

Murphy, P. E., Brown, T. G., and Moore, D. T. (2000). *Appl. Opt.* **39**, 2122–2129.

Murty, M. V. R. K. (1963). *J. Opt. Soc. Am.* **53**, 568–570.

Murty, M. V. R. K. (1964). *Appl. Opt.* **3**, 531–534.

Murty, M. V. R. K., and Hagerott, E. C. (1966). *Appl. Opt.* **5**, 615–619.

Murty, M. V. R. K., and Shukla, R. P. (1976). *Opt. Eng.* **15**, 461–463.

Naulleau, P. P., and Goldberg, K. A. (1999). *Appl. Opt.* **38**, 3523–3532.

Negro, J. E. (1984). *Appl. Opt.* **23**, 1921–1930.

Netterfield, R. P., Freund, C. H., Seckold, J. A., and Walsh, C. J. (1997). *Appl. Opt.* **36**, 4556–4561.

Nieuwenhuizen, H. (1970). *Mon. Not. Roy. Astron. Soc.* **150**, 325–335.

Ning, Y. N., and Jackson, D. A. (1992). *Proc. SPIE* **1795**, 165–172.

Nomura, T., Okuda, S., Kamiya, K., Tashiro, H., and Yoshikawa, K. (2002). *Appl. Opt.* **41**, 1954–1961.

North-Morris, M. B., VanDelden, J., and Wyant, J. C. (2002). *Appl. Opt.* **41**, 668–677.

Nugent, K. A. (1985). *Appl. Opt.* **24**, 3101–3105.

Oberparleiter, M., and Weinfurter, H. (2000). *Opt. Commun.* **183**, 133–137.

Ohman, Y. (1938). *Nature* **141**, 157–158, 291.

Ohtsuka, Y., and Itoh, K. (1979). *Appl. Opt.* **18**, 219–224.

Ohtsuka, Y., and Sasaki, I. (1974). *Opt. Commun.* **10**, 362–365.

Ohtsuka, Y., and Sasaki, I. (1977). *Opt. Commun.* **22**, 211–214.

Ohtsuka, Y., and Tsubokawa, M. (1984). *Opt. Laser Technol.* **16**, 25–29.

Okada, K., Sakuta, H., Ose, T., and Tsujiuchi, J. (1990). *Appl. Opt.* **29**, 3280–3285.

Okada, K., Sato, A., and Tsujiuchi, J. (1991). *Opt. Commun.* **84**, 118–124.

Okuda, S., Nomura, T., Kamiya, K., Miyashiro, H., Yoshikawa, K., and Tashiro, H. (2000). *Appl. Opt.* **39**, 5179–5186.

Olbright, G. R., and Peyghambarian, N. (1986). *Appl. Phys. Lett.* **48**, 1184–1186.

Olsson, A., and Tang, C. L. (1981). *Appl. Opt.* **20**, 3503–3507.

Omura, K., and Yatagai, T. (1988). *Appl. Opt.* **27**, 523–528.

Onodera, R., and Ishii, Y. (1994). *Appl. Opt.* **33**, 5052–5061.

Onodera, R., and Ishii, Y. (1995). *Appl. Opt.* **34**, 4740–4746.

Onodera, R., and Ishii, Y. (1999a). *Opt. Commun.* **167**, 47–51.

Onodera, R., and Ishii, Y. (1999b). *Opt. Eng.* **38**, 2045–2049.

Oreb, B. F., Farrant, D. I., Walsh, C. J., Forbes, G., and Fairman, P. S. (2000). *Appl. Opt.* **39**, 5161–5171.

Oslander, R., Champion, J., Kistenmacher, T., Wickenden, D., and Miragliotta, J. (March 2001). *SPIE's OE Magazine* 29–31.

Ou, Z. Y., Hong, C. K., and Mandel, L. (1987). *Opt. Commun.* **63**, 118–122.

Ou, Z. Y., and Mandel, L. (1988). *Phys. Rev. Lett.* **61**, 54–57.

Ou, Z. Y., Wang, L. J., Zou, X. Y., and Mandel, L. (1990a). *Phys. Rev. A* **41**, 566–568.

Ou, Z. Y., Zou, X. Y., Wang, L. J., and Mandel, L. (1990b). *Phys. Rev. Lett.* **65**, 321–324.

Ou, Z. Y., Zou, X. Y., Wang, L. J., and Mandel, L. (1990c). *Phys. Rev. A* **42**, 2957–2965.

Ovryn, B., and Andrews, J. H. (1999). *Appl. Opt.* **38**, 1959–1967.

Paez, G., and Strojnik, M. (1997). *Opt. Lett.* **22**, 1669–1671.

Paez, G., and Strojnik, M. (1998). *Opt. Lett.* **23**, 406–408.

Paez, G., and Strojnik, M. (1999). *J. Opt. Soc. Am. A* **16**, 475–480.

Paez, G., and Strojnik, M. (2000). *J. Opt. Soc. Am. A* **17**, 46–52.

Paez, G., Strojnik, M., and Torales, G. G. (2000). *Appl. Opt.* **39**, 5172–5178.

Panarella, E. (1973). *J. Phys., E: Sci. Instrum.* **6**, 523.

Pancharatnam, S. (1956). *Proc. Ind. Acad. Sci. A* **44**, 247–262.

Pancharatnam, S. (1963). *Proc. Ind. Acad. Sci. A* **57**, 231–243.

Panish, M. B., Hayashi, I., and Sumski, S. (1970). *Appl. Phys. Lett.* **16**, 326–327.

Papoulis, A. (1965). "Probability, Random Variables and Stochastic Processes." McGraw-Hill, New York.

Park, M.-C., and Kim, S.-W. (2000). *Opt. Eng.* **39**, 952–959.

Patel, N. S., Hall, K. L., and Rauschenbach, K. A. (1996). *Opt. Lett.* **21**, 1466–1468.

Patel, N. S., Hall, K. L., and Rauschenbach, K. A. (1998). *Appl. Opt.* **37**, 2831–2842.

Paulsson, L., Sjödahl, M., Kato, J., and Yamaguchi, I. (2000). *Appl. Opt.* **39**, 3285–3288.

Pearson, J. E., Bridges, W. B., Hansen, S., Nussmeier, T. A., and Pedinoff, M. E. (1976). *Appl. Opt.* **15**, 611–621.

Peck, E. R., and Obetz, S. W. (1953). *J. Opt. Soc. Am.* **43**, 505–509.

Pedretti, E., Labeyrie, A., Arnold, L., Thureau, N., Lardiere, O., Boccaletti, A., and Riaud, P. (2001). *Astron. Astrophys. Suppl. Ser.* **147**, 285–290.

Peek, Th. H., Bolwijn, P. T., and Alkemade, C. Th. J. (1967). *Am. J. Phys.* **35**, 820–831.

Peřina, J. (1971). "Coherence of Light." Van Nostrand Reinhold, New York.

Petley, B. W. (1983). *Nature* **303**, 373–376.

Pfleegor, R. L., and Mandel, L. (1967a). *Phys. Lett. A* **24**, 766–767.

Pfleegor, R. L., and Mandel, L. (1967b). *Phys. Rev.* **159**, 1084–1088.

Pfleegor, R. L., and Mandel, L. (1968). *J. Opt. Soc. Am.* **58**, 946–950.

Pförtner, A., and Schwider, J. (2001). *Appl. Opt.* **40**, 6223–6228.

Philbert, M. (1958). *Rev. Opt.* **37**, 598–608.

Phillion, D. W. (1997). *Appl. Opt.* **36**, 8098–8115.

Phoenix, S. J. D., Barnett, S. M., and Chefles, A. (2000). *J. Mod. Opt.* **47**, 507–516.

Pipkin, F. M. (1978) *In* "Advances in Atomic and Molecular Physics" (D. R. Bates and B. Bederson, eds.), Vol. XIV, pp. 281–341. Academic Press, New York.

Post, E. J. (1967). *Rev. Mod. Phys.* **39**, 475–493.

Potter, J., Ezbiri, A., and Tatam, R. P. (1997). *Opt. Commun.* **140**, 11–14.

Puschert, W. (1974). *Opt. Commun.* **10**, 357–361.

Qian, F., Wang, X., Wang, X., and Yang, B. (2001). *Opt. Laser Technol.* **33**, 479–486.

Quirrenbach, A. (2001). *Ann. Rev. Astron. Astrophys.* **39**, 353–401.

Raine, K. W., and Downs, M. J. (1978). *Opt. Acta* **25**, 549–558.

Rajbenbach, H., and Huignard, J. P. (1985). *Opt. Lett.* **10**, 137–139.

Ramachandran, G. N., and Ramaseshan, S. (1952). *J. Opt. Soc. Am.* **42**, 49–56.

Ramaseshan, S., and Nityananda, R. (1986). *Curr. Sci. (India)* **55**, 1225–1226.

Ramsay, J. V. (1966). *Appl. Opt.* **5**, 1297–1301.

Ramsay, J. V. (1969). *Appl. Opt.* **8**, 569–574.

Ramsay, J. V., Kobler, H., and Mugridge, E. G. V. (1970). *Solar Phys.* **12**, 492–501.

Rao, D. N., and Kumar, V. N. (1994). *J. Mod. Opt.* **41**, 1757–1763.

Rao, Y.-J., and Jackson, D. A. (1996). *Meas. Sci. Technol.* **7**, 981–999.

Rarity, J. G., and Tapster, P. R. (1989). *J. Opt. Soc. Am. B* **6**, 1221–1226.

Rarity, J. G., and Tapster, P. R. (1990). *Phys. Rev. Lett.* **64**, 2495–2498.

Rarity, J. G., and Tapster, P. R. (1992). *Phys. Rev. A* **45**, 2052–2056.

Rarity, J. G., Tapster, P. R., Jakeman, E., Larchuk, T., Campos, R. A., Teich, M. C., and Saleh, B. E. A. (1990). *Phys. Rev. Lett.* **65**, 1348–1351.

Räsänen, J., Abedin, K. M., Tenjimbayashi, K., Riju, T., Matsuda, K., and Peiponen, K.-E. (1997). *Opt. Commun.* **143**, 1–4.

Rashleigh, S. C. (1981). *Opt. Lett.* **6**, 19–21.

Recknagel, R.-J., and Notni, G. (1998). *Opt. Commun.* **148**, 122–128.

Reed, I. S. (1962). *IEEE Trans. Information Theory* **IT-8**, 194–195.

Řeháček, J., and Peřina, J. (1996). *Opt. Commun.* **125**, 82–89.

Reichert, J., Holzwarth, R., Udem, Th., and Hänsch, T. W. (1999). *Opt. Commun.* **172**, 59–68.

Restaino, S. R., Baker, J., Carreras, R. A., Fender, J. L., Loos, G. C., McBroom, R. J., Nahrstedt, D., and Payne, D. M. (1997). *Meas. Sci. Technol.* **8**, 1105–1111.

Restaino, S. R., McBroom, R. J., Baker, J., Carreras, R. A., and Loos, G. C. (1996). *Opt. Commun.* **130**, 231–234.

Rhodes, W. T., and Goodman, J. W. (1968). *J. Opt. Soc. Am.* **63**, 647–657.

Ribeiro, P. H. S., Pádua, S., Machado da Silva, J. C., and Barbosa, G. A. (1994). *Phys. Rev. A* **49**, 4176–4179.

Richardson, W. H., Machida, S., and Yamamoto, Y. (1991). *Phys. Rev. Lett.* **66**, 2867–2870.

Rimmer, M. P., King, D. M., and Fox, D. G. (1972). *Appl. Opt.* **11**, 2790–2796.

Rimmer, M. P., and Wyant, J. C. (1975). *Appl. Opt.* **14**, 142–150.

Roberts, R. B. (1975). *J. Phys. E: Sci. Instrum.* **8**, 600–602.

Robertson, D. I., Morrison, E., Hough, J., Kilbourn, S., Meers, B. J., Newton, G. P., Robertson, N. A., Strain, K. A., and Ward, H. (1995). *Rev. Sci. Instrum.* **66**, 4447–4452.

Robertson, J. G., and Tango, W. J. (eds.) (1994). "Very High Angular Resolution Imaging." *IAU Symposium 158.* Astronomical Society of the Pacific, San Francisco.

Robinson, D. W., and Reid, G. T. (1993). "Interferogram Analysis: Digital Processing Techniques for Fringe Pattern Measurement." IOP Publishing, London.

Rogstad, D. H. (1968). *Appl. Opt.* **7**, 585–588.

Roland, J. J., and Agrawal, G. P. (1981). *Opt. Laser Technol.* **13**, 239–244.

Ronchi, V. (1964). *Appl. Opt.* **3**, 437–451.

Rosenthal, A. H. (1962). *J. Opt. Soc. Am.* **52**, 1143–1148.

Ross, I. N., and Singh, S. (1983). *J. Phys. E: Sci. Instrum.* **16**, 745–746.

Rousselet-Perraut, K., Vakili, F., and Mourard, D. (1996). *Opt. Eng.* **35**, 2943–2955.

Rowley, W. R. C., and Wilson, D. C. (1972). *Appl. Opt.* **11**, 475–476.

Rudd, M. J. (1968). *J. Sci. Instrum. (J. Phys. E) Series 2* **1**, 723–726.

Rudenko, V. N., and Sazhin, M. V. (1980). *Sov. J. Quantum Electron.* **10**, 1366–1373.

Sakai, I. (1986). *Opt. Quant. Electron.* **18**, 279–289.

Sakai, I., Parry, G., and Youngquist, R. C. (1986). *Opt. Lett.* **11**, 183–185.

Saldner, H. O., and Huntley, J. M. (1997). *Opt. Lett.* **36**, 2770–2775.

Salimbeni, R., and Pole, R. V. (1980). *Opt. Lett.* **5**, 39–41.

Salit, M. L., Travis, J. C., and Winchester, M. R. (1996). *Appl. Opt.* **35**, 2960–2970.

Saltiel, S. M., Van Wonterghem, B., and Rentzepis, P. M. (1989). *Opt. Lett.* **14**, 183–185.

Sandercock, J. R. (1970). *Opt. Commun.* **2**, 73–76.

Sandercock, J. R. (1976). *J. Phys. E: Sci. Instrum.* **9**, 566–568.

Sanders, B. C., and Milburn, G. J. (1995). *Phys. Rev. Lett.* **75**, 2944–2947.

Sanders, G. A., Prentiss, M. G., and Ezekiel, S. (1981). *Opt. Lett.* **6**, 569–571.

Sandford, S. P., Luck, W. S., and Rohrbach, W. W. (1996). *Appl. Opt.* **35**, 2923–2926.

Sandoz, P. (1996). *J. Mod. Opt.* **43**, 1545–1554.

Sandoz, P., Calatroni, J., and Tribillon, G. (1999). *J. Mod. Opt.* **46**, 327–339.

Sandoz, P., Devillers, R., and Plata, A. (1997). *J. Mod. Opt.* **44**, 519–534.

Sandoz, P., Tribillon, G., and Perrin, H. (1996). *J. Mod. Opt.* **43**, 701–708.

Sasaki, O., Murata, N., and Suzuki, T. (2000). *Appl. Opt.* **39**, 4589–4592.

Sasaki, O., and Okazaki, H. (1986). *Appl. Opt.* **25**, 3137–3140.

Sasaki, O., Okazaki, H., and Sakai, M. (1987). *Appl. Opt.* **26**, 1089–1093.

Sasaki, O., Sasazaki, H., and Suzuki, T. (1991). *Appl. Opt.* **30**, 4040–4045.

Sasaki, O., Sato, T., Abe, T., Mizuguchi, T., and Niwayama, M. (1980). *Appl. Opt.* **19**, 1306–1308.

Sasaki, O., Takahashi, K., and Suzuki, T. (1990). *Opt. Eng.* **29**, 1511–1515.

Sasaki, O., Yoshida, T., and Suzuki, T. (1991). *Appl. Opt.* **30**, 3617–3621.

Sato, S., Ohashi, M., Fujimoto, M.-K., Fukushima, M., Waseda, K., Miyoki, S., Mavalvala, N., and Yamamoto, H. (2000). *Appl. Opt.* **39**, 4616–4620.

Saulson, P., and Cruise, M. (eds.) (2002). "Gravitational-Wave Detection." *Proc. SPIE* **4856**. SPIE, Bellingham, WA.
Saunders, J. B. (1955). *J. Opt. Soc. Am.* **45**, 133.
Schäfer, F. P. (1973). "Dye Lasers." Springer-Verlag, Berlin.
Schaham, M. (1982). *Proc. SPIE* **306**, 183–191.
Schawlow, A. L., and Townes, C. H. (1958). *Phys. Rev.* **112**, 1940–1949.
Schmidt, H., Salzmann, H., and Strowald, H. (1975). *Appl. Opt.* **14**, 2250–2251.
Schmit, J., and Creath, K. (1995). *Appl. Opt.* **34**, 3616–3619.
Schmit, J., and Creath, K. (1996). *Appl. Opt.* **35**, 5642–5649.
Schnatz, H., Lipphardt, B., Helmcke, J., Riehle, F., and Zinner, G. (1996). *Phys. Rev. Lett.* **76**, 18–21.
Schreiber, H., and Schwider, J. (1997). *Appl. Opt.* **36**, 5321–5324.
Schulz, G. (1993). *Appl. Opt.* **32**, 1055–1059.
Schulz, G., and Grzanna, J. (1992). *Appl. Opt.* **31**, 3767–3780.
Schulz, G., and Schwider, J. (1967). *Appl. Opt.* **6**, 1077–1084.
Schulz, G., and Schwider, J. (1976). *In* "Progress in Optics" (E. Wolf, ed.), Vol. XIII, pp. 95–167. North-Holland, Amsterdam.
Schwarz, J. P., Robertson, D. S., Niebauer, T. M., and Faller, J. E. (1999). *Meas. Sci. Technol.* **10**, 478–486.
Schwider, J. (1998). *Optik* **108**, 181–196.
Schwider, J., Burow, R., Elssner, K.-E., Grzanna, J., Spolacyzk, R., and Merkel, K. (1983). *Appl. Opt.* **22**, 3421–3432.
Schwider, J., Falkenstorfer, O., Schreiber, H., Zöller, A., and Streibl, N. (1993). *Opt. Eng.* **32**, 1883–1885.
Scott, R. M. (1969). *Appl. Opt.* **8**, 531–537.
Scully, M. O., and Druhl, K. (1982). *Phys. Rev. A* **25**, 2208–2213.
Scully, M. O., Englert, B.-G., and Walther, H. (1991). *Nature* **351**, 111–116.
Serabijn, E., and Weisstein, E. W. (1996). *Appl. Opt.* **35**, 2752–2763.
Serabyn, E., and Colavita, M. M. (2001). *Appl. Opt.* **40**, 1668–1671.
Serabyn, E., Wallace, J. K., Hardy, G. J., Schmidtlin, E. G. H., and Nguyen, H. T. (1999). *Appl. Opt.* **38**, 7128–7132.
Servin, M., Malacara, D., Malacara, Z., and Vlad, V. I. (1994). *Appl. Opt.* **33**, 4103–4108.
Servin, M., Malacara, D., and Marroquin, J. L. (1996). *Appl. Opt.* **35**, 4343–4348.
Shack, R. V., and Hopkins, G. W. (1979). *Opt. Eng.* **18**, 226–228.
Shaklan, S. B. (1990). *Opt. Eng.* **29**, 684–689.
Shaklan, S. B., and Roddier, F. (1987). *Appl. Opt.* **26**, 2159–2163.
Shankland, R. S. (1973). *Appl. Opt.* **12**, 2280–2287.
Shao, M. (ed.) (2002). "Interferometry in Space." *Proc. SPIE* **4852**. SPIE, Bellingham, WA.
Shao, M., and Colavita, M. M. (1992). *Ann. Rev. Astron. Astrophys.* **30**, 457–498.
Shao, M., and Staelin, D. H. (1977). *J. Opt. Soc. Am.* **67**, 81–86.
Sheem, S. K., and Giallorenzi, T. G. (1979). *Opt. Lett.* **4**, 29–31.
Shelby, R. M., Levenson, M. D., Perlmutter, S. H., de Voe, R. G., and Walls, D. F. (1986). *Phys. Rev. Lett.* **57**, 691–694.

Shen, W., Chang, M.-W., and Wan, D.-S. (1997). *Opt. Eng.* **36**, 905–913.

Shi, P., and Stijns, E. (1993). *Appl. Opt.* **32**, 44–51.

Shih, Y. H., Sergienko, A. V., and Rubin, M. H. (1993). *Phys. Rev. A* **47**, 1288–1293.

Shimizu, E. T. (1987). *Appl. Opt.* **26**, 4541–4544.

Shinohara, S., Mochizuki, A., Yoshida, H., and Sumi, M. (1986). *Appl. Opt.* **25**, 1417–1419.

Shukla, R. P., Dokhanian, M., George, M. C., and Venkateswarlu, P. (1991). *Opt. Eng.* **30**, 386–390.

Shukla, R. P., Moghbel, M., and Venkateswarlu, P. (1992). *Appl. Opt.* **31**, 4125–4131.

Shurcliff, W. A. (1962). "Polarized Light: Production and Use." Harvard University Press, Cambridge, MA.

Shurcliff, W. A., and Ballard, S. S. (1964). "Polarized Light." Van Nostrand, Princeton, NJ.

Siegman, A. E. (1966). *Proc. IEEE* **54**, 1350–1356.

Siegman, A. E. (1971). "An Introduction to Lasers and Masers." McGraw-Hill, New York.

Silvfast, W. T. (1996). "Laser Fundamentals." Cambridge University Press, Cambridge, UK.

Sliney, D., and Wolbarsht, M. (1980). "Safety with Lasers and Other Optical Sources: A Comprehensive Handbook." Plenum Press, New York.

Slusher, R. E., Hollberg, L. W., Yurke, B., Mertz, J. C., and Valley, J. F. (1985). *Phys. Rev. Lett.* **55**, 2409–2412.

Smartt, R. N. (1974). *Appl. Opt.* **13**, 1093–1099.

Smartt, R. N., and Hariharan, P. (1985). *Opt. Acta* **32**, 1475–1478.

Smartt, R. N., and Steel, W. H. (1975). *Japan. J. Appl. Phys.* **14**, Suppl. 14-1, 351–356.

Smith, P. W. (1965). *IEEE J. Quantum Electron.* **QE-1**, 343–348.

Smythe, R., and Moore, R. (1984). *Opt. Eng.* **23**, 361–364.

Snyder, J. J. (1977). *In* "Laser Spectroscopy III" (J. L. Hall and J. L. Carsten, eds.), pp. 419–420. Springer-Verlag, Berlin.

Snyder, J. J. (1981). *Proc. SPIE* **288**, 258–262.

Šolc, I. (1965). *J. Opt. Soc. Am.* **55**, 621–625.

Sommargren, G. E. (1981). *Appl. Opt.* **20**, 610–618.

Sommargren, G. E., and Thompson, B. J. (1973). *Appl. Opt.* **12**, 2130–2138.

Stahl, H. P., and Koliopoulos, C. L. (1987). *Appl. Opt.* **26**, 1127–1136.

Steel, W. H. (1962). *Opt. Acta* **9**, 111–119.

Steel, W. H. (1964a). *Opt. Acta* **11**, 9–19.

Steel, W. H. (1964b). *Opt. Acta* **11**, 211–217.

Steel, W. H. (1965). *In* "Progress in Optics" (E. Wolf, ed.), Vol. V, pp. 147–194. North-Holland, Amsterdam.

Steel, W. H. (1975). *Opt. Commun.* **14**, 108–109.

Steel, W. H. (1983). "Interferometry." Cambridge University Press, Cambridge, UK.

Steel, W. H., Smartt, R. N., and Giovanelli, R. G. (1961). *Austral. J. Phys.* **14**, 201–211.

Steinberg, A. M., and Chiao, R. Y. (1995). *Phys. Rev. A* **51**, 3525–3528.

Steinberg, A. M., Kwiat, P. G., and Chiao, R. Y. (1992a). *Phys. Rev. Lett.* **68**, 2421–2424.

Steinberg, A. M., Kwiat, P. G., and Chiao, R. Y. (1992b). *Phys. Rev. A* **45**, 6659–6665.

Steinberg, A. M., Kwiat, P. G., and Chiao, R. Y. (1993). *Phys. Rev. Lett.* **71**, 708–711.

Stetson, K. A. (1992). *Appl. Opt.* **31**, 5320–5325.

Stevenson, A. J., Gray, M. B., Bachor, H.-A., and McClelland, D. E. (1993). *Appl. Opt.* **32**, 3481–3493.

Stone, J. A., Stejskal, A., and Howard, L. (1999). *Appl. Opt.* **38**, 5981–5993.

Strain, K. A., and Meers, B. J. (1991). *Phys. Rev. Lett.* **66**, 1391–1394.

Streifer, W., Burnham, R. D., and Scifres, D. R. (1977). *IEEE J. Quantum Electron.* **QE-13**, 403–404.

Strong, J., and Vanasse, G. A. (1959). *J. Opt. Soc. Am.* **49**, 844–850.

Strong, J., and Vanasse, G. A. (1960). *J. Opt. Soc. Am.* **50**, 113–118.

Strong, J. D., and Vanasse, G. (1958). *J. Phys. Radium* **19**, 192–196.

Stumpf, K. D. (1979). *Opt. Eng.* **18**, 648–653.

Sudol, R., and Thompson, B. J. (1979). *Opt. Commun.* **31**, 105–110.

Surrel, Y. (1993). *Appl. Opt.* **32**, 3598–3600.

Surrel, Y. (1996). *Appl. Opt.* **35**, 51–60.

Surrel, Y. (1997a). *Appl. Opt.* **36**, 271–276.

Surrel, Y. (1997b). *Appl. Opt.* **36**, 805–807.

Surrel, Y. (1998). *Opt. Eng.* **37**, 2314–2319.

Sutton, E. C. (1979). *In* "High Angular Resolution Interferometry," Proceedings of Colloquium No. 50 of the IAU, pp. 16/1–16/14. University of Sydney, Sydney.

Sutton, E. C., Storey, J. W. V., Betz, A. L., Townes, C. H., and Spears, D. L. (1977). *Astrophys. J. Lett.* **217**, L97.

Suzuki, T., Hirabayashi, S., Sasaki, O., and Maruyama, T. (1999a). *Opt. Eng.* **38**, 543–548.

Suzuki, T., Kobayashi, K., and Sasaki, O. (2000). *Appl. Opt.* **39**, 2646–2652.

Suzuki, T., Matsuda, M., Sasaki, O., and Maruyama, T. (1999b). *Appl. Opt.* **38**, 7069–7075.

Suzuki, T., Muto, T., Sasaki, O., and Maruyama, T. (1997a). *Appl. Opt.* **36**, 6196–6201.

Suzuki, T., Okada, T., Sasaki, O., and Maruyama, T. (1997b). *Opt. Eng.* **36**, 2496–2502.

Suzuki, T., Sasaki, O., Higuchi, K., and Maruyama, T. (1989). *Appl. Opt.* **28**, 5270–5274.

Suzuki, T., Sasaki, O., Higuchi, K., and Maruyama, T. (1991). *Appl. Opt.* **30**, 3622–3626.

Suzuki, T., Sasaki, O., Higuchi, K., and Maruyama, T. (1992). *Appl. Opt.* **31**, 7242–7248.

Suzuki, T., Sasaki, O., and Maruyama, T. (1989). *Appl. Opt.* **28**, 4407–4410.

Suzuki, T., Sasaki, O., and Maruyama, T. (1996). *Opt. Eng.* **35**, 492–497.

Suzuki, T., Yazawa, T., and Sasaki, O. (2002). *Appl. Opt.* **41**, 1972–1976.

Suzuki, T., Zhao, X., and Sasaki, O. (2001). *Appl. Opt.* **40**, 2126–2131.

Svelto, O. (1989). "Principles of Lasers." Plenum Press, New York.

Swantner, W., and Chow, W. W (1994). *Appl. Opt.* **33**, 1832–1837.

Swart, P. L., and Spammer, S. J. (1996). *Opt. Eng.* **35**, 1054–1058.

Takahashi, N., Kakuma, S., and Ohba, R. (1996). *Opt. Eng.* **35**, 802–807.

Takeda, M. (1990). *Ind. Met.* **1**, 79–99.

Takeda, M., Ina, H., and Kobayashi, S. (1982). *J. Opt. Soc. Am.* **72**, 156–160.

Tango, W. J., and Davis, J. (1996). *Appl. Opt.* **35**, 621–623.

Tango, W. J., Davis, J., Lawson, P., and Booth, A. J. (1994). *Proc. SPIE* **2200**, 1267–1269.

Tango, W. J., and Twiss, R. Q. (1980). *In* "Progress in Optics" (E. Wolf, ed.), Vol. XVII, pp. 241–277. North-Holland, Amsterdam.

Tanner, L. H. (1966). *J. Sci. Instrum.* **43**, 878–886.

Tanner, L. H. (1967). *J. Sci. Instrum.* **44**, 1015–1017.

Tatsuno, K., and Arimoto, A. (1981). *Appl. Opt.* **20**, 3520–3525.

Tavrov, A., Bohr, R., Totzeck, M., Tiziani, H., and Takeda, M. (2002). *Opt. Lett.* **27**, 2070–2072.

Terrien, J. (1958). *J. Phys. Radium* **19**, 390–396.

Terrien, J. (1976). *Rep. Progr. Phys.* **39**, 1067–1108.

Tewari, S. P., and Hariharan, P. (1997). *J. Mod. Opt.* **44**, 543–553.

Texereau, J. (1963). *Appl. Opt.* **2**, 23–30.

Thompson, B. J., and Wolf, E. (1957). *J. Opt. Soc. Am.* **47**, 895–902.

Thompson, G. H. B. (1980). "Physics of Semiconductor Laser Devices." John Wiley, Chichester, UK.

Tilford, C. R. (1977). *Appl. Opt.* **16**, 1857–1860.

Timmermans, C. J., Schellekens, P. H. J., and Schram, D. C. (1978). *J. Phys. E: Sci. Instrum.* **11**, 1023–1026.

Tiwari, S. C. (1992). *J. Mod. Opt.* **39**, 1097–1105.

Tiziani, H. J., Rothe, A., and Maier, N. (1996). *Appl. Opt.* **35**, 3525–3533.

Tolansky, S. (1945). *Phil. Mag.* **36**, 225–236.

Tolansky, S. (1955). "An Introduction to Interferometry." Longmans, London.

Tolansky, S. (1961). "Surface Microtopography." Longmans, London.

Tomassini, P., Giuletti, A., Gizzi, L. A., Galimberti, M., Giuletti, D., Borghesi, M., and Willi, O. (2001). *Appl. Opt.* **40**, 6561–6568.

Tomita, A., and Chiao, R. Y. (1986). *Phys. Rev. Lett.* **57**, 937–940.

Tomita, Y., Yahalom, R., and Yariv, A. (1989). *Opt. Commun.* **73**, 413–418.

Townes, C. H. (1984). *J. Astrophys. Astron.* **5**, 111–130.

Toyama, K., Fesler, K. A., Kim, B. Y., and Shaw, H. J. (1991). *Opt. Lett.* **16**, 1207–1209.

Toyooka, S., Nishida, H., and Takezaki, J. (1989). *Opt. Eng.* **28**, 55–60.

Traub, W. A. (1986). *Appl. Opt.* **25**, 528–532.

Traub, W. A. (ed.) (2002). "Interferometry for Optical Astronomy II." *Proc. SPIE*, **4838**. SPIE, Bellingham, WA.

Traub, W. A., Winkel, R. J., Jr., and Goldman, A. (1996). *Appl. Opt.* **35**, 2732–2733.

Troup, G. J., and Turner, R. G. (1974). *Rep. Progr. Phys.* **37**, 771–816.

Truax, B. E., Demarest, F. C., and Sommargren, G. E. (1984). *Appl. Opt.* **23**, 67–73.

Tsuruta, T. (1963). *J. Opt. Soc. Am.* **53**, 1156–1161.

Tveten, A. B., Dandridge, A., Davis, C. M., and Giallorenzi, T. G. (1980). *Electron. Lett.* **16**, 854–856.

Twyman, F. (1957). "Prism and Lens Making." Hilger and Watts, London.

Tyson, R. K. (1991). "Principles of Adaptive Optics." Academic Press, San Diego.

Udem, Th., Reichert, J., Holzwarth, R., and Hänsch, T. W. (1999a). *Phys. Rev. Lett.* **82**, 3568–3571.

Udem, Th., Reichert, J., Holzwarth, R., and Hänsch, T. W. (1999b). *Opt. Lett.* **24**, 881–883.

Ulrich, R. (1980). *Opt. Lett.* **5**, 173–175.

Umeda, N., Tsukiji, M., and Takasaki, H. (1980). *Appl. Opt.* **19**, 442–450.

Urquhart, K. S., Lee, S. H., Guest, C. C., Feldman, M. R., and Farhoosh, H. (1989). *Appl. Opt.* **28**, 3387–3396.

Vacher, R., Sussner, H., and Schickfus, M. V. (1980). *Rev. Sci. Instrum.* **51**, 288–291.

Vali, V., and Bostrom, R. C. (1968). *Rev. Sci. Instrum.* **39**, 1304–1306.

Vali, V., and Shorthill, R. W. (1976). *Appl. Opt.* **15**, 1099–1100.

Vanasse, G. A., and Sakai, H. (1967). *In* "Progress in Optics" (E. Wolf, ed.), Vol. VI, pp. 261–330. North-Holland, Amsterdam.

van Cittert, P. H. (1934). *Physica* **1**, 201–210.

van Cittert, P. H. (1939). *Physica* **6**, 1129–1138.

Vandaele, A. C., and Carleer, M. (1999). *Appl. Opt.* **38**, 2630–2639.

van Wingerden, J., Frankena, H. J., and Smorenburg, C. (1991). *Appl. Opt.* **30**, 2718–2729.

Vaughan, J. M. (1989). "The Fabry–Perot Interferometer." Adam Hilger, Bristol.

Verma, K., and Han, B. (1991). *Appl. Opt.* **40**, 4981–4987.

Vittoz, B. (1956). *Rev. Opt.* **35**, 253–291, 468–480.

Voumard, C. (1977). *Opt. Lett.* **1**, 61–63.

Wallard, A. J. (1972). *J. Phys. E: Sci. Instrum.* **5**, 926–930.

Walls, D. F. (1977). *Am. J. Phys.* **45**, 952–956.

Walls, D. F. (1983). *Nature* **306**, 141–146.

Walsh, C. J. (1987). *Appl. Opt.* **26**, 1680–1687.

Walsh, C. J., and Brown, N. (1985). *Rev. Sci. Instrum.* **56**, 1582–1585.

Wang, X., Sasaki, O., Suzuki, T., and Maruyama, T. (2000). *Appl. Opt.* **39**, 4593–4597.

Wang, X., Sasaki, O., Takebayashi, Y., Suzuki, T., and Maruyama, T. (1994). *Opt. Eng.* **33**, 2670–2674.

Weber, J. (1969). *Phys. Rev. Lett.* **22**, 1320–1324.

Wei, C., Chen, M., and Wang, Z. (1999). *Opt. Eng.* **38**, 1357–1360.

Weigelt, G. P. (1978). *Appl. Opt.* **17**, 2660–2662.

Weinberg, F. J. (1963). "Optics of Flames." Butterworths, London.

Weiss, R. (1972). *Quart. Prog. Rep. Res. Lab. Electron. M.I.T.* **105**, 54–76.

Weiss, R. (1999). *Rev. Mod. Phys.* **71**, S187–S196.

White, A. D. (1967). *Appl. Opt.* **6**, 1138–1139.

White, A. D., and Rigden, J. D. (1962). *Proc. IRE (Correspondence)* **50**, 1697.

White, J. O., Cronin-Golomb, M., Fischer, B., and Yariv, A. (1982). *Appl. Phys. Lett.* **40**, 450–452.

Whitford, B. G. (1979). *Opt. Commun.* **31**, 363–365.

Willson, J. P., and Jones, R. E. (1983). *Opt. Lett.* **8**, 333–335.

Winter, J. G. (1984). *Opt. Acta* **31**, 823–830.

Wolf, E. (1954). *Proc. Roy. Soc. A* **225**, 96–111.

Wolf, E. (1955). *Proc. Roy. Soc. A* **230**, 246–265.

Wolf, E. (1981). *Opt. Commun.* **38**, 3–6.

Wolf, E. (1982). *J. Opt. Soc. Am.* **72**, 343–351.

Wolf, E. (1986). *J. Opt. Soc. Am. A* **3**, 76–85.

Wolf, E. (1987). *Phys. Rev. Lett.* **58**, 2646–2648.

Wolfelschneider, H., and Kist, R. (1984). *J. Opt. Commun.* **5**, 53–55.

Womack, K. H., Jonas, J. A., Koliopoulos, C., Underwood, K. L., Wyant, J. C., Loomis, J. S., and Hayslett, C. R. (1979). *Proc. SPIE* **192**, 134–139.

Wood, T. H. (1978). *Rev. Sci. Instrum.* **49**, 790–793.

Wright, O. B. (1991). *Opt. Lett.* **16**, 56–58.

Wu, L., Kimble, H. J., Hall, J. L., and Wu, H. (1986). *Phys. Rev. Lett.* **57**, 2520–2523.

Wyant, J. C. (1973). *Appl. Opt.* **12**, 2057–2060.

Wyant, J. C. (1975). *Appl. Opt.* **14**, 2622–2626.

Wyant, J. C. (1976). *Opt. Commun.* **19**, 120–121.

Wyant, J. C., and Bennett, V. P. (1972). *Appl. Opt.* **11**, 2833–2839.

Wyant, J. C., and Creath, K. (November 1985). *Laser Focus/Electro Optics*, 118–132.

Wyant, J. C., and O'Neill, P. K. (1974). *Appl. Opt.* **13**, 2762–2765.

Xiao, M., Wu, L.-A., and Kimble, H. J. (1987). *Phys. Rev. Lett.* **59**, 278–281.

Yablonovitch, E. (1993). *J. Opt. Soc. Am. B* **10**, 283–295.

Yamada, M., Ikeshima, H., and Takahashi, Y. (1980). *Rev. Sci. Instrum.* **51**, 431–434.

Yamaguchi, I., Liu, J., and Kato, J. (1996). *Opt. Eng.* **35**, 2930–2937.

Yamaguchi, I., Yamamoto, A., and Yano, M. (2000). *Opt. Eng.* **39**, 40–46.

Yamamoto, Y., Machida, S., Saito, S., Imoto, N., Yanagawa, T., Kitagawa, M., and Björk, G. (1990). *In* "Progress in Optics" (E. Wolf, ed.), Vol. XXVIII, pp. 87–179. North-Holland, Amsterdam.

Yang, T.-S., and Oh, J. H. (2001). *Opt. Eng.* **40**, 2771–2779.

Yarborough, J. M., and Hobart, J. (1973). *IEEE/OSA Conference on Laser Engineering and Applications, Washington, D.C.* (post-deadline paper).

Yariv, A. (1978). *IEEE J. Quantum Electron.*, **QE-14**, 650–660.

Yariv, A., and Winsor, H. (1980). *Opt. Lett.* **5**, 87–89.

Yaroslavskii, L. P., and Merzlyakov, N. S. (1980). "Methods of Digital Holography." Consultants Bureau, Plenum Publishing Company, New York.

Yatagai, T., and Kanou, T. (1984). *Opt. Eng.* **23**, 357–360.

Yeh, Y., and Cummins, H. Z. (1964). *Appl. Phys. Lett.* **4**, 176–178.

Yim, N.-B., Eom, C. I., and Kim, S.-W. (2000). *Meas. Sci. Technol.* **11**, 1131–1137.

Yokota, M., Asaka, A., and Yoshino, T. (2001). *Appl. Opt.* **40**, 5023–5027.

Yokoyama, S., Ohnishi, J., Iwasaki, S., Seta, K., Matsumoto, H., and Suzuki, N. (1999). *Meas. Sci. Technol.* **10**, 1233–1239.

Yokoyama, T., Araki, T., Yokoyama, S., and Suzuki, N. (2001). *Meas. Sci. Technol.* **12**, 157–162.

Yokozeki, S., and Suzuki, T. (1971). *Appl. Opt.* **10**, 1575–1580.

Yoshihara, K. (1968). *Japan J. Appl. Phys.* **7**, 529–535.

Yoshino, T., Kurosawa, K., Itoh, K., and Ose, T. (1982a). *IEEE Trans. Microwave Theory Tech.* **MTT-30**, 1612–1621.

Yoshino, T., Kurosawa, K., Itoh, K., and Ose, T. (1982b). *IEEE J. Quantum Electron.* **QE-18**, 1624–1633.

Yoshino, T., Nara, M., Mnatzakanian, S., Lee, B. S., and Strand, T. C. (1987). *Appl. Opt.* **26**, 892–897.

Yoshino, T., and Yamaguchi, H. (1998). *Opt. Lett.* **23**, 1576–1578.

Yurke, B., McCall, S. L., and Klauder, J. R. (1986). *Phys. Rev. A* **33**, 4033–4054.

Zajonc, A. G., Wang, L. J., Zou, X. Y., and Mandel, L. (1991). *Nature* **353**, 507–508.

Zernike, F. (1938). *Physica* **5**, 785–795.

Zernike, F. (1950). *J. Opt. Soc. Am.* **40**, 326–328.

Zhang, H., Lalor, M. J., and Burton, D. R. (1999). *Opt. Eng.* **38**, 1524–1533.

Zhao, B. (1997). *Meas. Sci. Technol.* **8**, 147–153.

Zhao, B., and Surrel, Y. (1997). *Appl. Opt.* **36**, 2070–2075.

Zhao, Y., Zhou, T., and Li, D. (1999). *Opt. Eng.* **38**, 246–249.

Zhou, W. (1984). *Opt. Commun.* **49**, 83–85.

Zhou, W. (1985). *Opt. Commun.* **53**, 74–76.

Zhu, Y., and Gemma, T. (2001). *Appl. Opt.* **36**, 2070–2075.

Zou, X. Y., Grayson, T. P., Barbosa, G. A., and Mandel, L. (1993). *Phys. Rev. A* **47**, 2293–2295.

Zou, X. Y., Grayson, T. P., and Mandel, L. (1992). *Phys. Rev. Lett.* **69**, 3041–3044.

Zou, X. Y., Wang, L. J., and Mandel, L. (1991). *Phys. Rev. Lett.* **67**, 318–321.

Zumberge, M. A., Rinker, R. L., and Faller, J. E. (1982). *Metrologia* **18**, 145–152.

Index

L

Lamb-dip stabilization, 88
lambda meter, 165
laser beams, 91
laser modes, 80
laser speckle, 91
laser-Doppler velocimetry, 5, 201
laser-feedback interferometers, 198
lasers, 4, 79
  beam expansion, 91
  beats, 5, 84
  confocal resonator, 81
  etalon stabilized, 86
  feedback effects, 86, 92
  frequency comparison, 84
  frequency measurements, 5, 111, 170
  frequency stabilization, 86, 111, 170
  frequency standards, 170
  frequency-doubling, 209
  generalized spherical resonator, 81
  for interferometry, 79
  longitudinal modes, 82
  measurement of line widths, 169
  modes, 80
  polarization-stabilized, 86
  safety precautions, 92
  single-frequency operation, 83
  spot size, 81
  stabilization by saturated absorption, 88
  stabilization by saturated fluorescence, 90
  stabilization on the Lamb dip, 88
  transverse Zeeman, 86
lateral shear
  with plane wavefronts, 144
  with spherical wavefronts, 146
lateral shearing interferometers, 125
  analysis of interferograms, 125
Lau effect, 129
length measurements, 105
  changes in length, 116
  exact fractions, 107

integral order, 106
  with lasers, 113
  sinusoidal wavelength scanning, 115
  two-wavelength interferometry, 113
  wavelength-scanning, 114
LIGO, 245
line standards, 105
linear systems
  two-dimensional, 289
Lippmann color photography, 33
liquid reference surface, 120
Lloyd's mirror, 13
localization of fringes, 18, 55
  in multiple-beam interference, 65
  secondary regions, 55
localized fringes, 21, 26, 28
low-coherence sensors, 194
Lummer–Gehrcke plate, 3, 158
Lyot filter, 163

M

Mach–Zehnder interferometer, 24, 26
  lateral shearing, 125
magnetometers, 206
measurement of the metre, 2, 109
metre
  definition, 2, 5, 111, 112
  measurement, 2, 109
  realization, 170
Michelson interferometer, 24
  double-passed, 77
  lateral shearing, 125
Michelson's stellar interferometer, 6, 221
Michelson–Morley experiment, 2, 7, 239
microscopy, 3, 143
Mirau interferometer, 143
modulation transfer function, 141
moire fringes, 129
monochromatic light waves, 10